VETERINARY SCIENCES AND MEDICINE SERIES

VETERINARY IMMUNOLOGY AND IMMUNOPATHOLOGY

VETERINARY SCIENCES AND MEDICINE SERIES

Rehabilitating the Athletic Horse
Hank W. Jann and Bud Fackelman (Editors)
2010. ISBN: 978-1-60876-672-7

Veterinary Immunology and Immunopathology
Leon Neumann and Sophie Meier (Editors)
2010. ISBN: 978-1-60876-342-9

VETERINARY SCIENCES AND MEDICINE SERIES

VETERINARY IMMUNOLOGY AND IMMUNOPATHOLOGY

LEON NEUMANN
AND
SOPHIE MEIER
EDITORS

Nova Science Publishers, Inc.
New York

For permission to use material from this book please contact us:
Telephone 631-231-7269; Fax 631-231-8175
Web Site: http://www.novapublishers.com

LIBRARY OF CONGRESS CATALOGING-IN-PUBLICATION DATA

Veterinary immunology and immunopathology / editors, Leon Neumann and Sophie Meier.
p. ; cm.
Includes bibliographical references and index.
ISBN 978-1-60876-342-9 (hardcover)
1. Veterinary immunology. 2. Immunopathology. I. Neumann, Leon. II. Meier, Sophie.
[DNLM: 1. Animal Diseases--immunology. 2. Immunity--physiology. 3. Neuroimmunomodulation. 4. Vertebrates--immunology. 5. Veterinary Medicine--methods. SF 757.2 V5855 2009]
SF757.2V486 2009
636.089'6079--dc22

2009045951

Published by Nova Science Publishers, Inc. ✝ New York

CONTENTS

PREFACE

Brucellosis is one of the most important zoonotic diseases worldwide. This book focuses on host natural resistance and innate immunity against Brucella infection. The impact of Johne's disease, a chronic, granulomatous enteritis of ruminants and some wild-type species is examined as well. In addition, the molecular and cellular mechanisms of DNA vaccines are addressed in this book, as well as the optimization strategies that are currently pursued to harness the therapeutic and commercial advantages of DNA vaccination. Moreover, immune responses may substantially differ between conventional laboratory antigens and microbial ones. This book focuses on the experimental pulmonary tuberculosis as a convenient in vivo model to study the disease, providing important advantages over other models. Other chapters examine the defense mechanisms of several flatfish species, which is a main priority to prevent economic losses. The role of the intestine in the immune response in all vertebrates is examined as well, including its cellular components.

Chapter 1 - As part of the mucosal immune system, the intestine plays a paramount role in the immune response in all vertebrates. The mucosal immunity includes inductive and effector tissues, and effector cells and molecules, all part of both innate and adaptive immunity and their connections. The digestive tract is the target organ of many parasites. In ectotherms, intestinal parasites include mainly flagellates, apicomplexans, myxozoans and helminths. The gut mucosal epithelium acts as a protective barrier against pathogens. Intestinal mucins and terminal carbohydrate residues are involved in the first contact pathogen-host, in some occasions through interaction with parasite lectins, as illustrated by mammalian and reptile *Entamoeba* spp. Mucins and their glycosylation patterns can change in reponse to intestinal parasites, including several piscine myxozoans and helminths. In addition, different pathogen recognizing receptors (PRRs) are sensed by the pathogen-associated molecular patterns (PAMPs) of invading pathogens. Although several PRRs have been thus far identified in ectotherms, little or no information is available on their presence in the intestine or on their involvement in parasite recognition. However, in mammals, several PRRs, mainly Toll-like receptors (TLRs) and nucleotide-binding oligomerization domains (NODs) play a significant role at the intestinal level by sensing pathogens and controlling inflammation and apoptosis, thus contributing to the maintenance of homeostasis. Several molecules and factors related to acute phase response (APR), such as AMPs, lysozyme, lectins, complement, cytokines and antiproteases can be activated at the gut level, in connection with innate and adaptive immune factors. Cellular components are also crucial in the intestinal immune response, and they include intestinal epithelial leucocytes (IECs), dendritic cells, macrophages, granulocytes and intraepithelial lymphocytes (IELs), all of them interacting at both the mucosal and systemic levels, to orchestrate an integrated immune response. The cross-talk between the different cellular types, involving also TLRs, AMPs, cytokines and other factors is well known in endotherms. The cellular reactions to several mammalian intestinal parasites are also well characterized. However, information is much scarcer for ectotherms. Although several types of immune cells (EGC/mast cells, other ganulocytes, rodlet cells (RCs), T and B cells, neuroendocrine cells) have been identified (mainly in teleosts), these cells are not well characterized and the knowledge of the different T cells subsets is in its infancy. In addition, though inflammatory reactions to several intestinal studied parasites have been reported, the cellular responses have been characterized for few species, mainly *Enteromyxum* spp. Available information points to the relevant role of RCs and EGCs/mast cells, but the cellular types have not been fully characterized, and the involvement of other granulocytes remains to be elucidated. In addition, the T and B cell responses and the role of secretory immunoglobulins are poorly known.

Chapter 2 - Johne's disease is a chronic, granulomatous enteritis of ruminants and some wild-type species that is caused by infection with *Mycobacterium avium* spp. *paratuberculosis* (MAP). This inflammatory bowel disease (IBD) is characterized by a lengthy yet variable asymptomatic subclinical phase in which infected animals can transmit MAP to their progeny and herd mates; this is followed by a clinical phase characterized by chronic diarrhea, wasting, and eventually death. The impact of Johne's disease is particularly evident in the dairy and beef industries, where significant losses due to reduced productivity and fertility, mortality, and premature culling have been reported. Treatment and preventative management strategies have been difficult to implement and are often ineffective for preventing this IBD due to the pathogenic nature of MAP, and the duration of its subclinical phase.

Individual resistance to Johne's disease is a heritable trait; therefore, it may be possible to selectively breed animals with enhanced disease resistance. To date, there have been few attempts to elucidate the genetic mechanisms involved in resistance to Johne's disease. Considering the polygenic nature of the disease, this task is inherently difficult, since there are likely dozens of loci scattered across the entire genome contributing small effects to disease resistance.

Insight into the genetics of Johne's disease may be gained from research on Crohn's disease, a human IBD that parallels Johne's disease in many respects. Extensive epidemiological studies that have been conducted and validated in a number of different human populations have identified numerous susceptibility loci for Crohn's disease. Variants in the corresponding bovine genes may also contribute to resistance to Johne's disease. Therefore, the focus of this review is to examine the bovine immune response during MAP infection, highlight pathological similarities between Johne's disease and Crohn's disease, and lastly, discuss the likelihood of comparable genetic mechanisms and susceptibility loci existing between these diseases.

Chapter 3 - Flatfish (*Pleuronectiformes* order) is a broad taxonomical group within teleosts, comprising 11 families and about 500 species worldwide. Farming of several flatfish species, such as Japanese flounder (*Paralichthys olivaceus*), Atlantic halibut (*Hippoglossus hippoglossus*), turbot (*Scophthalmus maximus*) and Senegalese sole (*Solea senegalensis*), has been improved in last years in diverse geographical areas; however, their cultivation can be restricted by the occurrence of viral diseases. Therefore, understanding defence mechanisms of these fish species is a main priority to prevent economic losses.

The innate immune system represents the first defence barrier against microbial pathogens. Interferons (IFNs) are key components of the innate immunity against viral infections. These molecules are cytokines that are divided into two categories: type I, (induced by viruses in most cells) and type II (produced by natural killer cells and T lymphocytes). Type I IFN promotes an antiviral state in neighbouring cells by inducing the expression of several antiviral proteins, such as the Mx proteins, whose antiviral mechanisms remain to be elucidated.

A variable number of Mx genes have been found in mammals, birds, fish and invertebrates. Regarding the *Pleuronectiformes* order, a single Mx gene has been reported in Japanese flounder and Senegalese sole, whereas two genes have been described in turbot and Atlantic halibut. These genes have been shown to be activated in response to viral factors such as dsRNA, ssRNA or DNA.

The full-length Mx cDNA from these fish species has been cloned, and the resulting sequences have been compared with those obtained from other species, showing the highest identity in the N-terminus. The variability recorded in the C-terminus may be related to the species-specificity of the Mx antiviral activity.

The Mx antiviral activity has been tested for several fish species using cells expressing recombinant Mx. The results derived from this kind of experiments show that Mx reduces the transcription of viral protein, which suggests that Mx may act blocking viral replication.

The understanding of the innate immune system, and particularly the IFN response, may help to develop new strategies and methodologies to control the diseases provoked by microbes and, consequently, to improve the culture conditions in the aquaculture farms. The antiviral activity of the Mx proteins described for several fish species seems to have a key role within the IFN type I response. In addition, Mx can be used as a model to study the pathogen-host interactions.

Chapter 4 - DNA vaccination is a promising immunology and biotechnology tool that effectively triggers the activation of a diversity of immune responses. DNA immunization has been shown to be an efficient way of generating antibodies, activating T helper cells and cytotoxic T lymphocytes, conferring protection against a wide range of infectious and non-infectious diseases. When compared with traditional vaccines, DNA vaccine technology is associated with simple and cost effective production protocols. The increase of the immunogenicity and potency of DNA vaccines is a necessary step for their successful application in larger animals and humans. A new generation of vaccines is being tested in ongoing clinical trials, using more efficient delivery devices and formulations. These efforts have resulted in the licensing of four DNA products for veterinary diseases which constitutes a huge step forward for the establishment of this immunization concept. In this chapter the molecular and cellular mechanisms of DNA vaccines will be addressed, as well as the optimization strategies that are currently pursued to harness the therapeutic and commercial advantages of DNA vaccination.

Chapter 5 - Brucellosis is one of the most important zoonotic diseases worldwide. Genetic resistance against brucellosis has been characterized in several host species. Recent studies have also shown that *Brucella* has evolved mechanisms for evading the host immune system. This chapter is focused on host natural resistance and innate immunity against *Brucella* infection.

Chapter 6 - Immune responses may substantially differ between conventional laboratory antigens and microbial ones. Pathogens like Mycobacterium tuberculosis (Mtb) have evolved complex mechanisms to evade, divert or subvert immune responses. Tuberculosis (TB) is a major cause of death worldwide whose etiology, *M. tuberculosis*, produces latent asymptomatic infections and chronic progressive diseases, being pulmonary TB its most important manifestation. TB immunity is mediated by Th1-type responses, nevertheless this response seems not enough to definitely eradicate the chronic infection, nor the latent bacilli.

The experimental pulmonary tuberculosis is a convenient *in vivo* model to study the disease because it provides important advantages. Mice infection is established through the airways and murine immune responses are well known. Different aspects of the disease can be explored using different mice strains with different susceptibilities. Another crucial advantage is the suitability to infect with different mycobacterial strains, including clinical isolates and mutant bacteria.

Mycobacteria induce strong Th1-type responses, however, infected immunocompetent hosts are somehow rendered unable to generate effective sterilizing immunity and bacilli can persist long periods in an ill-understood dormant state. Although the molecule Programmed Death-1 (PD1) belongs to the CD28 co-stimulatory family, unlike CD28, PD1 is an inhibitory receptor inducible upon activation of T, B and NKT cells, with two ligands described (PDL-1/PDL-2). The relevance of the PD1/PDL-1 pathway to induce and maintain immunologic tolerance is well documented, and dysfunction in this pathway is associated with autoimmunity. Recently, the PD1/PDL-1 pathway was shown as a suitable target for exploitation by chronic viral diseases provoking T cell exhaustion (a phenotype that can be reverted *in vivo* by blocking the PD1/PDL1 interaction). It is thus possible that other chronic infectious agents may use this strategy too.

The authors therefore evaluated the *in vivo* PD1 expression on CD4+ and CD8+ T cells, both in the TB target organ, the lung, and the mediastinal lymph nodes, along the infection (days 1-60) in Balb/c mice. Results revealed that compared to the less virulent *M. canetti*, airborne infection with the virulent strain *M. tuberculosis* H37Rv, triggered an increase in PD1+ T cells, especially in the lungs, although with different kinetics for CD4+ and CD8+ lymphocytes. Conceivably, virulent Mtb might elicit this inhibitory molecule in effector T cells to facilitate both the establishing and the persistence of the infection.

Chapter 7 - The link between stress, health and social factors has been the focus of many studies and the preferred subject of investigation of researchers from different disciplines. In particular rodent-based models modulating the social environment are becoming increasingly popular in the scientific literature as realistic models of human disease. A growing number of studies now prove that social factors in both early phases and adult life do have a widespread effect on neuroimmune functions. Accordingly, here we will provide an updated review of studies conducted prominently on laboratory rodents. However a number of studies have also been conducted in non-human primates, farm animals and birds. Results from different disciplines will be integrated to reach a clear conclusion: an allostatic overload, including neuroimmune abnormalities, is more likely to develop when unpredictable stressors of social nature, chronically induce physiological and behavioral adjustments that may 'wear and tear' the underlying physiological functions.

Chapter 8 - Multiple sclerosis (MS) is one of the most intensively studied immune-based inflammatory diseases (IMID) of the human central nervous system. Experimental autoimmune encephalomyelitis (EAE) is the elected animal model of MS in which immunopathogenic mechanisms are investigated and the efficacy of new therapies can be tested. However, of the many new therapies that showed promising effects in EAE models only very few were found effective in patients. One possible explanation is the immunological gap between the laboratory mouse and rat strains in which EAE is modeled and the MS patient. In this publication we discuss how EAE models in non-human primates can bridge this gap. Not only is the immune system of non-human primates more closely related to humans, but the EAE models in these species also offer unique possibilities for preclinical research.

In: Veterinary Immunology and Immunopathology
Editors: L. Nermann, S. Meier
ISBN: 978-1-60876-342-9

Chapter 1

MUCOSAL INTESTINAL IMMUNITY AND RESPONSE TO PARASITE INFECTIONS IN ECTOTHERMIC VERTEBRATES

Pilar Alvarez-Pellitero
Instituto de Acuicultura Torre de la Sal, Consejo Superior de Investigaciones Científicas, Torre de la Sal s/n, 12595 Ribera de Cabanes, Castellón, Spain

ABSTRACT

As part of the mucosal immune system, the intestine plays a paramount role in the immune response in all vertebrates. The mucosal immunity includes inductive and effector tissues, and effector cells and molecules, all part of both innate and adaptive immunity and their connections. The digestive tract is the target organ of many parasites. In ectotherms, intestinal parasites include mainly flagellates, apicomplexans, myxozoans and helminths. The gut mucosal epithelium acts as a protective barrier against pathogens. Intestinal mucins and terminal carbohydrate residues are involved in the first contact pathogen-host, in some occasions through interaction with parasite lectins, as illustrated by mammalian and reptile *Entamoeba* spp. Mucins and their glycosylation patterns can change in reponse to intestinal parasites, including several piscine myxozoans and helminths. In addition, different pathogen recognizing receptors (PRRs) are sensed by the pathogen-associated molecular patterns (PAMPs) of invading pathogens. Although several PRRs have been thus far identified in ectotherms, little or no information is available on their presence in the intestine or on their involvement in parasite recognition. However, in mammals, several PRRs, mainly Toll-like receptors (TLRs) and nucleotide-binding oligomerization domains (NODs) play a significant role at the intestinal level by sensing pathogens and controlling inflammation and apoptosis, thus contributing to the maintenance of homeostasis. Several molecules and factors related to acute phase response (APR), such as AMPs, lysozyme, lectins, complement, cytokines and antiproteases can be activated at the gut level, in connection with innate and adaptive immune factors. Cellular components are also crucial in the intestinal immune response, and they include intestinal epithelial leucocytes (IECs), dendritic cells, macrophages, granulocytes and intraepithelial lymphocytes (IELs), all of them interacting at both the mucosal and systemic levels, to orchestrate an integrated immune response. The cross-talk between the different cellular types, involving also TLRs, AMPs, cytokines and other factors is well known in endotherms. The cellular reactions to several mammalian intestinal parasites are also well characterized. However, information is much scarcer for ectotherms. Although several types of immune cells (EGC/mast cells, other ganulocytes, rodlet cells (RCs), T and B cells, neuroendocrine cells) have been identified (mainly in teleosts), these cells are not well characterized and the knowledge of the different T cells subsets is in its infancy. In addition, though inflammatory reactions to several intestinal studied parasites have been reported, the cellular responses have been characterized for few species, mainly *Enteromyxum* spp. Available information points to the relevant role of RCs and EGCs/mast cells, but the cellular types have not been fully characterized, and the involvement of other granulocytes remains to be elucidated. In addition, the T and B cell responses and the role of secretory immunoglobulins are poorly known.

Abbreviations

Ab	antibody
Ag	antigen
AMP	antimicrobial peptide
APC	antigen-presenting cell
APR	acute phase response;
CARD	caspase recruitment domain
CATERPILLER	transcription enhancer, R(purine)-binding, pyrin, lots of leucine repeats
CH	carbohydrate
CRD	carbohydrate recognition domain
DC	dendritic cell
EGC	eosinophilic granule cell
FAE	follicle-associated-epithelium
Gal	galactose
GalNAc	N-acetylgalactosamine
GALT	gut-associated lymphoid tissue
GlcNAc	N-acetylglucosamine
HK	head kidney
IECs	intestinal epithelial cells
IELs	intraepithelial lymphocytes
IL	interleukin
LFs	lymphoid follicles
LNs	lymph nodes
LP	lamina propria
LRR	leucine-rich repeat
MALT	mucosa-associated lymphoid tissue
MΦ	macrophage
MHC	major histocompatibility complex
MMCs	melano-macrophage centres
MUC	mucin
MyD88	myeloid differential primary response protein
NLR	NOD-like receptor

NOD	nucleotide-binding oligomerization domain
PAMPs	pathogen-associated molecular patterns
PP	Peyer's patches
PRR	pathogen recognizing receptor
RB	respiratory burst
RC	rodlet cell
SARM	Sterile-alpha and Armadillo motif protein
TCR	T cell receptor
TGF	transforming growth factor
TIR	Toll/interleukin-1 receptor
TLR	Toll-like receptor

1. INTRODUCTION

There is, currently, a general agreement as to the presence of the main immune mechanisms described for mammals in most ectothermic vertebrates, including the elements of both innate and adaptive immunity. As in other vertebrates, the innate immune system of ectotherms provides the first line of immune defence. Adaptive immunity relies on the generation of random and highly diverse repertoires of T- and B- lymphocyte receptors encoded by recombinant activation genes (RAGs) and contributes to a more specific and efficient response against infections (McGuinees et al., 2003; Medzhitov, 2007). There is, however, increasing evidence of the integration of the different immune mechanisms into a multilevel network, which challenges the artificial dichotomy between innate and adaptive systems (Flajnik and Du Pasquier, 2004).

Substantial progress has recently been made in the characterization of the immune mechanisms and pathways in ectotherms, especially in teleosts. Nevertheless, there are still important gaps in the knowledge of numerous immune mechanisms, and the available information varies according to the group and species. In the case of teleosts, where knowledge on genome and/or continuous cell lines is available, more information exists. This is the case of zebrafish *Danio rerio* (Trede et al., 2004; Meeker and Trede, 2008), pufferfish *Takifugu rubripes* (Bei et al., 2006) and Japanese flounder *Paralichthys olivaceus* (Aoki and Hirono, 2006). The channel catfish *Ictalurus punctatus* has played an important role as a model species for the study of comparative immunology (Clem et al., 1990). Recently, data on fish immune response are increasing for economically relevant species in aquaculture and fisheries, mainly salmonids (*Salmo salar*, *Oncorhynchus* spp.), cyprinids, flatfish (scophthalmids, pleuronectids), gadids, ictalurids, moronids or sparids. However, few data on immune responses to parasites are available (reviewed in Alvarez-Pellitero, 2008a). In the case of amphibians and reptiles, some studies are related to the evolution of certain immune mechanisms and pathways. Information on chondrichthyans, teleosts, amphibians and reptiles is included in the review of Zapata and Amemiya (2000). The genomic databases have greatly contributed to the comprehensive analysis of some immune factors in the anuran *Xenopus laevis* (Vasta et al., 2004). Some information on the urodelan axolotl (*Ambystoma mexicanum*) is also available, though the large genome size of this species makes physical analysis of its gene architecture a daunting task (Flajnik and Kasahara, 2001). Data on reptiles are scarcer and they concern mainly lizards and turtles (reviewed in Zapata and Amemiya, 2000).

All metazoans have mucosal epithelia, one of the most ancient and universal modules of the innate immunity. Together with the skin, the mucosal epithelia are the main interface between the host and the microbial world, acting as a protective barrier against pathogens (Medzhitov, 2007). The gut mucosa makes part of this mucosal immune system, which is more complex than the systemic counterpart, both in terms of anatomy (inductive and effector tissues) and effectors (cells and molecules) (Brandtzaeg et al., 2008). The different elements of the mucosal system make part of both innate and adaptive immunity and interact, not only at the mucosal level but also with the systemic immune system, to orchestrate an effective immune response. In mammals, and more specifically in humans, studies on intestinal mucosal immunity have substantially increased in the last years, due to its relevance in relation to certain inflammatory diseases, including inflammatory bowel disease, and some autoimmune diseases (Artis and Grencis, 2008). Mucosal immunity plays also a paramount role in relation to protection and mucosal vaccination (Holmgren and Czerkinsky, 2005).

In the present paper, the most relevant information on the immune response of ectotherms to intestinal parasites is reviewed. Most data on ectotherms are related to teleosts, but some data on amphibians and reptiles are also considered. Information on the involvement of the different mechanisms in the response to intestinal parasites is included, when available. As the knowledge of some elements of the intestinal mucosal immunity is limited for ectotherms, the main mucosal mechanisms are briefly described taking those of mammals as reference. Responses at the mucosal surface are firstly considered, following by other elements related mainly to innate immunity, though reference is made to the cross talk of innate and adaptive immunity. Finally, inflammation and cellular mechanisms are analyzed and selected examples of the response to parasite infections in both endotherms and ectotherms are provided.

2. Immune Mechanisms at the Mucosal Surface

2.1. Mucins

The mucosal surface of gastrointestinal tract is a complex organization of epithelium, immune cells and resident microbiota (McCraken and Lorenz, 2001). The intestinal epithelium is covered by a mucus layer, with mucins acting as the main structural component. Mucins are secreted by goblet cells and are mainly found at the periphery of epithelial cells and their extracellular environment or covering epithelial cells. Thus, they form a mesh-like structure that impedes the diffusion of offending macromolecules, constituting an immune defence barrier (Dharmani et al., 2009).

Mucins are subdivided into secretory and membrane-bound forms depending on their structure and location. They are large glycoproteins that range in size from 0.5 × 106 to 25 × 106 Da. The protein core of a typical mucin contains mucin domains that consist of tandem repeats rich in threonine, proline and/or serine, the hydroxyl residues of which are heavily substituted with O-linked oligosaccharides. The extensive O-glycosylation of mucins protects them from proteolytic ezymes and induces a relatively extended conformation. The carbohydrate (CH) chain is initiated with an N-acetylgalactosamine (GalNAc) linked to serine or threonine and is elongated by the formation of the so-called core structures followed by the backbone region (type-1 and type-2 chains). The chains are typically terminated by fucose, galactose (Gal), GalNAc or sialic acid residues in the peripheral region forming histo-blood-group antigens (Ags). Sulfation of Gal and N-acetylglucosamine (GlcNAc) residues causes further diversification. In addition to the O-linked glycans, mucins contain a small number of N-linked oligosaccharides, which have been implicated in folding, oligomerization (MUC2), or surface localization (MUC17). The N and C-terminal regions of mucin are poorly glycosylated compared to the mucin domains and they contain a wide range of amino acids, most notably cysteine residues (Moncada et al., 2003).

The process of glycosylation confers many of the general properties of mucins including high–charge density from sialic acid and sulfate residues, protease resistance and water holding capacity (Moncada and Chadee, 2002). The CH structures present in mucins are determined by expression of specific enzymes, the glycosyl-transfersases and glycosidases. Thus, mucin glycosylation is governed by genetics (due to polymorphisms in these enzymes), tissue-specific enzyme expression, and host and environmental factors influencing transferase expression (Linden et al., 2008). Glycosylation produce an abundant, diverse, and highly regulated repertoire of cellular glycans that are frequently attached to proteins and lipids. Glycans participate in many biological processes, including cell adhesion, molecular trafficking and clearance, receptor activation, signal transduction and endocytosis (reviewed in Ohtsubo and Marth, 2006). The CH structures present on mucosal surfaces vary between/within species and between and even within tissues, and also according to the cell lineage and developmental stage.

Both gel-forming and cell-surface mucins show constitutive and inducible gene expression in mucosal epithelial cells. The promoters of the MUC genes have generally not been fully characterized, although partial promoter characterization is available for some human MUCs. To date, at least 16 human mucins have been included in the family and their expression profile varies between tissues with the gastrointestinal tract showing the highest and most diverse expression. MUC1, the most extensively studied membrane-associated mucin, is the most ubiquitously expressed across all mucosal tissues. MUC1 associated with the cell surface is constantly internalized and recycled. During recycling, sialic acid is added to the premature form of MUC1. The cytoplasmic domains of the cell surface mucins are complex, often contain known phosphorylation motifs, and are highly conserved across species, suggesting important intracellular functions (reviewed in Linden et al., 2008).

Mucins and glycoconjugates are also present in piscine intestine. Neutral mucins and alkaline phosphatase activity in the digestive epithelium of fish could be positive correlated with absorption and transport of macromolecules (Stroband et al., 1979). Gut mucosubstances may also exert an osmotic function in fish, mainly in marine species (Smith, 1989). Glycoconjugate composition of mucous secretion in fish is remarkably different between species, and intraspecific variations associated to age, the region of the fish alimentary canal or other factors also occur (Domeneghini et al., 1998; Sarasquete et al., 2001). Differences in the pattern of glycoconjugate glycosylation in different cyprinid fish could be correlated with the feeding habitats (Fiertak and Kilarski, 2002). Variations with age were demonstrated in several piscine species, such as *Sparus aurata* (Domeneghini et al., 1998), *Umbrina cirrosa* (Parillo et al., 2002) and *Paralichthys dentatus* (Soffientino et al., 2006), using lectin histochemistry. Lectin binding has also been used to characterize the glycosydic pattern of intestinal cells of other piscine species (Imagawa et al., 1990; Scocco et al., 1997; Sarasquete et al., 2001; Pedini et al., 2002; Domeneghini et al., 2005).

Several studies have been also carried out on the glycoconjugates of the digestive tract of some amphibians (Fry and Kaltenbach, 1999; Ferri et al., 2001; Liquori et al. 2007) and reptiles (Madrid et al., 1989; Pérez-Tomás et al., 1990) using both classical histochemical and methods and lectin binding studies. Glycosylation patterns can also change between species.

2.2. Participation of Mucins and Carbohydrate Residues in the Interaction with Pathogens

2.2.1. Adhesion

Both the secreted and adherent mucosal barriers are constantly renewed and could potentially be rapidly adjusted to changes in the environment, for example, in response to microbial infection (Linden et al., 2008). The survival of many gastrointestinal pathogens lies in their ability to colonize the mucus layer. The first step in the establishment of these pathogens is the binding of the microorganisms to mucin. Once bound, pathogenic microorganisms must colonize or penetrate the mucus layer to survive in the gastrointestinal tract. According to Belley et al. (1999), the interactions between microorganisms and intestinal mucins can lead to four possible scenarios: 1) mucin adherence, followed by elimination of the pathogen through sloughing and peristalsis; 2) colonization, where the pathogen is confined to the mucus blanket and is denied access to the underlying epithelium; 3) colonization of the mucus layer with elaboration of pathogen virulence factors; 4) the mucus barrier is broached, invasion occurs and the invading pathogen gains access to the intestinal epithelium. Microbes use a wide variety of strategies to subvert or avoid the mucin barrier, including the production of enzymes capable of degrading mucin core proteins and mucin CHs (like the protozoan *Entamoeba hystolytica*, which cleaves the MUC2 mucin), and effective motility through mucus gels. Binding site of mucins has been shown to compete with those of the underlying epithelium, preventing attachment of pathogens to the mucosal surface. This mechanism of competitive inhibition protects de epithelium from pathogenic microorganims (parasites, bacteria and virus) while allowing the establishment of normal gut flora (reviewed in Moncada et al., 2003). Infection by pathogens is generally initiated by the specific recognition of host epithelia surfaces. Receptors present in the mucin layer can act as binding sites in the subsequent adhesion, which is essential for invasion. In their infection strategy, microorganisms often use sugar–binding proteins, such as lectins and adhesins to recognize and bind to host glycoconjugates where sialilated and fucosylated oligosaccharides are the major targets. The lectin/glycoconjugate interactions, characterized by their high specificity, are known to play a significant role in the adhesion of bacteria and parasites and in their interaction with the host. Binding is usually of low affinity but clustering of adhesions and receptors results in multivalent binding generating higher affinity (Imberty and Varrot, 2008). Fimbriae (or pili), outer membrane proteins, and cell wall components (e.g., lipopolisaccharides) may all function as adhesins.

One of the best characterized lectin/CH interactions is that of the enteric protozoan *Entamoeba hystolytica*. Colonization by this protozoan is mediated by binding Gal and GalNAc residues by the 170-kDa Gal/GalNAc lectin with high affinity. This Gal-lectin is not only involved in the host cell binding, but also in cytotoxicity and complement resistance, induction of encystation, and generation of the cyst wall. Thus, it plays a pivotal role in both differentiation and virulence, and is considered a promising subunit vaccine candidate (Petri et al., 2002; Gaucher and Chadee, 2003; Frederick and Petri, 2005). The surface of *Entamoeba invadens*, a reptilian parasite, was found to express a galectin which has a heterodimeric structure similar to that of *Entamoeba histolytica.* This galectin seem to monitor the concentrations of Gal ligands in the colon and contribute to stimuli that induce encystment (Eichinger, 2001).

Other enteric pathogen, *Cryptosporidium parvum*, also expresses a Gal/GalNAc-sporozoite surface lectin that is involved in the mechanism of parasite attachment to epithelial cells (Chen and LaRusso, 2000). In the case of *Giardia lamblia*, adherence to epithelium is inhibited by glucose and mannose, though the direct involvement of a surface lectin with specificities for both sugars has not been demonstrated (reviewed in Moncada et al., 2003). In the non-enteric parasite *Acanthamoeba castellani*, causing corneal keratitis, the mannose binding protein is a major virulence factor responsible for the parasite-induced target cell destruction (Garate et al. 2005). The blocking or inhibition of microbial lectins by suitable CHs or their analogs is the aim of anti-adhesion therapy for the prevention and treatment of infectious diseases (Sharon, 2006).

To the best of our knowledge, no information on lectins of piscine parasites is available. However, the interaction of parasite molecules with CH residues present in the fish intestine is likely to occur.

2.2.2. Changes in mucin patterns with infection

Microbial products can alter the production of mucins by mucosal epithelial cells. Mucin secretion is enhanced in response to probiotic bacteria. In addition, many enteric microbes and their toxins are known to have a potent secretagogue effect on goblet cells. This rapid mucin secretion can be an important mechanism of protection by eliminating the offending pathogens. In contrast, mucus depletion can occur in other cases (as with the pathogen *Helycobacter pilori*), which, in some instances, could be a deleterious side effect (Moncada et al., 2003; Linden et al., 2008).

In addition to regulation of their synthesis and release, mucins are regulated in terms of their glycosylation, which can change during mucosal infection/inflammation. This may be an important mechanism for unfavourably changing the niche occupied by mucosal pathogens or blocking the mechanisms that pathogens use to subvert the mucin barrier. Several immune factors also influence the secretion and glycosylation of mucins. As a matter of fact, mucin is one of the components involved in the relationship of the diverse elements of innate and adaptive immunity at the gut level. In the response to pathogens, the effects of adhesion on host cells can include altered morphology, fluid loss, up-regulation of adhesion molecules and apoptosis, and induction of cytokine release, with subsequent involvement of other immune mechanisms. Expression of cell-surface and gel-forming mucins can be up-regulated by inflammatory cytokines such as interleukin (IL)-1β, IL-4, IL-6, IL-9, IL-13, interferons, TNF-α, nitric oxide and other uncharacterized inflammatory factors. Responsiveness to these cytokines provides a link between mucins, innate mucosal immunity, and mucosal inflammatory responses. Pathogens also induce alterations in glycosylation that appears to be driven partly by CD4+ T cells, and also by T-cell independent mechanisms. Such changes in mucin glycosylation need to be considered as a component of the innate and adaptive immune responses to mucosal infection (reviewed in Linden et al., 2008).

Dynamic changes on mucins in intestinal parasitoses have been widely studied for nematode infections, although mostly described through differential histological staining of goblet cells (Karlsson et al., 2000). The use of some host/nematode models has allowed to conduct several studies on the mucin regulation and immune mechanisms involved in some nematodoses, such as those produced by *Nippostrongylus brasiliensis*, *Trichuris muris* and *Trichinella spiralis* (reviewed in Tse and Chadee, 1991; Else, 2005; Patel et al., 2009). Infections by *Nippostrongylus brasiliensis* produced increasing in acidic and sialilated mucins (Koninkx et al., 1988; Karlsson et al, 2000), probably related to the up-regulation of determined sialiltransferase sulfotransferase genes (Soga et al., 2008). Using lectin-binding assays and other methods, glycosylation alterations have also been reported (Ishikawa et al., 1993; Holmén et al. 2002). Such mucin and glycosylation alterations could contribute to the rejection of the parasite. Infection by *T. spiralis* leads to an enteropathy characterized by goblet cell and mast cell hyperplasia, up-regulation of MUC2 and MUC3, and enhanced production/expression of intestinal trefoil factor. T cell immune responses are also involved in such up-regulation (reviewed in Khan, 2008; Knight et al., 2008). Up-regulation of MUC-2 and its transcripts has been reported in the large intestine following the infection with *Trichuris muris* (reviewed in Artis and Grencis, 2008). There is a general consensus that acidic mucins, such as sialomucin and sulfomucin, play an important role in the protection of mucosa from infectious agents. This is illustrated by *Strongyloides venezuelensis* infections, in which sulphated glycoconjugates, either from mucosal mast cells or goblet cells, prevent the mucosal invasion by this nematode, and hence they have important effects on the longitudinal distribution of its females (Maruyama et al. 2000, 2002).

Data on intestinal trematodes were reviewed by Toledo et al. (2006). Goblet cell hyperplasia and the associated increased mucus production play an important role in the expulsion of some species, such as *Neodiplostomum seoulense*, *Echinostoma spp.* and *Gymnophaloides seoei*. The distribution and intensity of glycoconjugates also changed in *Echinostoma*-infected hosts.

Parasite infections can also produce changes in the lectin-binding patterns to terminal CH residues. The coccidian *Isospora suis* induces a decrease in the expression of some α-GalNAc and fucose terminals in the small intestine of infected piglets with respect to controls, whereas the enhanced expression of glucose/mannose terminals could favour the intestinal colonization by *Escherichia coli* (Choi et al., 2002).

Limited knowledge on the involvement of mucin and terminal CH residues in the interaction of pathogens with ectothermic vertebrate hosts is available and it is mainly related to teleosts (Table 1). Some data concern bacteria or bacterial products but little information is available for parasites. The oral administration of endotoxin LPS to common carp induced a decrease in intestinal mature luminal mucins linked to a new biosynthesis of intracellular mucin glycoproteins. The increasing secretion of mucins can induce a flushing-out effect, with in combination with altered glycosylation patterns provide a removal mechanism or bacterial noxes. Consequently, pseudofeces of fish, which are commonly sign of intestinal parasitic infections, may be also interpreted as an elimination mechanism of bacterial noxes (Neuhaus et al., 2007). Peroral application of two strains of *Aeromonas hydrophyla* altered intestinal secretion and composition of intestinal mucus glycoproteins in carp. *A. hydrophyla* 60 apparently could be eliminated very quickly from the intestinal tract due to an increased mucus secretion. In contrast, *A. hydrophyla* 38 seemed to destroy the mucus layer, leaving the intestinal mucosa insufficiently protected against further pathogens. In addition, an increased glycosylation with fucose and neuraminic acid was observed (Schoers et al., 2009).

As for piscine parasites, some information is available on the enteric myxozoans *Enteromyxum* spp. In *Sparus aurata* parasitized by *E. leei*, the number of goblet cells was considerably and significantly reduced in heavily infected fish (Fleurance et al., 2008). Similarly, in *E. scophthalmi*-infected turbot, the number of goblet cells decreased dramatically in advanced infections, whereas and increase occurred in mild infections (Bermúdez et al., 2009a, b). Glucose/mannose, fucose and GlcNAc residues, were less abundant in both infected hosts with respect to uninfected fish. In contrast, D-Gal and D-GalNAc moities increased in *E. scophthalmi*-infected turbot. N-acetyl-neuraminic acid, which was abundant in uninfected *Sparus aurata*, decreased dramatically in *E. leei*-infected fish, in accordance with a decrease in acidic mucins. This influence of *Enteromyxum* spp. infection on the intestinal CH pattern of infected fish suggests the involvement of such terminal CH residues in the parasite invasion and host response (Redondo and Alvarez-Pellitero, 2009a, b).

Table 1. Changes in goblet cells, mucins and/or carbohydrate residues in response to pathogens or pathogen products in the intestine of teleosts. Gal: Galactose. GalNAc: N-acetylgalactosamine. GlcNAc:N-acetylglucosamine. Neu-Nac: N-acetylneuraminic acid

Pathogen or their products	Fish	Effect	Reference
LPS	*Cyprinus carpio*	↓ mature luminal mucins ↑intracellular mucins altered glycosylation patterns	Neuhaus et al., 2007
Bacteria *Aeromonas hydrophyla*	*C. carpio*	Strain 60: ↑ mucus secretion Strain 38 : destruction mucus layer ; ↑ glycosylation	Schoers et al., 2009
Myxozoa *Enteromyxum leei*	*Sparus aurata*	↓↓ goblet cells in heavily infected fish ↓ acidic mucins, ↓↓↓ NeuNAc, ↓↓ fucose, ↓GlcNAc and glucose/mannose residues	Fleurance et al., 2008 Redondo and Alvarez-Pellitero, 2009a, b
E. scophthalmi	*Psetta maxima*	↓ fucose and GlcNAc, ↓↓ glucose/mannose, ↑ Gal and GalNAc residues ↑ goblet cells in mild infections, ↓ in severe infections	Redondo and Alvarez-Pellitero, 2009a, b Bermúdez et al., 2009a, b
Trematoda *Deropristis inflata*	*Anguilla anguilla*	↑ number mucous cells and acidic mucins	Dezfuli et al., 1997
Cestoda *Cyatocephalus truncatus* and Acanthocephala *Pomphorhynchus laevis*	*Salmo trutta*	↑mucous cells with acidic or mixed glycoconjugates ↑ thickness of mucous gel	Bosi et al., 2005a

Some data are also known for helminths. Eels (*Anguilla anguilla*) infected by the digenetic trematode *Deropristis inflata* presented a mild catarrhal enteritis formed by leucocytes, mucus and epithelial cell debris. The number of mucous cells increased in the parasitized segments of alimentary canal in comparison to control fish, including acidic mucin secreting cells (Dezfuli et al., 1997). Bosi et al. (2005) also reported a significant increase in the number of mucous cells containing acid or mixed glycoconjugates in *Salmo trutta* naturally parasitized by the acanthocephalan *Pomphorhynchus laevis* and the cestode *Cyatocephalus truncatus*, with respect to uninfected fish. In addition, when intestinal helminths were present, the thickness of the adherent mucous gel increased.

2.3. Pathogen Recognizing Receptors (PRRs)

2.3.1. Toll-like receptors (TLRs)

Once the pathogen is able to overcome the luminal mucin barrier, the second host defence comes into play at the level of epithelium and lamina propria (LP), in which several elements of innate and adaptive immunity can interact. The innate immune system plays a crucial role, as it sense the invading pathogens by recognizing pathogen-associated molecular patterns (PAMPs) via pathogen recognizing receptors (PRRs). PRRs, with a broad specificity, are germline encoded and have evolved to recognise PAMPs (Medzhitov and Janeway, 2001; Janeway and Medzhitov, 2002). A unifying feature of PRRs is their highly conserved structures, which are invariant between microorganisms of a given class. The different PRRs are involved in performing specific tasks, including opsonization, activation of complement cascade, phagocytosis, etc. (Pasare and Medzhitov, 2004).

There are several functionally distinct classes of PRR, but the best characterized are the TLRs, which are type-I transmembrane glycoproteins composed of extracellular, transmembrane and intracellular signaling domains (Gay and Gangloff, 2007). The extracellular component is characterized by a leucine-rich repeat (LRR) motif and the intracellular domain is a Toll/interleukin-1 receptor (TIR). The TIR domain of TLRs has been associated with the signaling cascade leading to the nuclear translocation of the transcription factor-kappa B (NF-κB), involved in the initiation of both innate and adaptive immune responses (Leulier and Lemaitre, 2008). Members of the TLR family contribute both to cell-cell interaction and to signalling. Toll, the founding member of the family, was initially implicated in the establishment of dorsoventral polarity in the early *Drosophila melanogaster* embryo. Mammalian TLRs have essential roles in the direct recognition of infectious agents (reviewed in Leulier and Lemaitre, 2008).

Apart from the specificity of TLRs towards different PAMPs (Akira and Takeda, 2004; Belinda et al. 2008), TLRs can differentially utilize the TIR domain-containing adaptor proteins to optimise the immune responses against microorganisms. Five intracellular TLR adaptor proteins harbouring the TIR domain have been identified: myeloid differential primary response protein (MyD88), MYD-88-adaptor-like (Mal), TRIF-related adaptor molecule (TRAM or TICAM-2) and sterile-alpha and Armadillo motif protein (SARM) (see Belinda et al., 2008).

The results of vertebrate and invertebrate genome projects have suggested that the TLR pattern recognition system is conserved across mammalian groups and probably most vertebrate species, including ectotherms, whereas invertebrates harbour TLR-like proteins with no homology to human TLRs. Ortologous of the mammalian-type TLR (M-type TLR) members are present in gnathostomes, including fish, amphibians, reptiles and mammals, and even in lampreys. M-type family members are mostly related to microbial pattern recognition for immunity (Roach et al., 2005; Ishii et al., 2007a). The vertebrate TLRs provide an example of a multi-gene family in which gene duplication has been followed by extensive changes in certain aspects of gene function while others have been conserved throughout vertebrate history. The TLR and the intracellular TIR domain have ancient evolutionary origins (Hughes and Piontskivska, 2008). The evolutionary conservation of key components of the TLR-signalling pathway that act as switches for differential gene induction, including MYD88, TIRAP, TRIF, TRAF6, IRF3 and IRF7, has also been demonstrated in teleosts (reviewed in Purcell et al., 2006). SARM is also functionally conserved from arthropod to human (Belinda et al., 2008).

At least eleven members of TLR family have been identified so far in mammals (reviewed in Kawai and Akira, 2005; Akira et al. 2006). However, nearly all vertebrate TLRs belong to one of the six major families (TLR1-6). According to their functional diversification, each TLR recognizes different PAMPs, and even self-molecules. Selective pressure presumably for maintenance of specific PAMP recognition has dominated the TLR2 subfamily (for lypopeptide), the TLR3 family (for double-stranded RNA), the TLR4 family (for bacterial lipopolysaccharide, LPS), the TLR5 family (for flagellin), and the TLR7-9 subfamilies (for nucleic acid and heme motifs). TLR7 and TLR8 seem to be responsible for virus detection. TLR9 mediates recognition of CpG DNA motifs found in bacteria and DNA viruses, and other non-DNA pathogenic components. TL3, TLR7 and TLR9, all of which recognize nucleic acids, are not expressed on the cell surface but are exclusively expressed in the endosomal compartments. The TLR family specific for lipopetide PAMPs includes TLR1, TLR2, TLR6, TLR10 and TLR14. The remaining major family, including the TLR11-13, TLR21-23 subfamilies, is represented in humans only by a pesudogene (reviewed in Roach et al., 2005). Different TLR ligands, such as TLR2, TLR4, TLR9 (reviewed in Gazinelli and Denkers, 2006) and TLR11 (Yarovinsky and Sher, 2006) are also involved in the recognition of protozoan PAMPs.

It appears that, with few exceptions, vertebrates have at least one member gene representative from each of the six major TLR families. Most of the mammalian TLR signalling molecules or TIR domains have been identified in ectothermic vertebrates. One occasional exception is the lack of TLR4 in some fish species (Roach et al., 2005; Tsujita et al., 2006). This fact, together wit the identification of a negative regulator of TLR signaling in zebra fish explains the resistance of fish to endotoxic shock and supports the idea that the TLR4 receptor complex for LPS recognition arose after the divergence of fish and tetrapods (Sepulcre et al., 2009). Several TLRs are known in the anuran amphibian *Xenopus laevis*. Phylogenetic analysis showed that the repertoire of the frog TLRs consisted of both fish- and mammalian-type TLRs (Ishii et al., 2007b). Amphibians and fish have a putatively soluble form of the TLR5 gene (*TLRS5*). Most TLR11 subclades have representatives from fish and frogs. The TLR16 subfamily, molecularly distant from all other TLRs and found only in *Xenopus*, may belong to the TLR11 family (Roach et al., 2005).

2.3.2. Other PRRs

Two additional families of innate receptors join the TLRs as pathogen sensors. They are intracellular receptors including nucleotide-binding domain (NOD)-like receptors (NLRs) and

Rig-I (retinoic-acid-inducible gene I)-like proteins (RLR). NLRs make part of a large family of receptors recently named the caspase recruitment domain (CARD), transcription enhancer, R(purine)-binding, pyrin, lots of leucine repeats (CATER-PILLER) (or NOD- LRR proteins). In humans, there are over 20 members of this family, including NOD1 (CARD4) and NOD2 (CARD15) (reviewed in Sanderson and Walker, 2007). The role of C-type lectins, another important group of PRRs is commented below (section 3.1.4).

In teleosts, enhanced TLR expression induced by pathogens has been demonstrated on some occasions, but almost no information on the fish TLR function has been reported (reviewed in Alvarez-Pellitero, 2008a). Studies of Tsujita et al. (2006) suggested that the fish soluble TLR5 serves as an adjuvant augmenting flagellin-TLR5 mediated NF-κB activation even in human. Evidence for a positive selection (acting on the structural components, not the interactive sites) of LRRs in the TLR9 gene of several teleosts has been obtained and may indicate their adaptation to different oligodeoxynucleotids present in different bacterial species (Chen et al., 2008).

2.3.3. PRRs in gut epithelia

The ability of intestinal epithelial cells (IECs) to recognize and respond to high diversity of microorganisms present in the intestinal lumen relies upon a set of PRRs that recognize conserved diverse pathogen motifs. The main receptors involved in the response of mucosal epithelia are TLRs and NODs (Dahan et al., 2007). Transmembrane signaling occurs through TLRs, whereas cytoplasmic receptors are generally members of NOD-LRR (CATERPILLER) family.

TLR expression is tightly controlled in the intestinal epithelium, as it plays a role in maintaining epithelial barrier function in response to enteric pathogens and parasites. Initial studies demonstrated that normal epithelium has very low expression of TLRs, which makes the mucosal immune response tolerant of commensal Ags, whereas pathologic conditions such as a very high antigenic dose or the presence of regulatory cytokines might stimulate expression of the TLR, leading to amplified innate immune response (Moncada et al., 2003). TLRs can participate in several immune mechanisms at the intestinal level. Expression of antimicrobial peptides (AMPs) by leukocytes, hepatocytes and epithelial cells can be induced by activation of TLRs (reviewed in Canny and Levy, 2008). At sites of organized mucosal lymphoid tissues, the intestinal epithelium is dramatically modified to fulfill a different function in which TLR are also involved. In mammals, mucosal lymphoid follicles (LFs), whether solitary or clustered as in Peyer's patches (PP), are covered by a specialized follicle-associated-epithelium (FAE) containing M cells (see below, section 5.1). TLR-mediated signals regulate the gate keeping functions of the FAE to promote Ag capture by dendritic cells (DCs) in organized mucosal lymphoid tissues (reviewed in Chabot et al., 2006) (see below, section 5.3.1). In addition, TLR signalling is important for the maturation of Ag-presenting cells (APCs) that is necessary for the activation of T cells. Recent findings lead to the hypothesis that TLR signals help to maintain the epithelial barrier and promote normal growth and differentiation of IECs (reviewed in Pancer and Cooper, 2006). TLRs have also essential roles in the inflammatory response. The members of NOD-LRR family are also involved in the response to pathogens and in inflammation at intestine as well.

Recent data in mammals indicate that stimulation of TLRs by intestinal microbiota supports colonic homeostasis. Intestinal villus and crypt IECs express several TLRs in their apical poles in vivo (Didierlaurent et al., 2002; Lee et al., 2008). A polarizing-tolerizing ability of IECs has been described and it represents a unique mechanism that aids to maintain the epithelial barrier and restrains inflammatory responses in a bacteria-rich environment (Cario and Podolsky, 2005). Several members of the TLR family can induce apoptosis or programmed cell death and thus they may have the capacity to cross talk with the death signaling pathways. At least TLR2, TLR3 and TLR4 have been involved in apoptotic signaling (Salaun et al., 2007; Gay and Gangloff, 2007). The apoptotic mechanisms and their connection at the intestinal level are commented below (section 4.3).

In relation to adaptive immunity, a member NOD-LRR family, the class II transactivator, regulates MHCII expression (Sanderson and Walker, 2007). NODs are also critical for the regulation of bacterial immunity within the intestine by controlling the expression of the AMPs cryptdins (reviewed in Pancer and Cooper, 2006)

3. Other Innate Defences at the Mucosal Intestinal System

3.1. Acute Phase Proteins

The acute phase response (APR) is a pervasive physiological response of the body to injury, trauma or infection. In its broadest context, the APR involves changes in at least the hepatic, neuroendocrine, haematopoietic, musculo-skeletal and immune systems, and is induced by pro-inflammatory cytokines such as IL-1, IL-6 and TNFα (Bayne and Gerwick, 2001). Several factors directly or indirectly related to APR, such as antimicrobial peptides, lysozyme, anti-proteases, complement, and lectins, are commented in this section, with special reference to their presence in the mucosal intestinal tissues.

3.1.1. Antimicrobial peptides (AMPs)

AMPs are widely expressed and rapidly induced at epithelial surface and circulating immune cells to repeal assault from diverse infectious agents, including bacteria, viruses, fungi and parasites, and thus they are likely to play a key role in host protection and homeostasis (Dommet et al., 2005; Lai and Gallo, 2009). AMPs can be expressed at the intestinal level. Some of them are constitutively expressed, but also the production of AMPs by epithelial cells can be regulated by cytokines and factors of the innate and adaptive immune systems in response to pathogen encounter. Besides their direct antimicrobial action, AMPs also function as opsonins, chemokines and modulators of cytokine production that in turn regulates innate immune responses. Recent observations have shown that some AMPs have a broad impact in host defence, since they can both trigger and tune immunomodulatory processes, such as chemotaxis, phagocytosis, cytokine production, reactive oxygen species production and DC maturation (Hölzl et al., 2008).

AMPs constitute a phylogenetically ancient form of innate immunity and are widespread in nature as defence mechanisms in plant and animals (Zasloff, 2002). The molecules referred as AMPs are extremely diverse. So far, more than 1200 AMPs of different origins have been identified or predicted (for a list see the Antimicrobial Peptide Database, APD: http://aps.unmc.edu/AP/main.php). Most of them maintain certain common features, such as being small (12-50 aminoacids), containing positive charge and an amphipatic structure. Based on their structure, AMPs can be divided in several categories, but their functions are unknown in many cases, so they are similar only by their capacity to killing or inhibit the growth of microbes (Lai and Gallo, 2009).

Kolls et al. (2008) distinguishes the following AMPs as key antimicrobial products in epithelial-cell defence: classical cationic AMPs (defensins, cathelicidins), the best characterized and conserved across many phyla; S100 family proteins; elastase-inhibitors (elafin, SLPI); peptidoglycan recognition proteins (PGLYRPs); calcium-dependent lectins (C-type lectins) (collectins, MBL); iron metabolism proteins (lactoferrin, hepcidin, lipocalin), and chemokines.

Cationic AMPs bind to negatively charged structures on microbial surfaces. The best known mammalian AMPs are defensins and cathelicidins that function via membrane perturbation. Others act catalytically (phospholipase A2) and others deprive microorganisms of important nutrients (e.g. lactoferrin that binds iron). A number of these peptides are constitutively produced providing resident protection, whereas others are induced during infection and inflammation. In addition to directly eradicate microorganisms, several AMPs have been shown to be multifunctional with additional immunomodulatory functions, including acting as chemoattractants (reviewed in Dommett et al., 2005; Canny and Levy, 2008). Defensins can act as mediators of cytotoxic CD8+ lymphocytes (Ross and Pawlina, 2006). Among cationic AMPs, the bactericidal/permeability-increasing protein (BPI) is a polypeptide of neutrophil primary (azurofilic) granules, which is also expressed in human monocytes, eosinophils, fibroblasts and epithelial cells. BPI and its homologous at mucosal sites have potent antimicrobial and endotoxin-neutralising activities, and they are so potential therapeutics as anti-infective agents (reviewed in Canny and Levy, 2008).

Increasing numbers of AMPs have been identified from teleosts in recent years, most of them cationic peptides (see Plouffe et al., 2005; Mulero et al., 2008). They include oncorhycins, histone or histone-like molecules, pardaxins, misgurin, piscidins, pleurocidins, moronecidins, chrisophins and several hepcidin-related AMPs, and most of them were isolated or expressed in skin, gills and intestine. Two cathelicidin genes have been found in *Oncorhynchus mykiss* and *Salmo salar* (Chang et al., 2006). Three distinct transcripts of NK-lysine (a new AMP class recently discovered in mammalians with 78 amino acid residues) exist in channel catfish and they seem to be similarly arranged in zebra fish (Wang et al. 2006). New findings of hepcidins have been reported recently (Martin-Antonio et al., 2009; Cho et al., 2009). In addition, Zou et al. (2007), discovered β-defensin homologues in *Danio rerio*, *Takifugu rubripes* and *Tetraodon nigroviridis*, and Ullal et al. (2008) found expression of AMPS derived from haemoglobin in the epithelium of *Ictalurus punctatus*.

Teleost AMPs can be activated in response to infections, mainly bacteria, as hepcidins (Chen et al., 2007; Cuesta et al., 2008; Cho et al., 2009), cathelicidins (Maier et al., 2008) and β-defensin (Zou et al., 2007). Up-regulation of the haemoglobin-derived AMP was observed in the gill epithelium of channel catfish in response to the ciliate parasite *Ichthyophthyrius multifiliis* (Ullal et al. 2008). Piscidines, α-helical haemolytic AMPs were detected in EGCs (named as mast cells) and rodlet cells (see below, section 5.4) of nine Perciform species. Piscidin-immunoreactive cells were most common at sites of pathogen entry, including the skin gill and gastrointestinal tract (Silphaduang and Noga, 2001; Silphaduang et al., 2006). A role of piscidins in killing bacteria was suggested by the delivery of these AMPs to bacteria-containing phagosome of granulocytes upon phagocytosis (Mulero et al., 2008). Some histone-like proteins isolated from fish tissues may be important for defence as they are able to kill the dinoflagellate *Amyloodinium ocellatum* (Noga et al., 2001, 2002).

The skin of anuran amphibians is an extraordinarily rich source of AMPs, to the extent that a large part of the reported molecules has this origin. The first AMP was found in the skin of *Bombina variegata* some 30 years ago. The subsequent discovery of magainins in skin secretions of *Xenopus laevis* was decisive. Acting as wide-spectrum microbicides against a variety of bacteria, protozoa and fungi, amphibian peptides have stimulated increasing interest because of their unique characteristics and potential therapeutic usefulness (Rinaldi, 2002). According to Rollins-Smith et al. (2002), AMPs in the skin can operate as a first line of defence against the organisms associated with global amphibian declines, such as the pathogenic chytrid fungus *Batrachochytrium dendorabditis*.

Members of the S100 protein family have been implicated in different regulatory activities, such as protein phosphorylation, the dynamics of cytoskeleton constituents, Ca^{2+} homeostasis, enzyme activities, transcription factors, cell growth and differentiation, and in the inflammatory response (reviewed in Donato, 2003). A S100 protein has recently been identified in rodlet cells (RCs) (see below, section 5.4) and it is considered to be linked to the regulation of RC activity of *Psetta maxima* (Vigliano et al., 2009).

3.1.2. Lysozyme

Lysozyme is another important defence molecule of the innate immune system, playing a role in mediating protection against microbial invasion and occurring also at the mucosal surfaces. Lysozyme is widely distributed in bacteriophages, plants, invertebrates and vertebrates (Jolles and Jolles, 1984). It is a mucolytic enzyme produced by leucocytes, especially monocytes, MΦs and neutrophils and is found in tissues, body fluids and secretions, such as mucus and saliva. Fish lysozyme possesses lytic activity against bacteria and can activate complement and phagocytes. It is present in mucus, lymphoid tissue, plasma and other fluids and is also expressed in a wide variety of tissues Lysozyme activity has been detected in the intestine of several teleosts, such as *Katsuwonus pelamis*, *Seriola quinqueradiata*, *Oncorhynchus mykiss*, *Salmo salar*, *Plecoglossus altivelis*, *Anguilla japonica* and *Gadus morhua* (reviewed in Saurabh and Sahoo, 2008).

Table 2. Changes in acute phase proteins or related factors in response to piscine intestinal parasites. α-2M: α-2 macroglobulin. GRP-75: mortalin

Parasite	Fish	Effect	Reference
Myxozoa *Ceratomyxa shasta*	*Oncorhynchus spp.*	↑ serum lysozyme levels	Foott et al., 2004
Enteromyxum leei	*Sparus aurata*	↓ serum lysozyme levels ↑ expression α-2M and GRP-75 ↑ serum complement	Sitjà-Bobadilla et al., 2008 Cuesta et al. 2006c
E. leei	*Diplodus puntazzo*	↑ serum antiproteases	Muñoz et al. 2006
E. scophthalmi	*Psetta maxima*	Initial ↓ and further ↑ serum lysozyme levels ↑ serum antiproteases oscillation serum complement	Sitjà-Bobadilla et al., 2006

Available data on changes in lysozyme in piscine intestinal parasitoses are related to its content in serum (Table 2). An increase in serum lysozyme level was detected in fish infected by *Ceratomyxa shasta* (Foott et al., 2004), whereas a decrease followed by an increase occurred in *E. scophthalmi* exposed turbot (Sitjà-Bobadilla et al., 2006). The low serum lysozyme levels found in *D. puntazzo* could be involved in the higher susceptibility of this species to *E. leei* (Alvarez-Pellitero et al., 2008). In *S. aurata*, lysozyme was significantly depleted after chronic exposure to *E. leei* (Sitjà-Bobadilla et al., 2008).

3.1.3. Anti-proteases

Host protease inhibitors modulate protease activities and control a variety of critical protease-mediated processes, including the resistance to invasion by infectious agents. Thus, anti-proteases can interact with innate and adaptive mechanisms (Dalmo et al, 2007). Variations in antiproteases, mainly α-2-macroglobulin (α-2M) have been observed in several parasite infections (Table 2). Information on piscine intestinal parasites concerns the enteric *Enteromyxum* spp. In *E. scophthalmi*-exposed turbot and *E. leei*-exposed *D. puntazzo*, total antiproteases and α-2M levels oscillated after exposure, but they were generally higher than in control fish (Sitjà-Bobadilla et al., 2006; Muñoz et al., 2007). In the intestine of *Sparus aurata* chronically-exposed to *E. leei*, up-regulation of α-2M and the heat shock protein GRP-75 (mortalin) were observed using rtqPCR (Sitjà-Bobadilla et al., 2008).

3.1.4. Lectins

Lectins are proteins or glycoproteins derived from both animal and plant materials that recognize and reversibly bind to specific sugar moieties or certain glycosidic linkages of polysaccharides, glycoproteins and glycolipids (Sharon and Liss, 1989). The role of lectins as cell recognition molecules of microorganisms and of animal cells is now widely accepted. In mammals and other vertebrates, lectins are cell surface constituent strategically located to serve as recognition molecules in a variety of systems, whereas others perform intracellular functions. Thus, they control the biosynthesis of glycoproteins and their intracellular trafficking, participate in cell signaling, and play key roles in cell interactions in the immune system (Sharon, 2008).

Among the lectins making part of the innate immune system, the C-type lectins can act as PRR in the recognition of different PAMPs. Such lectins are characterized by C-type lectin receptors (CLRs), proteins that contain CH recognition domains (CRD) (Mc Greal et al., 2004). The best characterized of these C-type lectins is the mannose-binding lectin (MBL), which is present on phagocytes, plasma and mucosal surfaces. It plays an essential role as initiator of the primary immune response and participates in inflammation (Petersen et al., 2001, Turner, 2003; Klein, 2005; Arnold et al., 2006). Lectins have been found in different tissues of several fish species, though few are known from the intestine. The C-type lectin, galectin and pentraxin repertoires are highly diversified in teleosts, but ficolines and P-type lectins remain to be identified (Vasta et al., 2004; Russell and Lumsden, 2005). MBL or its homologous are present in teleosts (Vitved, 2003; Nikolakopoulou and Zarkadis, 2006) and can recognize bacterial pathogens (Ewart et al. 1999). This lectin was found in intestine and other organs of *Cyprinus carpio* and *Carassius auratus* and a role as possible component of complement system has been suggested (Vitved et al., 2003). The pufflectin, another mannose-binding lectin, is present in the mucosal tissues of skin and digestive tract. The galectin congerin, located at skin and mucosal tissues (Nakamura et al., 2001) can play immune functions in the Japanese conger eel (*Conger myriaster*) intestinal lumen (Nakamura et al., 2007).

Another interesting group of lectins are the intelectins, firstly identified in mouse as homologous of the *Xenopus laevis* oocyte lectin (Komiya et al., 1998). In mammals, intelectin is a Ca2+-dependent, D-galactosyl-specific lectin expressed in Paneth and goblet cells of the small intestine and proposed to have antimicrobial actions via their ability to bind a variety of non-mammalian sugars or to lactoferrin, as it is structurally identical to the intestinal lactoferrin receptor present in the enterocyte brush border. Intelectins are especially abundant in the brush border, but are also present in secretory granules (Wrackmeyer et al., 2006; reviewed in Nair et al., 2006; Artis and Grencis, 2008). Recently, intelectins have been also found in several fish species and some of them can interact with bacteria. The complete sequence of the intelectin gene has been reported for grasscarp, *Ctenopharyngodon idella* (Chang and Nie, 2007), and the lectin was also histochemically detected in several tissues, including intestine. In *Oncorhynchus mykiss*, intelectin was identified as one of the acute phase response genes (see Takano et al., 2008). In the same host, Russell et al. (2008) identified and cloned an intelectin that binds to GlcNAc and mannose. This intelectin recognizes bacteria and chitinous targets and is expressed in intestine and other tissues. Two intelectin genes were reported in channel catfish (*Ictalurus punctatus*) and blue catfish (*Ictalurus furcatus*), with catfish intelectin-2 being strongly induced following intraperitoneal injection with *Edwarsiella ictaluri* (Takano et al., 2008). Lin et al. (2009), have recently characterized seven zebrafish intelectins (zINTLs), of which zINTL2 was distributed predominantly in the gut and up-regulated upon *Aeromonas salmonicida* infection, whereas zINTL1, expressed in the gut and other organs, was down-regulated by this bacteria. A comparative analysis with other species showed that intelectin family may be a deuterostome specific gene family. Though their sequences are highly conserved, their expression patterns, quaternary structures and glycosylation vary considerably between species.

A new family of lectins specific for fucose (F-type lectins) was identified in a variety of taxa from prokaryotes to amphibians. This lectin family shares a characteristic sequence motif in the CRD and a novel structural fold. Studies on eel fucolectins demonstrated that they are secretory proteins with unique structures among lectins, exhibiting similarities to frog pentraxin (Honda et al. 2000). Other F-type lectins have been characterized in the serum of several fish species (see Vasta et al., 2004 and Cammarata et al., 2007). F-lectin analogues are also found in higher vertebrates and may have originated by CRD that later evolved new specificities (reviewed in Vasta et al., 2004).

The pentraxins C-reactive protein and serum amyloid P, as well as transferrin and trombin have been identified in teleosts (reviewed in Bayne and Gerwick, 2001). Some of them are involved in fish responses to parasites (reviewed in Alvarez-Pellitero, 2008a), but no information on intestinal parasites is available.

In amphibians, lectin repertoires are equally diversified, including C-type lectins, the oocyte cortical lectin XL35, ficolins, galectins and pentraxins, and the *Xenopus* spp. genomic databases have greatly contributed to their comprehensive analysis. Prototype galectins have been characterized in toads, frogs and salamanders (*Bufo arenarum*, *Rana catesbiana*, *Xenopus laevis* and *Ambystoma mexicanum*) (reviewed in Vasta et al., 2004).

Biochemical studies on lectins of reptile origin are not well advanced and most information is related with their presence in venoms, as a galectin from *Crotalus atrox* (Hirabayashi et al., 1991). Other galectins were purified from the liver of the lizard *Sceloporus spinosus* and are related to hepatic functions (Fenton et al., 2004).

Several transmembrane or soluble lectin domains in the host can interact with mannose, fucose or Gal structures in different parasites and can thus play a crucial role in the host/parasite interaction and invasion, and also in the immune evasion. The formation of mixed molecular receptor assemblages might further extend the PAMP profile recognition (Cambi and Figdor, 2005). MBL is known to be involved in the interaction and response to some parasites (Gruden-Movsesijan and Milosavljevic, 2006; Klabunde et al. 2000; Hokke and Yazdanbakhsh, 2005). Intelectins have been also proposed to have a role in the response to parasite infections. Intelectin-1 and intelectin-2 are up-regulated in mice infected with *Trichuris muris* and *Trichinella spiralis*, respectively (Pemberton et al., 2004; reviewed in Artis and Grencis, 2008). Lectins binding to O-linked glycoconjugates containing the Tn (GalNAcα1→Ser/Thr) or T [Gal(β1→3)GalNAcα1→Ser/Thr)] Ags mediate inhibition of sporozoite infectivity of *Cryptosporidium parvum in vitro* (Gut and Nelson, 1999). Similarly, Chen and LaRusso (2000) and Stein et al. (2006) demonstrated that Gal/GalNAc epitopes of glycoproteins on the epithelial apical membrane are involved in the attachment of *C. parvum* to cultured cells. Thus, the lectins recognising these CH residues could, potentially, lead to invasion-inhibiting treatments for cryptosporidiosis (Gut and Nelson, 1999). A clinical application of such use of lectins and lectin- carbohydrate binding is illustrated by the case of giardiasis. The lectin WGA blocks parasite growth by binding to GlcNAc residues on the surface membrane of *Giardia lamblia*, which raises the possibility of influence the course of giardiasis using dietary lectins (Ortega-Barria et al., 1994; Grant et al., 2001).

Lectins may also act as significant defensive factor in fish (Tsutsui et al., 2005), as specific lectin-carbohydrate interaction has been proven for several fish parasites. Specific binding of pufflectin to the monogenean *Heterobothrium okamotoi* has been reported (Tsutsui et al., 2003), but no data are known for intestinal parasites. Several CH residues have been detected by lectin histochemistry in the myxozoans *Enteromyxum scophthalmi* (Redondo et al., 2008) and *E. leei* (Redondo and Alvarez-Pellitero, 2009c). However, their specific involvement in host-parasite interactions remains to be demonstrated.

3.1.5. Cytokines, chemokines

Cytokines and chemokines make part of the complex components and pathways related to both innate and adaptive immune responses. The intestinal epithelial and endothelial cells of endotherms secrete several cytokines including IL-1, IL-2, IL-3, IL-4, IL-6, IL-8, IL-13, IFN-γ, and TNF. In addition, the cytokine family linked to the third lineage of regulatory T cells, Th-17 is specially related to the mucosal immunity and inflammation. This IL-17 family includes six members (IL-17A-F), IL-21, IL-22, IL-23. Although the focus of most studies on IL-17 production has been on CD4+δβ T cells, γδ T cells (see below, section 5.6.1) have also been shown to be a potent source of IL-17, which they express in the LP (reviewed in Roark et al., 2008). The linking of the IL-23/Th17 axis to chronic intestinal inflammation could have a beneficial role in intestinal protection and homeostasis (Maloy and Kullberg, 2008). IL-22 is one of the key cytokines that mediate the cross talk that enhance the innate responses of epithelial cells and restore the integrity of the tissue barrier. Thus, it may play a protective role in the gut (Ouyang and Valdez, 2008).

The main inflammatory cytokines TNFα, IL-1β and IL-6 elicited after injury, and those further released in the downstream cascade, are present in teleosts (reviewed in Secombes et al., 1999, 2001; Huising et al., 2004, Whyte et al.; 2007, Randelli et al., 2008; Alvarez-Pellitero, 2008a). Information on their presence in the intestine is more limited, though IL-17 and IL-22 are known from teleosts (Gunimaladevi et al., 2006; Igawa et al., 2006). In addition, some cytokine genes are expressed in fish in response to different stimuli. Genes of both pro-inflammatory (IL-1β and TNF-α1) and anti-inflammatory (IL-10 and TGF-β) cytokines were involved in the enteritis process produced by soybean meal (SBM) in *Cyprinus carpio* (Urán et al., 2008). TGF-β and IFN-γ inducible lysosomal thiol reductase (GILT) were downregulated in *Salmo salar* with SBM enteropathy (Lilleeng et al., 2009). Infection of *Oncorhynchus mykiss* with *Aeromonas salmonicida* induced up-regulation of proinflammatory cytokine expression (IL-1β, IL-8, TNF- α and IFN-γ) in the proximal intestine, whereas TGF-β was decreased in the distal intestine (Mulder et al., 2007). Isolated IECs (IELs were also present in the used IEC-rich fraction) from rainbow trout showed up-regulation of IL-1β and TNF-α2 after stimulation with live *Aeromonas salmonicida*. In addition, TNF was detected by immunohistochemistry in IECs of trout intra-anally challenged with live bacteria. Killed bacteria could not up-regulate the expression of these cytokines, and thus the effect is caused by the adhesion of the bacteria (Komatsu et al., 2009).

Few data on the effect of intestinal parasitoses on cytokine expression in ectothermic vertebrates are available (Table 3). In the intestine of *S. aurata* chronically-exposed to *E. leei*, a down-regulation of IL-1β and TNF-α were observed using rtqPCR, but no significant changes were seen in head kidney (HK) (Sitjà-Bobadilla et al., 2008). However, up-regulation of IL-1β was detected in HK soon after exposure, using semiquantitative PCR (Cuesta et al., 2006a).

3.1.6. Complement

Complement is amongst the main mechanisms involved in the initiation of the innate response and further mounting of an adaptive response. Complement includes a combination of three pathways, the alternative, lectin and classical. The classical pathway is initiated by a complex between an Ag and an antibody (Ab). It is triggered by binding of the Fc portion of the IgG to the C1q component of the C1 complex. In the alternative pathway, the spontaneous activation of C3 is amplified upon the covalent binding of C3(H_2O) to various microbial surfaces (Boshra and Sunyer, 2006). The lectin pathway requires the interaction of lectins such as MBL and ficolins with sugar moieties found on the surface of microbes (Fujita et al., 2004). Among other functions in innate response and in its connection with adaptive response, complement plays important roles in the killing of pathogens, through opsonization and activation of phagocytes, and in inflammation.

The complement cascade is part of the phyllogenetically ancient innate immune response and is crucial to our natural ability to ward off infection (Gasque, 2004). With the emergence of Igs in cartilaginous fish, the classical and lytic pathways first appeared. The rest of piokiloterm species, from teleosts to reptilians, seem to contain a well developed complement system resembling that of homoeothermic vertebrates. However, some differences exist. Several species of poikiloterms possess multiple forms of complement components (C3 and factor B) that are structurally and functionally more diverse than those of higher vertebrates. In teleosts, the genes encoding the component C3 can have several isoforms in a single species (reviewed in Sunyer and Lambris, 1998; Holland and Lambris, 2002; Boshra and Sunyer, 2006; Whyte, 2007).

Table 3. Changes in cytokines in piscine intestine (intest.) or head kidney (HK) in response to food stimulus or intestinal pathogens. SBM: soybean meal. GILT: IFN- γ inducible lysosomal thiol reductase. IECs: intestinal epithelial cells; IELs: intraepithelial lymphocytes. IHC: immunohistochemistry.

Food content or pathogen	Fish	Effect	Reference
SBM	*Cyprinus carpio*	Enteritis involving IL-10, IL-1β, TNF-α1 and TGF-β genes	Urán et al., 2008
	Salmo salar	↓ TGF-β and GILT expression	Lilleeng et al., 2009
Bacteria *Aeromonas salmonicida*	*Oncorhynchus mykiss*	↑ IL-1β, TNF-α1 and IFN-γ expression in proximal intest. ; ↓ TGF-β expression in distal intest.	Mulder et al., 2007
		↑ IL-1β and TNF-α2 expression in isolated IECs (and possibly IELs) TNF detection by IHC in IECs after intra-anal challenge	Komatsu et al., 2009
Myxozoa *Enteromyxum leei*	*Sparus aurata*	↓ IL-1β and TNF-α1 expression in chronic infection	Sitjà-Bobadilla et al., 2008
		HK: initial↑ IL-1β expression	Cuesta et al., 2006a

Most data on the implication of complement in piscine parasitic diseases were obtained by measuring the alternative activity in serum (Table 2). In the case of intestinal parasites, such alternative activity increased in serum of *Sparus aurata* exposed to *E. leei* (Cuesta et al., 2006a) and oscillated over the post-exposure period in turbot exposed to *E. scophathlmi* (Sitjà-Bobadilla et al., 2006). Variations in gene expression of some complement factors of teleosts in response to several parasites have been reported, but no data are available for intestinal parasites (reviewed in Alvarez-Pellitero, 2008a).

4. Inflammation and Some Related Pathways

4.1. Inflammation

A combination of innate mechanisms (with the contribution of some adaptive ones) leads to an inflammatory reaction, in which TLRs and some NLRs have essential roles. TLRs activate tissue-resident macrophages (MΦs) to produce inflammatory cytokines, which activate hepatocytes to produce acute phase proteins (APRs) (collectins, pentraxins). Such APRs, in turn, activate complement and opsonize pathogens for phagocytosis by MΦs and neutrophils. Besides such indirect induction of antimicrobial response, TLRs also directly trigger such a response by inducing MΦs to produce AMPs and nitric-oxide synthase and also inducing other oxidative mechanisms (reviewed in Medzhitov, 2007). This connection between TLR and inflammation has special relevance in the mucosal intestinal tissue, in which TLR expression is carefully regulated to shut down a proinflammatory response to commensal organisms. Among NLRs, NOD1 and NOD2 act as bacterial detectors. Members of NOD-LRR members can interact to increase inflammatory and apoptotic responses, or counterbalance TLR activity (Sanderson and Walker, 2007). Important interactions also occur between TLRs and certain NLRs for inducing the pro-inflammatory IL-1β, and further priming of a multisubunit complex called the "inflammasome", in which caspases are activated to respond to bacterial products and products of damaged cells (Creagh and O'Neill, 2006; Pétrilli et al, 2007). Recently, Eisenbarth and Flavell (2009) reviewed aspects of the innate-adaptive immunity cross talk and the role NLR-based activation in the inflammasome.

4.2. Phagocytes and Oxidative Mechanisms

Phagocytosis is one of the main mechanisms involved in the host protective responses leading to the clearance of pathogens. As in mammals, the leucocytes involved in phagocytosis in teleosts include mainly neutrophils, acidophilic granulocytes and monocyte-macrophages (Dalmo et al., 1997; Sepulcre et al., 2002). Inducible antimicrobial responses such as the respiratory burst (RB) and nitric oxide (NO) have been demonstrated in fish phagocytes, and display biochemical and physiological similarities to homologous responses induced in mammalian phagocytes (Neumann et al., 2001; Whyte, 2007). Phagocytic response as well as the role of phagocytosis and other indices of phagocyte activity, i.e. oxidative mechanisms, as defence mechanisms for the elimination of piscine parasites has frequently been reported (reviewed in Alvarez-Pellitero, 2008a). The available information on piscine intestinal parasites is presented below (section 5.7.2)

4.3. Apoptosis and Anoikis

Apoptosis or programmed cell death is crucial for maintaining social order among the cells comprising metazoans and is also an important effector mechanism of innate and adaptive host responses to pathogens (Schaumburg et al., 2006). Apoptotic processes can be divided into intrinsic and extrinsic pathways. Different factors and routes can participate in the induction of apoptosis. Detailed descriptions of the apoptotic pathways are presented in the reviews of James and Green (2004) and Schaumburg et al. (2006). Reference to factors involved in apoptosis thus far identified in teleosts can be found in Alvarez-Pellitero (2008a).

The caspase pathway is one of the main routes involved in apoptosis or programmed cell death (Schaumburg et al., 2006). Several caspases have been characterized in teleosts, including pro-inflammatory caspases, such as caspase-1 (López-Castejón et al., 2008) and caspases related to apoptosis, such as caspase-9 (Reis et al., 2007a) or caspase-3 (Reis et al., 2007b). Members of the TLR family can also induce apoptosis (Salaun et al., 2007) (see above, section 2.3.3).

One of the main sites in which apoptosis can occur is the intestine, with involvement of different elements of innate and adaptive immunity and in relation to inflammatory response. An effective immune response depends on the proper balance between proliferation and death of activated T cells. This is particularly important in the intestine, where mucosal T cells are subjected to a high antigenic pressure. Apoptosis can proceed via an active or passive form. Passive cell death is Ag-independent, whereas active apoptosis requires engagement of T cell receptor (TCR), triggering of death receptors, and activation of a proteolytic cascade initiating with caspase 8. In the human intestine, the LP T cells play a dual role in providing protection against pathogens, dietary and enteric flora Ags while limiting its potentially damaging immune response. This daunting task of adaptive immunity is complemented by that of other cells (MΦs, eosinophils, mast cells, and NK cells) mediating innate immunity. This establishes a controlled inflammatory status termed "physiological intestinal inflammation" (reviewed in Levine and Fiocchi, 2001). Apoptosis is differentially regulated in the intestine by the type and quantity of absorbed Ags as well as the characteristics of local T cells. DCs play a central role in the regulation of immune responses by processing and presenting Ags to naïve cells (see below, section 5.3.1). It has been proposed that after the initial interaction between DCs and T cells, T-cell-induced DC apoptosis serves as a down-regulatory mechanism that prevents the otherwise continuous activation of T cells. The reciprocal regulation between DCs and T cells in the gastrointestinal immune system is under strict control compared to the systemic immune system and could contribute to the generation of fine balanced responses to intestinal Ags (Nicoletti et al., 2006). Thus, defective apoptosis of lamina propria T cells is a component of chronic intestinal inflammation (reviewed in Levine and Fiocchi, 2001).

Multiple lines of evidenced indicate that necrosis, the other cell death process, can be a programmed event, termed necroptosis (Galluzi and Kroemer, 2008). Hitomi et al. (2008) have elucidated the connection between necrosis and apoptosis and implicated necroptosis as a critical regulatory pathway for innate immunity. The molecular signaling pathway is dependent on the RIP1 kinase and is activated in apoptotic-deficient conditions. Caspase-8 mediated degradation of RIP1 may represent one of the major molecular switches between apoptosis and necrosis (Galluzi and Kroemer, 2008). The relevance of this relationship in vivo and the major therapeutic implications, including the intestinal level, remain to be elucidated.

Apoptosis has also been described in ectothermic vertebrates. The factors involved in apoptosis in teleosts were reviewed in Alvarez-Pellitero (2008a). In amphibians, Ducoroy et al. (1999) reported apoptosis as a natural current process during lymphocyte development in thymus and spleen of the urodele axolotl (*Ambitoma mexicanum*). Natural apoptosis can be amplified by superantigen (*Staphylococcus aureus* enterotoxin B).

A special type of apoptosis termed anoikis can occur in the epithelia. Anoikis is a Greek word meaning "homelessness" that is used to designate apoptosis induced by loss of cell adhesion or inappropriate cell adhesion. Anoikis is in fact apoptosis with its characteristic features, and constitutes an important mechanism for preventing detached cells surviving in inappropriate places and growing displastically. Epithelial cells are particularly sensitive to anoikis, which induces their detachment and this situation occurs frequently in the intestine, in occasions in response to pathogens. The cell-extracellular matrix complexes depending on integrins are involved in anoikis (reviewed in Valentijn et al., 2004), but other mechanisms related to the disassembly of adherens junctions have recently been reported (see Suzanne and Steller, 2009).

Apoptosis plays an important role in the host response to infections. Parasitic protozoa and helminths are able to modulate host apoptotic pathways to survive within their hosts, and hence apoptosis is an important mechanism for immune evasion (James and Green, 2004). The intestinal *Cryptosporidium parvum* induces apoptosis in cultured human epithelial cells (McCole et al., 2000). The involvement of caspase activation was demonstrated, as well as the ability of *C. parvum* to suppress the apoptotic response via up-regulation of survivin (Liu et al., 2008). Apoptosis is also involved in the response of fish to some parasites (reviewed in Alvarez-Pellitero, 2008a), but few data are available on intestinal pathogens. The epithelial form of apoptosis (anoikis) has also been described. Detachment-induced apoptosis of enterocytes was observed in sea bass (*Dicentrarchus labrax*) suffering a septicemia by *Photobacterium damselae* subsp. *piscicida*, or following injection of its exotoxin. Linings of enterocytes detached from intestine mucosa dissociate into isolated enterocytes which undergo caspase 3-mediated anoikis. The enterocyte anoikis terminates by secondary necrosis and lysis due to the lack of elimination by phagocytosis of apoptosing enterocytes (do Vale et al., 2007). Anoikis seems also occur in infections by the myxozoan *E. scophtahlmi* (see below, section 5.7.2.3)

Inflammatory reactions involving cellular responses either accompanied by the formation of granulomas or not, have been reported for numerous piscine parasites belonging to different groups (reviewed in Alvarez-Pellitero et al., 2008a), including intestinal parasites. The available information on inflammmatory/apoptotic responses to intestinal parasites of ectotherms is presented in the section 5.7.2, in which the cellular reaction to these infections is described.

5. Cellular Components in the Intestinal Mucosal System

5.1. The Structure of Intestine

The mammalian intestine is covered by columnar epithelia whose surface has unique projections, villi, which optimize interaction between the contents of the lumen and the cellular binding. Intestinal glands, the crypts of Lieberkühn, open at the base of the villi. Stem cells are located at the base of the glands (Cross and Mercer, 1993). Mature cells in the mucosa of the small intestine include enterocytes, goblet cells, Paneth cells, enteroendocrine cells and M cells. Different types of leucocytes are also present in the intestinal epithelium (Ross and Pawlina, 2006). In addition, almost all activities involved in the vertebrate gut function are mediated by the enteric nervous system, which controls the motility, exocrine and endocrine secretions, and microcirculation of the gastrointestinal tract; it is also involved in regulating immune and inflammatory processes (reviewed in Goyal and Hirano, 1996). A loose connective tissue layer, the LP underlies the epithelial cells. This layer contains the blood and lymph vessels that collect processed nutrients, and many cells that protect this region from foreign invasion from the lumen (Cross and Mercer, 1993).

The main elements of the mammalian mucosal immune system were depicted by Brandtzaeg et al. (2008), which also recommended a terminology and abbreviations. The different components of mucosal immunity include inductive and effector tissues, and effector cells and molecules, all making part of both innate and adaptive immunity and their connections. The inductive sites for mucosal immunity are constituted by organized mucosa-associated lymphoid tissue (MALT), as well as local/regional mucosa-draining lymph nodes (LNs). MALT is subdivided according to anatomical regions. In mammals, the gut-associated lymphoid tissue (GALT) comprises Peyer's patches (PP) and the isolated lymphoid follicles (ILFs). PP in the small intestine are believed to be a main source of conventional (B2) surface

(s) IgA-expressing primed (memory/effector) and class-switched mucosal B cells. PP also contain M cells, located at the FAE. The role of these tissues is in the induction phase of intestinal immune responses, as they represent sites where Ags are taken and presented to B and T lymphocytes. The effector sites of the intestine are the mucosal epithelia and the underlying LP. Many different cell types are present in these effector sites in the normal intestine, including activated T cells, plasma cells, mast cells, DCs and MΦs (Platt and Mowat, 2008; Brandtzaeg et al., 2008).

Lymphoid aggregates occur in the LP of Chondrichthyes. In teleosts, the GALT consists principally of different sized lymphocytes, plasma cells and MΦs, as well as several types of granulocytes, PAS positive cells and eosinophil granular cells (Zapata and Amemiya, 2000). Gut intraepithelial lymphocytes (IELs) are largely considered T cells, whereas lymphoid cells in the LP are mainly B lymphocytes (Zapata et al., 2006). GALT in fish lacks PP and M cells (Buddington et al., 1997), however, some evidence exists that the distal intestine contains functional analogues of mammalian M cells (see below, section 5.2). The fish hindgut seems to have an inducing function, as all cells necessary for the mucosal immune response are present in this second segment: B cells, T cells, plasma cells and Ig-binding and Ag-presenting MΦs (Rombout et al., 1993a). MΦ aggregates, also known as melano-macrophage centres (MMCs) are present in HK and spleen of many fish. In higher teleosts MMCs often exist as complex discrete centres, containing lymphocytes and MΦs (Agius and Roberts, 2003). MMCs may also develop in association with chronic inflammatory lesions in other organs, such as the intestinal LP. It has been hypothesized that teleost MMCs play a role analogous to the germinal centres of LNs in mammals. This hypothesis was confirmed by the labelling of free melano-macrophages and MMCs from the kidney and spleen of three teleost species by CNA-42, an Ab usually employed for labelling follicular DCs of higher vertebrates (Vigliano et al. 2006). Whether or not the intestinal MMCs could play a role similar to that of LNs in mammals, remains to be elucidated.

In amphibians, lymphoid aggregates appear also in the gut. In anurans, development of GALT is greater than in urodeles and lymphoid tissue occurs throughout the gut from the small intestine to cloaca. Several reptile species contain numerous densely cell populated lymphoid aggregates throughout the gut. In fact, the development of reptilian GALT is greater than in other ectothermic vertebrates (reviewed in Zapata and Amemiya, 2000).

5.2. Role of Intestinal Epithelial Cells (IECs) in Immune Responses

IECs exhibit numerous physical and biochemical adaptations to maintain barrier function, including expression of elaborate tight junctions, actin-rich microvillar extensions, and secretion of mucin and polysaccharide-rich glycocalyx. IECs can express a battery of AMPs (see above section 3.1.1), including defensins, cathelicidins and calprotectins (reviewed in Artis and Grencis, 2008). In addition to barrier function, IECs can influence innate and adaptive immune response. They express PRRs, including TLRs and NODs (see above section 2.3) that enable microbial recognition. Besides the involvement of TLRs, there are other mechanisms whereby IECs act as a sentinel for mucosal immune responses. Ligation of TLRs and NODs results in activation of innate responses, including the induction of expression of proinflammatory cytokines and chemokines. A wide array of cytokines has been shown to be secreted by IECs, including IL-8, IL-7, IL-15, granulocyte-macrophage colony-stimulating factor, IL-6 and transforming growth factor-β (TGF- β). These cytokines can play inflammatory or regulatory roles, depending upon the local conditions (Dahan et al., 2007).

IECs can control the uptake, transmission and presentation of luminal Ags through a diverse set of pathways (Shao et al., 2001). In addition to the transfer of information by IECs to MALT indirectly via the secretion of chemokines, IECs can transmit information from luminal microorganisms and Ags by direct processing and presenting them to Ag-specific lymphocytes. It is well documented that IECs can act as non-professional Ag-presenting cells (APCs), as they express major histocompatibility complex (MHC) class I and class II molecules and all the machinery required for Ag processing and presentation, though they lack de classical co-stimulatory molecules on the surface (Hershberg and Mayer, 2000; Dahan et al., 2007). Epithelial Ag processing occurs in a highly polarized fashion with apical Ags being stored, processed and presented exclusively basolaterally (reviewed in Dahan et al., 2007). In addition, activation of NF-κB expression in IECs enhances the production of factors that activate DCs (reviewed in Coombes and Powrie, 2008).

Another cellular type acting as a principal site for the sampling of gut luminal Ags in mammals is the highly specialized M cell. M cells locate at the FAE of PP and are characterized by an irregular brush border and reduced glycocalyx. Thus, they facilitate the transport of macromolecules to the underlying lymphoid tissue, which contains several types of professional APCs, including MΦs, DCs, and B cells. Up to now, M cells have not been found in fish. However, enterocytes of the distal segment of carp gut seem to be specialized in the uptake and processing of Ags from the lumen and their transport to lymphoid cells and MΦs present in the epithelium and LP; hence they may be functional analogues of mammalian M cells (Rombout et al., 1989, 1993a). The finding of intestinal villous UEA-1 (*Ulex europeus* agglutinin-1)$^{+}$ cells, - which serve as an Ag gateway for the sampling of gut bacteria and subsequent induction of Ag-specific responses in a PP-independent manner - (Jang et al., 2004), support that M-like type cells could also occur in fish, in spite of lacking PP.

5.3. Antigen Presenting Cells and major histocompatibility complex (MHC)

5.3.1. Macrophages and dendritic cells (DCs)

The intestine and associated lymphoid tissues are home to an extensive network of innate immune cells with Ag-presenting function, including MΦs, conventional CD11 c^{hi} DCs and plasmacytoid DCs (pDCs) (a subset of DCs with microscopic appearance of plasmablasts) (Coombes y Powrie, 2008). They play an important role in the recognition of pathogens by acting as APCs.

In mammals, MΦ are located in the LP just below the epithelium and in the subepithelial dome region of PP. Murine and human intestinal MΦ express a number of characteristic markers similar or different to those found in other MΦ populations. Ag acquisition could occur through breaches of intestinal barrier or indirectly, following uptake and transfer from IECs. Although intestinal MΦ are highly phagocytic with prominent phagocytic vacuoles, secondary lysosomes and pseudopodia, as well as expressing some class II MHC, they express only low levels of co-stimulatory molecules such as CD40, CD80 and CD86. Thus, intestinal MΦ appear to be held in a state of partial activation, in which they can clear large numbers of bacteria without provoking inflammation, playing thus a major role to maintain homeostasis (see also above the role of TLRs, section 2.3.1). MΦ recognize pathogens through PRRs, mainly TLRs. At difference with their counterparts in other tissues, intestinal MΦ are unable to respond fully after recognition of microbes via TLR and fail to produce some inflammatory cytokines (reviewed in Platt and Mowat, 2008).

DCs, which are known to be unique in their ability to bind and activate naïve T cells, are prominently localized to the mucosal surfaces, both at sites of Ag uptake and within inductive lymphoid tissues. Mucosal DCs, through their ability to orchestrate protective immunity and immune tolerance in the host, have a key role in shaping the intestinal immune response, and the decision between tolerance and immunity in the intestine (reviewed in Coombes and Powrie, 2008). In addition, subpopulations of mucosal DCs have unique functions when compared to DCs from non-mucosal sites, both regards to the induction of regulatory T cells in a resting state and in the processing and presentation of pathogens following mucosal infection. DC subsets, primarily defined by the expression of cell surface markers, localize to discrete regions within primary inductive tissues, like PP and mesenteric LNs and effector sites, like the LP. In these tissues, they interact with T and B cells to differing degrees and may have unique roles in driving regulatory, effector and memory T cell responses (reviewed in Johansson and Kelsall, 2005). DCs and MΦs are also involved in inflammation/apoptosis (Nicoletti et al, 2006) (see above section 4).

APCs are less studied in ectothermic vertebrates. Ag-presenting MΦs are present in the teleost intestine (Rombout et al., 1993a). In fugu, neutrophils and monocyte/macrophages from an inflammatory site expressed CD8α, similar to mammalian DCs (Araki et al., 2008). In the amphibian *Rana pipiens*, DCs of the epidermis nictitant membrane and cornea showed co-expression of ATPase and MHC class II molecules (Castell et al., 1999). Evidence for the expression of non-specific esterase reactivity in such cells led to Castell-Rodríguez et al. (2001) to consider them as the amphibian counterpart of Langerhans cells.

5.3.2. Major histocompatibility complex (MHC)

The major histocompatibility (MH) receptors are immunoglobulin superfamily member proteins that interact with T cells through a specific TCR in order to initiate immune responses. Recent discoveries have demonstrated the role of MH receptors in the integration of innate and adaptive responses (reviewed in Dixon and Steht, 2001). In most vertebrates the MHC comprises several loci which fall into three subgroups, class I, II and III. Proteins encoded by the MHC class I and II loci form complexes with antigenic peptides and present them at the cell surface to the T cells (usually class I for intracellular pathogens and class II for extracellular pathogens). The class III region encompasses may genes and although some of them are involved in immunity, they are not directly involved in the presentation of Ags (Campbell and Trowsdale, 1993). The class I gene family is divided into the classical class Ia genes and the non-classical Ib genes. Class Ia genes are very polymorphic, with most of the polymorphisms occurring within the peptide binding region, which is the area that recognizes Ags. Class Ib genes have limited or no polymorphism and their function remains largely unknown (reviewed in Flajnik and Kasahara, 2001). Both types of MH receptors, class I and class II receptors, are present in teleosts, and polymorphism has been found in some species (Chistiakov et al., 2007). MHCs have been characterized in some amphibians and differences between species as well as changes in the expression of class I and class II genes occurring in metamorphosis have been found (Flajnik and Kasahara, 2001; Teacher et al., 2009)). Some information has recently been obtained on the MHC of reptiles. MHCI genes in *Sphenodon* sp. are highly polymorphic and highly divergent from other vertebrate MHC genes (Miller et al., 2006). MHC of non classical class I class II have been characterized from marine iguana *Amblyrhynchus cristatus* (Glaberman et al., 2008, 2009).

To the best of our knowledge, scarce information is available for MHC in ectotherms in relation to intestinal immunity or intestinal parasites, though some data on expression in certain tissues or cells have been published. Granulocytes of *S. aurata* express MHC II genes, suggesting a role as APCs (Cuesta et al., 2006b). A highly regulated expression of MH II in *S. salar* inoculated with *Neoparamoeba* sp. was suggested by the numerous positive cells detected within amoebic gill disease lesions, and thus such cells could contribute to Ag presentation (Morrison et al., 2006). Association of disease resistance to MHC polymorphism has been suggested for amoebic gill disease in salmonids (Wynne et al., 2007) and *Aeromonas salmonicida* infections in *Xenopus laevis* tadpoles (Barribeau et al., 2008). Wegner et al. (2003) found a consistent relationship between parasite diversity among different habitats and MHC diversity in wild *Gasterosteus aculeatus*.

5.4. Eosinophilic Granule Cells, Mast Cells, Rodlet Cells

In mammals, eosinophils and mast cells are mainly tissular, and may be abundant in the intestine. Eosinophils are characterized by large specific granules containing several cationic proteins. It is likely that the major function of eosinophils is the destruction of parasite helminths. Eosinophils possess Fc receptors that bind to Ab-coated parasites. Once bound, eosinophils degranulate and release their contents to the worm cuticula, including products of the RB (Tizard, 1995). Mast cells are large cells with conspicuous granules distributed throughout the body in connective tissues. Mast cells are a heterogeneous population that can adapt to changes in the environment and thus they can have different staining properties. They contain histamine, proteases and proteoglycans, whereas leukotrienes and cytokines are produced at activation. Mast cells are mainly activated via the Ab IgE, which is formed by plasma cells during initial exposure to certain Ags, and binds (by the Fc portion) to receptors on the mast cell. When mast cells are activated, they release the contents of their granules thus attracting eosinophils to the sites where a more effective defence is required (Cross and Mercer, 1993). Mast cells make innumerable contributions to both protective and pathologic immune responses. They can modulate the DC migration and maturation, can interact with T cells and are a normal component of most adaptive immune responses, thus illustrating the now established concept that innate and adaptive immunity are not complete independent entities (Sayed and Brown, 2007).

Other granular intestinal cells are the Paneth cells, located at the base of crypts of Lieberkühn in the small intestine of many mammalian species. They are recognized by the large apical secretory granules that are rich in cryptdins (isoforms of α-defensins) and other AMPs, such as lysozyme (Ouellette, 1999). Paneth cells can also produce Igs and are capable of phagocitaze certain bacteria and protozoans (Ross and Pawlina, 2006).

Mast cells or their equivalents have been reported in all classes of vertebrates, including ectothermic vertebrates (see Mulero et al., 2007a). However, the terminology and nature of fish eosinophilic/mast cells have been controversial and still remain confusing. The term eosinophilic granule cells (EGCs) was introduced by Roberts et al. (1971) to designate mononuclear eosinophilic granule-containing cells distributed in the connective tissues of various teleosts. EGCs are common in the intestine of salmonids and other fish. They constitute a heterogeneous cell population, with marked staining diversity and cytoplasmic granules described as either basophilic or eosinophilic (Reite, 1998; Reite and Evensen, 2006; Rocha and Chiarini-Garcia, 2007). Thus, EGCs have been suggested to be mast cell analogous or equivalent (Reite, 1998), similar to intestinal Panneth cells (Sveinbjorsson et al., 1996) or representing a cell type originating from the evolutionary precursors of both Panneth cells and mast cells (Paulsen et al., 2001). Important differences can also occur between fish species, and thus, the characterization of these cells using specific markers is needed (Sveinbjorsson et al.; 1996, Reite and Evensen, 2006; Leknes, 2007). Some piscine EGCs produce lysozyme (Sveinbjorsson et al., 1996) or AMPS, such as piscidins (Silphaduang and Noga, 2001; Silphaduang et al., 2006) (see above section 3.1). Recently, Mulero et al. (2007a) found histamine in Perciform fish cells containing eosinophilic granules, though not showing the metachromasia staining characteristics observed with toluidine blue in the mammalian mast cells. These authors referred to these granulocytes, present at the connective tissue of the gills and intestine, as mast cells. These cells were not stained by the G7 mAb, which is specific to *Sparus aurata* acidophilic granulocytes (AGs) (Sepulcre et al., 2002). In contrast, such AGs, which were also present in the intestine, did not contain histamine (Mulero et al., 2007a). During the ontogeny of lymphomyeloid organs, these AGs appeared earlier in the posterior intestine and in the blood than in the kidney, suggesting an important role in the immune surveillance of the gut (Mulero et al., 2007b). In the LP of *Oncorhynchus mykiss*, besides EGCs, another cell type with characteristics of mast cells was reported (Sire and Vernier, 1995). Grove et al. (2006) described different leucocyte populations in the LP of the posterior intestine of halibut (*Hippoglossus hippoglossus*), including non-specific esterase+ and acid phosphatase+ MΦs but no further information on their characteristics was provided. Intestinal piscine EGCs can participate in the immune response. In *Oncorhynchus mykiss*, they can internalize heterologous proteins when these are administered anally (Dorin et al., 1993). In *Cyprinus carpio*, the contribution of basophilic granulocytes (morphologically homologous of salmon EGCs) and IELs (mainly T cells) to the enteritis produced by soybean meal was demonstrated (Urán et al., 2008).

Another enigmatic cellular type in teleosts is the rodlet cell (RC), commonly associated with the epithelial tissues of virtually every fish species with a high variability in abundance and distribution. The function of RCs has been controversial, and they were even initially considered to be parasites. However, recent findings have confirmed their endogenous origin and it is thought that they are a type of inflammatory cell akin to other piscine inflammatory cells (reviewed in Manera and Dezfuli, 2004; Reite, 1998; Reite and Evensen, 2006). Their early appearance during carp ontogeny supports this hypothesis (Mazon et al 2006). Scarce information in relation to the precise RCs function is available. Using lectin histochemistry, several terminal CH residues have been characterized, though differences between fish species occur (Imagawa et al., 1990; Redondo et al., 2008; Redondo and Alvarez-Pellitero, 2009a,b). AMPs, such as piscidins, have also been detected in the RCs of some fish (Silphaduang et al., 2006). Vigliano et al. (2009) found expression of the S100 protein in the RCs of *Psetta maxima* in association with the RC development. Their findings suggest the probable relation of S100 with the release of rodlets due to the contractile capability of the fibrous layer, in which such protein was identified (see above section 3.1.1). Some parasites, including protozoans, myxozoans and helminths, seem to induce the recruitment of RC or changes in their abundance and distribution, as well as the releasing of rodlets (Manera and Dezfuli, 2004). The RC changes produced by piscine intestinal parasites are commented below (section 5.7.2).

5.5. Neuroendocrine Cells

Several lines of evidence support a possible role for enteric nervous system in the pathogenesis of inflammatory bowel disease (Goyal and Hirano, 1996). It is also known that such system could coordinate a multi-system response aimed at eradication of parasites, and subsequent intestinal recovery (McKay and Fairwather, 1997). The available information on the changes of neuromodualtors in fish infected by intestinal parasites is presented below (section 5.7.2).

5.6. Conventional and Innate-like Lymphocytes

Adaptive immunity arose early in vertebrate evolution, between the divergence of cyclostomes (lampreys) and cartilaginous fish (sharks). All jawed vertebrates possess the genetic elements essential for the function of the adaptive/combinatorial immune response (Marchalonis et al. 2006). The combinatorial immune system consists of Ag-recognizing lymphocytes, immunoglobulins (Abs and Ig-family TCR), MHC products, and recombination-activating (RAG) 1 & 2 genes. The overall shape of the molecules and the recombination mechanisms that create junctional diversity in TCRs and Igs are similar in fish and mammals (Du Pasquier, 2001).

According to Medzhitov (2007), there are two types of lymphocytes that express Ag receptors: conventional lymphocytes and innate-like lymphocytes, that is B1 cells, marginal-zone B cells, natural-killer T cells and subsets of γδ T cells. For these innate-like lymphocytes, the diversity of Ag receptors is restricted and not entirely random. In fact, B1 cells can be activated directly by PRRs and are programmed to produce Abs with a wide specificity for common bacterial Ags. Innate T-like cells recognize microbial Ags presented by non-classical MHC molecules.

In the case of conventional lymphocytes (mostly αβ T cells and B2 cells), Ag receptors are essentially assembled randomly (not predetermined). T and B types of conventional lymphocytes have been demonstrated in teleosts, although analysis of lymphocyte subsets is only just beginning in fish (Fischer et al. 2006).

5.6.1. T cells and T cell receptors (TCRs)

In vertebrates, the T cell has a membrane bound TCR that is responsible for Ag recognition in T-cell mediated immune responses. There are two types of T-cell populations based upon their heterodimeric receptors (αβ or γδ). The receptor polypeptide chains are derived from variable (V), joining (J), diversity (D) and constant (C) gene segments. Two types of conventional αβT cells are recognised: T helper (Th) cells, which are marked by the co-receptor CD4 on the cell surface; and cytotoxic T cells, which express CD8. These cells recognize antigenic peptides bound to the MHC class II and class I molecules, respectively.

The effector Th cells are CD4+ and can be divided into three main types. Th-1 cells secrete IFN-γ and TNF-β, which allows these cells to be particularly effective in protecting against intracellular infections. Th2 cells secrete IL-4, IL-5, IL-10 and IL-13 and activate B cells, which are adapted for defence against parasites that are vulnerable to IL-4 switched IgE production, IL-5-induced eosinophilia, and IL-3 and IL-4-stimulated mast cell proliferation and degranulation. A third subset of CD4 effector cells, known as Th17 cells, was recently discovered. These cells secrete ILs of the IL-17 group (reviewed in Kaiko et al., 2007) (see above, section 3.1.5). Another non-effector cell type is the regulatory T (Treg) cell. Treg cells secrete IL-10 and TGF-β, which modulate helper T-cell activity and suppress some of their functions. Specialized subsets of Treg contribute to the regulatory network that control parasitic infections (Belkaid et al. 2006). In teleosts, besides the better known Th1 cytokines, several cytokines related to Th17 activity (see above, section 3.1.5) and possibly with Th2 responses (reviewed in Alvarez-Pellitero, 2008) have been identified in fish. In addition, IL-10 and TGF-β, both secreted by Treg cells, are expressed in the intestine of some teleosts (see above, section 3.1.5) and could be involved in the recovery of inflammation (Mulder et al., 2007, Urán et al., 2008).

In mammals, the intestinal epithelium and underlying LP contains a large number of T cells localized in the PPs and solitary isolated lymphoid structures, or diffusely throughout the intestinal LP and overlying single layered epithelium. LP lymphocytes are primarily conventional $CD4^+TCR\ \alpha\beta^+$ or $CD8\alpha\beta^+\ TCR\ \alpha\beta^+$ T cells that display a previously activated or memory phenotype and enter the intestinal mucosa subsequently to their priming and activation in secondary lymphoid organs (reviewed in Agace, 2008). In addition to the αβ conventional T cells, there are unconventional T cell populations, mainly represented by T cells that express a γδ receptor, constituting a major population of lymphocytes in epithelium-rich tissues. Thus, such cells bearing γδ TCRs could act as the first line of defence against invading pathogens and they may perform immunoregulatory functions, modulating the functions of αβ T cells. In addition, γδ T cells may help preserve the integrity of damaged epithelial surfaces by providing localized delivery of an epithelial growth factor (Chen et al., 2002). It has also been shown that human γδ T cells can participate in the induction of adaptive immunity by acting as professional Ag-presenting cells (see André et al., 2007). Besides $CD4^+$ and $CD8^+$ lymphocytes in LP, intraepithelial lymphocytes (IELs), are primarily $CD8^+$ cells and have been classified into two major subgroups. The first is conventional $CD8\alpha\beta^+TCR\alpha\beta^+$ T cells and the second group is $CD8\beta^-CD8\alpha\alpha^+$ IELs that express either $TCR\alpha\beta^+$ or $TCR\gamma\delta^+$ (Agace, 2006). Virtually all γδ-IELs and many αβ-IELs in the mouse small intestine are known to express CD8αα homodimers (Ishikawa et al., 2007). Recent data indicate an extra-thymic origin of the CD8αα intraepithelial T-lymphocyte effector population in mammals (Ishikawa et al., 2007; Rocha, 2007). A similar process could take place in bony fish, as suggested by a very early (prethymic) appearance of T-like cells in the gut (Rombout et al., 2005), and in axolotl, in which the TR delta repertoires of thymectomized individuals appear different than those of normal axolotls (André et al., 2007). However, the ontogeny of these "unconventional" $CD8\alpha\alpha^+$ IELs continues to be the subject of much debate. According to Cheroutre and Lambolez (2008), some gut T cells showing an unconventional $TCR\gamma\delta^+$ phenotype and postulated to develop locally in the intestine require a thymic education that directs lineage commitment and functional differentiation.

The IEC/T cell relationship plays an important role in the immune regulation of the gut (Shao et al. 2001). Various phenotypically distinct subpopulations of CD4+ and CD8+ T cells interact with IECs within the mucosa. The cross talk occurring between the Ag-rich lumen and the lymphocyte-rich LP serves to maintain intestinal homeostasis. IECs have the capacity to talk to LP lymphocytes, activating populations of unique regulatory T cells. These cells have the capacity to talk back to the epithelium, influencing epithelial growth and differentiation (reviewed in Dahan et al., 2007).

Studies on TCR structure and expression in condrichtians, teleosts and amphibians have demonstrated the conservation of the structure and of the potential diversity of TCR molecules (Charlemagne et al., 1998). Presently, the availability of some tools has allowed the detection of some specific T cell markers in teleosts (reviewed in Randelli et al., 2008). Genes corresponding to the TCRα, β, γ, and δ subunits have been characterized. Other putative T cell markers in fish, such as CD3, CD4 and CD8α are known for a few species (Hordvik et al., 2004; Taylor et al., 2005; Randelli et al. 2008; Araki et al., 2008). T cells have also been characterized from the gut of rainbow trout, as they express transcripts of T cell marker homologs of CD8, CD4, CD28, CD3ε, TCRζ, TCRγ, and TCRβ and lacked IgM. Thus, a highly diverse αβ TCR repertoire is maintained in fish IEL in absence of PP and mesenteric LNs (Bernard et al., 2006). In *Salmo salar*, the TCR γ gene has extensive diversity and is mainly expressed in the same tissues that TCRδ genes are expressed (including intestine), which suggest that Atlantic salmon have a functional γδ T-cell immune component (Yazawa et al., 2008). TCRγ and CD3 γ/δ chains were cloned and characterized in *Cyprinus carpio* (Shang et al., 2008). In *Salmo salar* suffering enteropathy induced by soybean meal, CD3ε+ T cells were demonstrated in the mixed inflammatory infiltrate, as well as expression of CD3pp, CD4 and CD8 (Bakke-McKellep et al., 2007).

TCRs have recently been characterized in some amphibians. TCRα and TCRβ-like chains were detected in *Xenopus laevis* and axolotl (*Ambystoma mexicanum*). The genes encoding the axolotl TCRαβ have been cloned and the TCRδ has been characterized (André et al. 2007).

5.6.2. B cells and immunoglobulins

B cells are characterized by the expression of the B cell receptor, a surface immunoglobulin (Ig) receptor. The expression of surface Ig related receptors has been conserved in phylogenetically distinct species as a critical checkpoint in B cell development (Pike and Ratcliffe, 2002). Ig classes (in mammals, IgM, IgA, IgD, IgG, IgE9 are defined by the isotypes of heavy (H) chains (μ, α, δ, ε). Each isotype is in turn distinguished by unique structures in its constant region domain. These different structures confer distinctive functions on the Ig classes. Each isotype is encoded by a distinct gene and multiple heavy chain isoforms can be produced by alternative pathways of RNA processing, such as the secreted (sIg) and membrane (mIg) forms of all H chains. Allelic variation in the constant (C) regions gives rise to allotypes. The different types of light (L) chains (κ and λ in mammals) are also typically referred as isotypes. However, the classification in isotypes for non mammalian vertebrates is not clear (Bengtén et al., 2000).

Two Ig classes are found in secretions, including the fluid intestinal. IgE is associated with immunity to parasites and type I hypersensitivities. IgA, however, appears to have evolved specifically for the purpose of protecting body surfaces. IgA is the predominant Ig isotype participating in mucosal immunity in mammals and as such can be considered an important element of the first line of defence. In the intestine, IgA is produced in amounts that exceed all other classes combined (Tizard, 1995). IgA exists in two forms, secretory and serum. The secretory form is typically a dimeric molecule, whereas serum IgA is predominantly monomeric. IgA dimers (with Cα2 domain of one molecule bound by a J chain to the Cα3 domain of the other molecule) bind a receptor synthesized by IECs and hepatocytes termed secretory component, to produce secretory IgA (Tizard, 1995). In mammals, the majority of IgA synthesized is of the secretory form (Bengtén et al., 2000). To date the only unequivocal homologous of mammalian IgA are found among the avian species, with αH chain. While α homologous have not been demonstrated in ectothermic vertebrates, it is clear that these animals possess functional analogous of IgA. IgM molecules in teleosts and *Xenopus*, IgX in *Xenopus* and IgY in axolotls all are found in secretions and hence are likely involved in mucosal immunity (Bengtén et al., 2000). The genes encoding the axolotl B cell receptor have been recently cloned (André et al. 2007).

IgA is synthesized by plasma cells located in the submucosa in response to local Ag stimulation. Some of this IgA binds to a receptor for polymeric immunoglobulin (pIgR) on the submucosal surface of epithelial cells. The complex IgA-pIgR is then endocytosed and transported across the epithelial cell; the endosome fuses with the plasma membrane and exposes the IgA to the intestinal lumen, and the IgA with the receptor peptide (secretory component, SC) still attached is released (Tizard, 1995). IgA can bind to viral proteins inside the cell and prevent their replication. IgA also binds to Ags that have penetrated to the submucosa. Suzuki and Fagarasan (2008) reviewed the IgA synthesis in the gut in host-bacterial interaction. The major sites of generation of intestinal IgA are represented by the organized lymphoid structures present in the wall of small and large intestine. In PP, most of the germinal centre B cells preferentially switch from IgM+ to IgA+. PP DCs critically contribute to creating a special gut microenvironment, due to their location at the subepithelial dome where they directly access luminal bacteria. Indeed, the epithelium that covers PP contains M cells, considered to be the major site of Ag entry to the mucosa. PP T cells are also conditioned by gut DCs to produce cytokines with facilitate preferential class switch recombination from IgM to IgA. LP is a site where T-independent IgA responses are triggered by intestinal bacteria. Not only IgA+ B cells but also activated IgM+ B cells from the PP migrate to the LP. The pIgRs in chicken and *Xenopus* have been cloned (Hamuro et al., 2007).

IgG and IgE isotypes, found thus far exclusively in the mammals, have been extensively studied due to their clinical importance. IgY is a low molecular weight, monomeric Ab of wide distribution. It is found in amphibians, reptiles and birds (reviewed in Warr et al. 1995 and Bengtén et al., 2000). Other classes in cartilaginous fish are reviewed in Bengtén et al. (2000).

The Ab repertoire in teleosts is more limited than in mammals. (Marchalonis et al., 2006). The most prevalent immunoglobulin in serum of teleosts is an IgM tetramer. Some teleosts have the monomer (H2L2) of IgM in serum, and a dimmer has been observed in secretions (Lobb and Clem, 1981). Other Ig classes have been found in fish, namely IgD, IgT and IgZ (reviewed in Randelli et al., 2008). The number of VH families varies tremendously between different teleost species. In addition, a mechanism to generate structural diversity in the tetrameric Ab, through a random polymerisation of constituent monomeric, has been suggested for salmonids to compensate the reduced isotypy (Kaattari et al., 1999).

Although fish do not have IgA or a J chain, there are indications that fish produce mucosal antibodies in both cutaneous and gut mucus. Differences in serum and mucus IgM have been reported based on their reaction with Mabs, and different IgM isotypes have been suggested (Rombout et al., 1993b). Fish mucosal Abs have been mainly found in the skin and few data are related to intestine. IgM+ positive cells were detected in the posterior intestine of halibut (*Hippoglossus hippoglossus*) and were more common in the epithelium than in the LP (Grove et al., 2006). Fournier Betz et al. (2000) found also IgM+ cells in the intestine of turbot. In addition, the gene encoding for pIgR in carp has been isolated and sequenced, and it is expressed in all lymphoid tissues, including a gut subpopulation of lymphoid cells in epithelium and LP (Rombout et al., 2008). PIgR-like sequences were also discovered in *Danio rerio* and *Takifugu rubripes* (Hamuro et al., 2007).

As indicated above, innate-like lymphocytes include B1 cells and marginal-zone B cells. A phagocytic function of innate-like B lymphocytes has been described. The recent identification of phagocytic B lymphocytes in teleosts and amphibians (*Xenopus laevis*) supports the idea of an evolutionary relationship between B-1 lymphocytes and MΦs of mammals, and that B cells might have evolved from ancient phagocytic cell (Li et al., 2006; see also Alvarez-Pellitero et al., 2008).

The few data available on B cell responses to parasites in teleosts are presented below (section 5.7.2).

5.7. Inflammatory and cellular responses to intestinal parasite infections

5.7.1. Examples from endothermic vertebrates

Most available data are related to mammalian intestinal parasites, mainly diplomonad flagellates, apicomplexans, and helminths (trematodes and nematodes).

5.7.1.1. Diplomonadida

Eckmann (2003) reviewed the mucosal defences against the intestinal flagellate *Giardia lamblia*. Several studies support the role of B cells and secretory IgA antibodies in anti-giardial defence. However, B cell-independent mechanisms can also contribute to the eradication of the parasite, such as nitric oxide and Paneth cell α-defensins and lactoferrin.

5.7.1.2. Apicomplexa

The involvement of gut IELs in the immune response to coccidia was reviewed by McDonald (1999). T cells (particularly those with the αβ form of TCR) have effector roles in immunity to these intracellular infections. TCRγδ+ T cells might also increase during mucosal infection (e.g. with *Eimeria* or *Cryptosporidium parvum*). Further data are available on T cell responses to *Eimeria* spp. infections. CD4+ cells may be particularly involved in primary infections of rats with *E. separata*, whereas immunity to a challenge infection seems to be mediated predominantly by CD8+ and TCRαβ^{+} T cells (Shi et al., 2001). Infection of rabbit with *E. intestinalis* induced strong response of mucosal T lymphocytes and low systemic responses after primary infection, at difference with protected rabbits. These facts suggest that protection of rabbits against *E. intestinalis* infection is due to an effective mucosal immune response, and that systemic responses, which increase after successive infections, are only reflections of repeated encounters with parasite Ags (Renaux et al., 2003). In chickens, recent evidence implicates cell-mediated immunity as the major factor conferring resistance to coccidiosis by *Eimeria* spp., whereas humoral immunity plays only a minor role in protection against this disease (Yun et al., 2000).

Cell-mediated immunity is believed to be the primary method for host elimination of *Cryptosporidium* spp. infections. IELs, mainly $CD4^{+}$T cells, and IFN-γ are critical for parasite elimination, though $CD8^{+}$ T cells are also involved (reviewed in McDonald, 1999; Thompson et al., 2005). Guk et al (2003) found that $CD4^{+}$ LP lymphocytes are important to the protective response to a primary infection with *C. parvum*, through the induction of IFN-γ and IgA, whereas IELs were more involved in protection against a challenge infection, through enhanced cytotoxicity.

5.7.1.3. Helminths

Eosinophils and mast cells have been reported to be involved in the immune response to several parasite diseases, but the specific function of these cells in the host response remains unclear in many cases (Klion y Nutman, 2004). Various lines of evidence suggest that esosinophils play a role in the elimination of helminth parasites. Eosinophilic infiltration is a characteristic feature in the intestinal mucosa of mice infected by the trematode *Echinostoma* spp. and peripheral eosinophilia was also observed in some cases (reviewed in Toledo et al. 2006). Davies et al. (2005) demonstrated intestinal esosinophilia and eosinopoietic responses in response to schistosome eggs and parasites. Mucosal mast cells are also important effector cells against intestinal helminths (Nawa et al., 1994). Worm expulsion concomitant with intestinal mastocytosis has been observed in infections by the trematodes *Neodiplostomum seoulense*, *Echinostoma trivolvi* and *Metagonimus yokogawai*. Systemic and intestinal Ab responses are also involved in several intestinal trematode infections in mammals (reviewed in Toledo et al., 2006).

Recent studies using mainly two murine models, *Trichuris muris* and *Trichinella spiralis*, have identified IECs as key populations in anti-nematode immune response. The robust innate immune IEC responses during nematode infection can influence both the initiation and effector phases of the anti-parasite response (reviewed in Artis and Grencis, 2008). CD4+ Th1 cells expressing IFN-γ promote parasite persistence and host susceptibility, while resistance to infection is dependent on CD4+Th2 cells and Th2 associated cytokines, including IL-4, IL-9, IL-13, IL-25 and IL-36. In addition, Treg cells are induced as part of the estrategy for immune evasion to control the pathology associated with the infection throufh cytokines such as IL-10 or TGF-β. Readers are referred to this review of Artis and Grencis (2008) for detailed analysis of the role of IECs in the response and their relationship with other immune epithelial components: MΦ-IECs and DCs interactions, role of goblet cells, T cells and cytokines and chemokines. In a similar way, Patel et al (2009) reviewed the effector mechanisms at the host-parasite interface during the immune response to tissue-dwelling intestinal nematode parasites. The response is characterized by elevations of IL-4 and IL-13 and increased number of CD4+ T cells, granulocytes and MΦs. Kamal et al. (2001) reported a thymic-independent population of mucosal T cells that modulates innate small intestine defense in mice by increasing the number of Paneth and intermediate cells in response to *T. spiralis* infections.

5.7.2. Ectothermic vertebrates

Until the last two decades, the information on the cellular response to the piscine parasites had been limited to the description of the histopathologial lesions produced by the parasites, and changes in leucocytes in blood or infected sites. Inflammatory reactions involving cellular responses either accompanied by the formation of granulomas or not, have been reported for numerous piscine parasites belonging to different groups. More specific data on the involvement of different factors related to inflammation, including cellular components (phagocytes, other leucocytes) and complement in the response against parasites are available in some cases (reviewed in Alvarez-Pellitero, 2008a). Few data on intestinal parasitoses are known and in many cases are related to systemic responses whereas cellular changes at the mucosal level have rarely been described. Several intestinal *Entamoeba* spp. (Amebozoa) are known from fish (see Lom and Dyková 1992), but no information on their efect on the gut has been provided. The remaining data concern diplomonad flagellates, Apicomplexa and Myxozoa (Table 4).

5.7.2.1. Diplomonadida

Intestine is the primary site of infection of *Hexamita/Spironucleus* flagellates, though diplomonads can invade the gall bladder. The anterior intestine and the pyloric region are the preferred locations for *H. salmonis* in farmed rainbow trout fry, but in heavily infected fish parasites occur through the entire length of the gut (Uldal and Buchmann, 1996). Catarrhal enteritis is the main pathology of the typical disease initially described as hexamitosis/spironuleosis, though severe damage with inflammatory and cellular reactions are produced in systemic infections of several salmonids. In contrast, *S. barkhanus* produced slight damage in *Salvelinus alpinus*, which could be related to a cellular response (Sterud et al. 2003) and the clearance of parasites by phagocytosis (reviewed in Alvarez-Pellitero, 2008b).

5.7.2.2. Apicomplexa

Amongst apicomplexans, several *Cryptosporidium* spp. have been found in ectothermic vertebrates, though in many cases they are unnamed species (see reviews of Thompson et al., 2005; Ryan, 2009). The information on the mechanisms involved in the mucosal response to these infections is poor and mainly deals with the description of histopathological effect. Piscine *Cryptosporidium* spp. produce variable degrees of histopathological damage, mainly in juvenile fish. Distortion of the normal mucosal architecture with distension of epithelial cells, which can lead to sloughing of epithelial cell remnants and even detachment of intestinal mucosa, is the most frequently described damage. An inflammatory reaction with lecucocyte infiltration has been also reported for some species, such as *C. scophthalmi*, parasitizing the intestine of *Psetta maxima* (Alvarez-Pellitero et al., 2004), an intestinal species from barramundi *Lates calcarifer* (Glazebrook and Campbell, 1987) and a gastric species from *Sciaenus ocellatus* (Camus and López, 1996). In this gastric cryptosporidiosis the submucosa and LP were infiltrated with EGCs, lymphocytes and MΦs. In *Sparus aurata* parasitized by the gastric *C. molnari*, some cells, probably MΦs, engulfed parasite stages and debris (Alvarez-Pellitero and Sitjà-Bobadilla, 2002). Increasing in RCs was observed in infections by *C. molnari* in *Sparus aurata* (Alvarez-Pellitero and Sitjà-Bobadilla, 2002) and by *C. scophthalmi* in *Psetta maxima* (Alvarez-Pellitero et al., 20

Table 4. Inflammatory and cellular reactions, and phagocytic and oxidative activities in response to parasites of the digestive tract in ectotherms. HK: head kidney; GPX-1: gluthatione peroxidase-1. [1]All hosts are teleosts, except those marked with (A): amphibians and (R): reptiles

Parasite	Host[1]	Effects in serum and non-digestive sites	Effects at intestine	References
Diplomonadida *Hexamita/ Spironucleus*	*Oncorhynchus mykiss* and other fish species		Enteritis	Uldal and Buchmann, 1996; Alvarez-Pellitero, 2008b
S. barkhanus	*Salvelinus alpinus*		Cellular response	Sterud et al., 2003
Apicomplexa *Cryptosporidium molnari*	*Sparus aurata*		Presence of MΦs	Alvarez-Pellitero and Sitjà-Bobadilla, 2002
C. scophthalmi	*Psetta maxima*		Inflammatory reaction with leucocyte infiltration	Alvarez-Pellitero et al., 2004
Cryptosporidium sp.	*Lates calcarifer*		Inflammatory reaction with leucocyte infiltration	Glazebrook and Campbell, 1987
Cryptosporidium sp.	*Sciaenus ocellatus*		EGC, lymphocyte and MΦ infiltration at LP	Camus and López, 1996
C. fragile	*Duttaphrynus malanostictus* (A)		Mucosal thickening	Jirků et al. (2008)
Cryptosporidium sp.	*Xenopus laevis* (A)		Proliferative gastritis	Green et al., 2003
Cryptosporidium sp.	*Thamnophis sirtalis* (R)		Heterophylic and lymphocytic inflammatory infiltrates in LP	Brower et al., 2001
Cryptosporidium sp.	*Eublepharis macularius* (R)		Hyperplasia and mononuclear inflammation	Terrel et al., 2003
Cryptosporidium sp.	*Testudo kleinmani* (R)		Heterophils, lymphocytes and MΦ in LP	Graczyk et al., 1998
Goussia carpelli	*Cyprinus carpio*	HK:↑ phagocytic activity and RB	Granulocyte, lymphocyte and MΦ infiltration	Jandrysek et al., 1994; Steinhagen and Hespe, 1997
G. subepithelialis	*C. carpio*	↑ phagocytic activity of blood neutrophils	Inflammatory reaction	Lom and Dykova, 1992, Studnicka and Siwicki, 1990
Myxozoa *Ceratomyxa shasta*	*Oncorhynchus spp.*		Inflammatory response, lymphocyte infiltration, LP thickening	Bartholomew et al., 1989

Table 4. (Continued)

Parasite	Host[1]	Effects in serum and non-digestive sites	Effects at intestine	References
Enteromyxum leei	*Sparus aurata*	HK leucoc.:↓ PO, phagocyt., RB; ↑↑↑ cytotox. act. Serum: ↑ ↓ PO		Cuesta et al., 2006a, b
		↑ RB blood leukocytes	↓ GPx-1 expression	Sitjà-Bobadilla et al. 2008
			EGCs, MΦ infiltration, LP thickening	Fleurance et al., 2008
		HK: ↑ caspase positive cells	↑ caspase positive cells	Estensoro et al., 2009
E. leei	*Diplodus puntazzo*	↑ serum NO levels		Golomazou et al., 2006
		↑ serum PO level		Muñoz et al., 2007
		↑ blood leukoc., ↑RB, MMC at certain infection points	↓ EGC1 and ↑ EGC2; lymphoc. infiltration: initial ↑ followed by ↓	Alvarez Pellitero et al., 2008
E. scophthalmi	*Psetta maxima* *P. maxima*	Blood: initial ↑ granulocytes and leucoc. RB; parasites in MΦ HK, spleen: ↑ apoptotic cells and MMCs; haematop. deplet.	Leucocyte infiltration of LP	Sitjà-Bobadilla et al., 2006 Redondo et al. , 2004
			Changes in neuromodulators	Bermúdez et al., 2007
			↑ RCs in low, mild infections; ↓ in severe infect.	Bermúdez et al., 2009a, b
		HK, spleen: caspase-3 apoptosis	Caspase-3. apoptosis	Losada et al., 2009
		HK, spleen: ↓IgM+ cells	↑ IgM+ cells, mainly in the LP	Bérmudez et al., 2006

Table 4. (Continued)

Parasite	Host[1]	Effects in serum and non-digestive sites	Effects at intestine	References
Cestodes *Cyatocephalus truncatus*	*Salmo trutta*		Changes in neuromodulators	Dezfuli et al., 2000
Diphyllobothrium dendriticum	*Coregonus lavaretus*		↑ mast cells /EGCs, MMCs	Dezfuli et al., 2007a
Eubothrium crassum	*Oncorhynchus mykiss*		Changes in neuromodulators	Bosi et al., 2005b
Nematodes *Anisakis simplex*	*Platichthys flessus*		RCs with discharge activity	Dezfuli et al., 2007b
Acanthocephalans *Acanthocephalus clavula*	*Anguilla anguilla* *Oncorhynchus mykiss*		↑ RCs	Dezfuli et al., 1998
Dentitruncus truttae	*Salmo trutta*		↑ mast cells, RCs Changes in neuromodulators	Dezfuli et al., 2008
Phomphorhynchus laevis	*Salmo trutta*		Changes in neuromodulators	Dezfuli et al., 2002
P. laevis	*Leuciscus cephalus*		Enteritis, EGCs in inflamm. tissue Changes in neuromodulators	Dezfuli et al., 2002a Bosi et al., 2005c

In the amphibian *Xenopus laevis*, *Cryptosporidium* sp. produced a proliferative gastritis (Green et al., 2003). Jirků et al. (2008) observed mild mucosal thickening in the anuran *Duttaphrynus malanostictus* infected by *Cyrptosporidium fragile*. An inflammatory reaction has been reported in several reptilians cryptosporidioses. In the garter snake (*Thamnophis sirtalis*), the enteritis associated with cryptosporidiosis was characterized by dense heterophylic and lymphocytic inflammatory infiltrates throughout the LP, with intestinal necrosis (Brower et al., 2001). Geckos (*Eublepharis macularius*) showed hyperplasia and mononuclear inflammation associated to *Cryptosporidium* infection of small intestine (Terrel et al., 2003). The intestinal LP and mucosa of the Egyptian tortoise *Testudo kleinmani* infected by *Cryptosporidium* sp. were infiltrated by heterophils, lymphocytes and MΦs (Graczyk et al., 1998).

Some data on intestinal piscine *Goussia* spp. are available. *G. carpelli* and *G. subepithelialis* cause epithelial detachment and inflammatory changes in epithelium and LP of carps (reviewed in Lom and Dykova (1992). Infiltration of granulocytes, lymphocytes and MΦs in infected carps (Jendrysek et al., 1994) as well as a response of HK phagocytes, were described in *G. carpelli*-infected carps (Steinhagen and Hespe, 1997). In addition, spontaneous infection by *G. subepithelialis* produced increased phagocytic activity of blood neutrophils (Studnicka and Siwicki, 1990).

5.7.2.3. Myxozoa

Infections by *Ceratomyxa shasta* in susceptible *Oncorhynchus* spp. *induced* inflammatory response with lymphocyte infiltration of the submucosa, as well as thickening of the LP. The mucosa was extensively destroyed at posterior intestine (Bartholomew et al. 1989; see also Lom and Dyková, 1992).

The remaining available information on inflammatory/cellular responses to Myxozoa is related mainly to *Enteromyxum* spp. infections. Inflammatory reaction with leucocytic infiltration was produced by both *Enteromyxum scophthalmi* (Bermúdez et al., 2009a, b) and *E. leei* (Fleurance et al., 2008), especially in severely parasitized fish. Sitjà-Bobadilla et al. (2006) reported higher abundance of MMCs in HK and spleen of *E. scophthalmi*-infected turbot, which also showed melanization of spleen MMC associated to the presence of parasite stages or debris engulfed in MΦs. MMCs can also increase in the intestinal LP of parasitized fish (Bérmudez, Quiroga and Alvarez-Pellitero, unplublished observations). The existence of apoptosis suggested by Sitjà-Bobadilla et al. (2006), seemed to be confirmed by the detection of caspase-3 positive cells, which appeared within the intestinal epithelium in the initial stages of the infection and in higher number in the inflammatory infiltrates of the LP-submucosa and in lymphohaemotopoietic organs as the lesions became severe (Losada et al., 2009). Caspase-3 positive cells increased also in *E.leei*-infected *Sparus aurata*, though more markedly in the HK than in the intestine (Estensoro et al., 2009).

In relation to oxidative mechanisms, gluthatione peroxidase-1 (GPx-1) gene expression was down-regulated in the intestine of *S. aurata* chronically-exposed to *E. leei* (Sitjà-Bobadilla et al., 2008). The remaining data concern changes in the blood or hematopoietic organs. The RB of circulating leucocytes increased in turbot soon after exposure to *E. scophthalmi*, in congruence with the rise in the percentages of granulocytic cells in blood (Sitjà.Bobadilla et al., 2006) and with the presence of blood MΦs harbouring stages of the myxosporean (Redondo et al., 2004). The leucocyte PO content, phagocytosis and RB activity of HK leukocytes decreased significantly in E. *leei*-exposed *S. aurata*, whereas the cytotoxic activity was up to 2-3 times higher than in control fish (Cuesta et al., 2006c). Serum peroxidase (PO) levels increased in such *E. leei*-exposed fish soon after infection but a subsequent fall occurred at later post-exposure times (Cuesta et al., 2006a). The RB of blood leucocytes increased in *S. aurata* chronically-exposed to *E. leei* (Sitjà-Bobadilla et al., 2008). In *E. leei*-exposed *D. puntazzo*, an increase in RB of circulating leucocytes was detected at certain infection points (Alvarez-Pellitero et al., 2008). Increasing in serum NO and PO levels have also been reported (Golomazou et al., 2006; Muñoz et al., 2007).

Some information is also available on the cellular reaction at intestine. In *E.leei*-infected *Sparus aurata*, numerous mononuclear and EGCs were present in the LP, wich in other cases appeared particularly thick and massively infiltrated mainly by MΦs (Fleurance et al., 2008). Alvarez-Pellitero et al. (2008) found two apparently different types of EGCs involved in the response of *Diplodus puntazzo* to the myxozoan *E. leei*. EGC1 (very similar to the EGC of salmonids), degranulated in infected intestines and their number decreased with the progression of the infection, in parallel with an increase in EGC2-type cells. In addition, recruitment of lymphocytes to the intestinal mucosa occured from the beginning of the infection, when still no appreciable changes in blood lymphocytes were appreciated but decreased at the last sampling (Alvarez-Pellitero et al., 2008). In turbot experimentally infected with *E. scophthalmi*, IgM+ cells increased in the intestine, the target organ (mainly in the LP), but decreased in kidney and spleen (Bermúdez et al., 2006).

Other intestinal cell type affected by enteromyxoses is the rodlet cell. Infections by the myxozoan *Enteromyxum scophtalmi* in turbot produce an initial increase in RCs in low and mild infections, followed by a decrease in severe disease (Bermúdez et al., 2009a, b).

The response of enteric neuroendocrine system to enteromyxoses was studied in *E. scophtalmi*-infected turbot (Bermúdez et al., 2007) and *E. leei*-infected *Sparus aurata* (Estensoro et al., 2009). The changes induced by the infections in the presence of neuromodulators may cause alterations in gut motility, electrolit and fluid secretion, and vascular and immune functions.

5.7.2.4. Helminths

Inflammatory and cellular reactions can occur in fish in response to intestinal helminthes. *Pomphorhynchus laevis* elicited enteritis in infected *Leuciscus cephalus* and EGCs vere identified within the inflammatory tissue (Dezfuli et al., 2002a). A larg number of mast cells/EGCs were seen within and around the cysts produced by *Diphyllobothirum dendriticum* in the digestive tract of *Coregonus lavaretus*. MMCs were also observed in close proximity to the encysted parasite (Dezfuli et al., 2007a). *Salmo trutta* parasitized by the intestinal acanthocephalan *Dentitruncus truttae* had larger numbers of mast cells than uninfected fish and their migration to the site of infection was suggested. Intense degranulation was encountered in all intestinal layers (Dezfuli et al., 2008). Changes in RCs can also occur in these infections. RCs were more numerous in the intestine of *Anguilla anguilla* and *Oncorhynchus mykiss* infected by the acanthocepahlan *Acanthocephalus clavula* and of *Salmo trutta* parasitized by *Dentitruncus truttae*, respectively (Dezfuli et al., 1998, 2008). RCs with discharge activity were found in the gut epithelia of several *Platichthys flesus* that harbored *Anisakis simplex* larvae on the outer surface of intestine (Dezfuli et al. 2007b).

The abundance and distribution of different neuromodulators in the digestive tract can change in fish parasitized by several helminthes with respect to uninfected fish. Infections of *Salmo trutta* by *Cyatopcephalus truncatus*, *Pomphorhynchus laevis* and *Dentitruncus truttae* (Dezfuli et al., 2000, 2002b, 2008), *Oncorhynchus mykiss* by *Eubothrium crassum* (Bosi et al., 2005b), and *Leuciscus cephalus* by *Pomphorhynchus laevis* (Bosi et al., 2005c) are examples of such a situation. Changes mostly involved neuromodulators that are known to be implicated in gut motility, digestive/absorptive processes, as well as the immune and inflammatory responses.

6. CONCLUSION

The knowledge of the mucosal system of mammals has substantially increased in recent years, mainly due to the relevance of mucosal immunity in the development of new strategies for mucosal vaccination and novel approaches for the treatment of inflammatory diseases. In contrast, limited information is available for ectotherms. Although the main mechanisms of mucosal immunity seem to be present in the intestine of ectotherms, there are important gaps in the knowledge of many factors and pathways. The information on their involvement in the response to parasites is even more limited. Changes in mucin and glycosylation patterns have been reported for several piscine parasites, such as the myxozoans *Enteromyxum* spp. and some helminths. However, no data on their specific involvement in the host-parasite interaction are available, in contrast with mammalian parasites, such as *Entamoeba* spp. or *Cryptosporidium* spp. Several TLRs and some other PRRs have been characterized in ectotherms, but no information on their expression in the intestine is known. Thus, studies on these receptors and their relationship with the mucin patterns, and with other effector molecules and intestinal cells are needed. APR related factors are also relevant at the mucosal level. Amongst them, AMPs have acquired paramount importance in antimicrobial therapies and an increasing number of these molecules have been characterized recently. Several AMPs and other APR related factors, such as lysozyme, antiproteases or cytokines, exist in teleosts, though their involvement in the response to parasites has been demonstrated in few cases. Amphibian AMPs, mainly from skin, have acquired relevance due to their potential as therapeutics. Several lectins have also been reported in ectotherms, especially in teleosts and some are present in the intestine. Lectins and other APRs could be involved in the responses to infections and hence their functional aspects deserve more attention. In the enteric infection produced by *E. leei* in *S. aurata*, cytokines and other factors can be down- or upper- regulated in the intestine, the target organ.

Cellular components (IECs, DCs, mast cells, macrophages, granulocytes, T and B lymphocytes) are also crucial in the intestinal immune response. The cross-talk between such cellular types— involving also mucins, TLRs, AMPs, cytokines and other factors — is well known in endotherms, and contributes to the control of inflammation/apoptosis and to the maintenance of homeostasis. In contrast, the leucocyte types of the teleost gut are poorly known, and their characterization is hampered by the lack of specific tools or markers. Inflammatory and cellular reactions to some parasitoses have been described, but the involved cells have been characterized in few cases. More studies are deserved on the piscine inflammatory/apoptotic pathways and their relation to homeostasis, a field of knowledge well studied in mammals. The role of IECs in intestinal mucosal immunity and in the apoptotic response to parasite infections at the epithelial level (anoikis)—thus far found only in enteromyxoses— requires more attention. Although different populations of leucocytes are present in the teleost gut, and the presence of mast cells has been demonstrated, the different types of EGCs and other granulocytes are not fully characterized, and the information on their functional aspects is very limited, mainly in relation to parasitoses. The changes in RCs in piscine intestinal infections confirm its character of immune cells, but more information is needed on their precise role in the response. Expression of cytokines and mediators in the different cellular types is scarcely known for ectotherms. Studies on the piscine MΦ aggregates (or MMCs), which are present also at the LP, and their possible role as analogous to the germinal centres of LNs in mammals is also required. Information on T cells is even scarcer. Although some T cell types have been characterized in ectotherms (mainly in teleosts), little information is available on the presence of the cells equivalent to the Th1, Th2, Th17 and Treg subsets—well established in mammals — in the piscine intestine. Several cytokines related to Th17 and Treg activities have been identified in the gut of some fish, but not in intestinal parasite infections. IgM+ cells have been described in some fish species, but information on their changes in intestinal infections is very scarce. A better characterization of the secretory Igs is also necessary.

Comprehensive studies taking into account the different factors of both innate and adaptive mechanisms at the mucosal level and in relation to systemic immunity are necessary to fully characterize the immune response to ectotherm intestinal parasites.

ACKNOWLEDGMENTS

Part of the information included in the present review has been generated through the research project AGL2006-13158-C02-01 funded by the Spanish *Ministerio de Educación y Ciencia*.

REFERENCES

Agace, W. W. (2008). T-Cell recruitment to the intestinal mucosa. *Cell, 29*, 514-522.

Agius, C. & Roberts, R. J. (2003). Melano-macrophage centres and their role in fish pathology. *J. Fish Dis., 26*, 499-509.

Akira, S. & Takeda, K. (2004). Toll-like receptor signalling. *Nat. Rev. Immunol., 4*, 499-511.

Akira, S., Uematsu, S., Takeuchi, O. (2006). Pathogen recognition and innate immunity. *Cell, 124*, 783-801.

Alvarez-Pellitero, P. (2008a). Fish immunity and parasite infections. From innate immunity to immunoprophylactic prospects. *Vet Immunol. Immunopathol., 126*, 171-198.

Alvarez-Pellitero, P. (2008b). Diseases caused by flagellates. In:, Eiras, J., Segner K., Wahli, T., Kapoor, B. G. (Eds.). *Fish Diseases*, Science Publishers, New Hampshire, 421-515.

Alvarez-Pellitero, P., Sitjà-Bobadilla, A. (2002). *Cryptosporidium molnari* n. sp. (Apicomplexa: Cryptosporidiidae) infecting two marine fish species, *Sparus aurata* L., and *Dicentrarchus labrax* L.. *Int. J. Parasitol., 32*, 1007-1021.

Alvarez-Pellitero, P., Quiroga, M. I., Sitjà-Bobadilla, A., Rodondo, M. J., Palenzuela, O., Padrós, F., Vázquez, S. & Nieto, J. M. (2004). *Cryptosporidium scophthalmi* n. sp. (Apicomplexa: Cryptosporidiidae) from cultured turbot *Scophthalmus maximus*. Light and electron microscope description and histopathological study. *Dis. Aquat. Org., 62*, 133-145.

Alvarez-Pellitero, P., Palenzuela, O. & Sitjà-Bobadilla, A. (2008). Histopathology and cellular response in *Enteromyxum leei* (Myxozoa) infections of *Diplodus puntazzo* (Telesotei). *Parasitol. Int., 57*, 110-120.

André, S., Kerfourn, F., Affaticati, P., Guerci, A., Ravassard, P. & Fellah, J. S. (2007). Highly restricted diversity of TCR delta chains of the amphibian axolotl (*Ambystoma mexicanum*) in peripheral tisssues. *Eur. J. Immunol., 37*, 1621-1623.

Aoki, T. & Hirono, I. (2006). Immune relevant genes of Japanese flounder, *Paralichthys olivaceus*. *Comp. Biochem. Physiol., D 1*, 115-121.

Araki, K., Akatsu, K., Suetake, H., Kikuchi, K. & Suzuki, Y. (2008). Characterization of $CD8^+$ leukocytes in fugu (*Takifugu rubripes*) with antiserum against fugu CD8α. *Dev. Comp. Immunol., 32*, 850-858.

Arnold, J. N., Dwek, R. A., Rudd, P. M. & Sim, R. B. 2006. Mannan binding lectin and its interaction with immunoglobulins in health and disease. *Immunol. Lett. 106,* 103-110.

Artis, D., Grencis, R.K. 2008. The intestinal epithelium: sensors to effectors in nematode infection. *Immunology, 1,* 252-264.

Bakke-McKellep, A. M., Froystad, M. K., Lilleeng, E., Dapra, F., Refstie, S., Krogdahl, Å. & Landsverk, T. (2007). Response to soy: T-cell-like reactivity in the intestine of Atlantic salmon, *Salmo salar* L. *J. Fish Dis.*, *30,* 13-25.

Barribaeu, S. M., Villinger, J. & Waldman, B. (2008). Major histocompatibility complex based resistance to a common bacterial pathogen of amphibians. *PloS One, 3,* e2692, doi: 10.371/journal.pone.0002692.

Bartholomew, J. L., Smith, C. E., Rohovec, J. S., Fryer, J. L. (1989). Characterization of a host response to the myxosporean parasite *Ceratomyxa shasta* (Noble), by histology, scanning electron microscopy and immunological techniques. *J. Fish Dis.*, *12,* 509-522.

Bayne, C. J. & Gerwick, L. (2001). The acute phase response and innate immunity of fish. *Dev. Comp. Immunol.*, *25,* 725-743.

Bei, J. X., Suetake, H., Araki, K., Yoshiura, Y., Lin, H. R. & Suzuki, Y. (2006). Two interleukin (IL)-15 homologues in fish from two distinct origins. *Mol. Immunol.*, *43,* 860-869.

Belinda, L. W. C., Wei, W. X., Hanh, B. T. H., Lei, L. X. B., Lei, L. X., Bow, H. & Ling, D. J. (2007). SARM: a novel Toll-like receptor adaptor, is functionally conserved from arthropod to human. *Mol. Immunol.*, *45,* 1732-1742.

Belkaid, Y., Sun, C. M., Bouladoux, N. (2006). Parasites and immunoregulatory T cells. *Curr. Op. Immunol.*, *18,* 406-412.

Belley, A., Keller, K., Gottke, M., Chadee, K. & Goettke, M. (1999). Intestinal mucins in colonization and host defense against pathogens. *Am. J. Trop. Med. Hyg.*, *60,* 10-15.

Bengtén, E., Wilson, M., Miller, N.W., Clem, L. W., Pilstrom, L. & Warr, G. W. (2000). Immunoblobulin isotypes: Structure, function, and genetics. In: Origin and evolution of the vertebrate immune system. *(Series: Current Topics in Microbiology and Immunology)*, *248,* 189-219.

Bermúdez, R., Faílde, L. D., Losada, A. P., Redondo, M. J., Álvarez-Pellitero, P. & Quiroga, M. I. (2009a). Morphopathological and ultrastructural studies on enteromyxosis of turbot (*Psetta maxima* L.). *14th EAFP International Conference. Prague*, 14-19 September, 2009, *Book of Abstracts*, 17-18.

Bermúdez, R., Losada, A. P., Vázquez, S., Redondo, M. J., Álvarez-Pellitero, P. & Quiroga, M. I. (2009b).Light and electrón microscopic studies on turbot (*Psetta maxima* L.) infected with *Enteromyxum scophthalmi*: histopathology of turbot enteromyxosis. *Dis. Aquat. Organ.* (in press).

Bermúdez, R., Vigliano, F. A., Marcaccini, A., Sitjà-Bobadilla, A., Quiroga, M. I. & Nieto, J. M. (2006). Response of Ig-positive cells to *Enteromyxum scophthalmi* (Myxozoa) experimental infection in turbot, *Scophthalmus maximus* (L.): A histopathological and immunohistochemical study. *Fish Shellfish Immunol.*, *21,* 501-512.

Bermúdez, R., Vigliano, F., Quiroga, M. I., Nieto, J. M., Bosi, G., Domeneghini, C. (2007). Immunohistochemical study on the neuroendocrine system of the digestive tract of turbot, *Scophthalmus maximus* (L.), infected by *Enteromyxum scophthalmi* (Myxozoa). *Fish Shellfish Immunol.*, *22,* 252-263.

Bernard, D., Six, A., Rigottier-Gois, L., Messiaen, S., Chilmonczyk, S., Quillet, E., Boudinot, P. & Benmansour, A. (2006). Phenotypic and functional similarity of gut intraepithelial and systemic T cells in a teleost fish. *J. Immunol.*, *176,* 3942-3949.

Bosi, G., Arrighi, S., Di Giancamillo, A. & Domeneghini, C. (2005a). Histochemistry of glycoconjugates in mucous cells of *Salmo trutta* uninfected and naturally parasitized with intestinal helminths. *Dis. Aquat. Organ., 64,* 45-51.

Bosi, G., Domeneghini, C., Arrighi, S., Giari, L., Simoni, E. & Dezfuli, B. S. (2005c). Response of the gut neuroendocrine system of *Leuciscus cephalus* to the presence of *Pomphorhynchus laevis* Müller, 1776 (Acanthocephala). *Cell. Mol. Biol., 20,* 509-518.

Bosi, G., Shinn, A. P., Giari, L., Simoni, E., Oironi, F. & Dezfuli, B. S. (2005b). Changes in the neuromodulators of the diffuse edocrine system of the alimentary canal of farmed rainbow trout, *Oncorhynchus mykiss*, naturally infected with *Eubothrium crassum* (Cestoda). *J. Fish Dis., 28,* 703-711.

Boshra, H., Li, J. & Sunyer, J. O. (2006). Recent advances on the complement system of teleost fish. *Fish Shellfish Immunol., 20,* 239-262.

Brandtzaeg, P., Kiyono, H., Pabst, R. & Russell, M.W. (2008). Terminology: nomenclature of mucosa-associated lymphoid tissue. *Mucosal Immunol., 1,* 31-37.

Brower, A. I. & Cranfield, M. R. (2001). *Cryptosporidium* sp.-associated enteritis without gastritis in rough green snakes (*Opheodrys aestivus*) and a common garter snake (*Thamnophis sirtalis*). *J. Zoo Wildl. Med., 32,* 101-105.

Buddington, R. K., Krogdahl, Å. & Backe-McKellep A. M. (1997). The intestines of carnivorous fish: structure and functions and the relation to diet. *Acta Physiol. Scand., 161,* 67-80.

Cammarata, M., Benenati, G., Odom, E. W., Salerno, G., Vizzini, A., Vasta, G. R. & Parrinello, N. (2007). Isolation and characterization of a fish F-type lectin from gilt head bream (*Sparus aurata*) serum. *Biochim. Biophys. Acta., 1770,* 150-155.

Cambi, A. & Figdor, C. G. (2005). Levels of complexity in pathogen recognition by C-type lectins. *Current Opin. Immunol., 17,* 345-351.

Campbell, R. D. & Trowsdale, J. (1993). Map of the human MHC. *Immunol. Today*, 14, 349-352.

Camus, A. C. & López, M. K. (1996). Gastric cryptosporidiois in juvenile red drum. *J. Aquat. Anim. Health., 8,* 167-172.

Canny, G. & Levy, O. (2008). Bactericidal/permeability increasing protein (BPI) and BPI homologs at mucosal sites. *Trends Immunol., 29,* 541-547.

Cario, E. & Podolsky, D. K. (2005). Intestinal epithelial TOLLerance versus inTOLLerance of commensals. *Mol. Immunol., 42,* 887-893.

Castell, A. E., Hernández, A., Sampedro, E. A., Herrera, M. A., Alvarez, S. J. & Rondán, A. (1999). ATPase and MHC class II molecules co-exprssion in *Rana pipiens* dendritic cells. *Dev. Comp. Immunol., 23,* 473-485.

Castell-Rodríguez, A. E., Sampedro-Carrillo, E. A., Herrera-Enriquez, M. A. & Rondán-Zárate, A. (2001). Non-specific esterasae positive dendritic cells in epithelia of the frog *Rana pipiens. Histoch. J., 33,* 311-316.

Chabot, S., Wagner, J. S., Farrant, S. & Neutra M. R. (2006). TLRs regulate the gatekeeping functions of the intestinal follicle-associated epithelium. *J. Immunol., 176,* 4275-4283.

Chang, C. I., Zhang, Y. A., Yi, X., Nie, P. & Secombes, C. J. (2006). Two cathelicidin genes are present in both rainbow trout (*Oncorhynchus mykiss*) and Atlantic salmon (*Salmo salar*). *Antimicrob. Agents Chemother., 50,* 185-195.

Chang, M. X. & Nie, P. (2007). Intelectin gene from the grass carp *Ctenopharyngodon idella*: cDNA cloning, tissue expression, and immunohistochemical localization. *Fish Shellfish Immunol., 23,* 128-140.

Charlemagne, J., Fellah, J. S., De Guerra, A., Kerfourn, F. & Partula, S. (1998). T-cell receptors in ectothermic vertebrates. *Immunol. Rev., 166,* 87-102.

Chen, J. S. C., Wang, T. Y., Tzeng, T. D., Wang, C. Y. & Wang, D. (2008). Evidence for a positive selection in the TLR9 gene of teleosts. *Fish Shellfish Immunol., 24,* 234-242.

Chen, S. L., Li, W., Meng, L., Sha, Z. X., Wang, Z. J. & Ren, G. C. (2007). Molecular cloning and expression analysis of a hepcidin antimicrobial peptide gene from turbot (*Scophthalmus maximus*). *Fish Shellfish Immunol., 22,* 172-181.

Chen, Y., Chou, K., Fuchs, E., Havran, W. L. & Boismenu, R. (2002). Protection of the intestinal mucosa by intraepithelial γδ T cells. *Proc. Nat. Acad. Sci., 99,* 14338-14343.

Chen, X. M. & LaRusso, N. F. (2000). Mechanisms of attachment and internalization of *Cryptosporidium parvum* to biliary and intestinal epithelial cells. *Gastroenterology., 118,* 368-379.

Cheroutre, H. & Lambolez, F. (2008). The thymus chapter in the life of gut-specific intraepithelial lymphocytes. *Curr. Op. Immunol., 20,* 185-191.

Chistiakov, D. A., Hellemans, B. & Volckaert, F. A. M. (2007). Review on the immunology of European sea bass *Dicentrarchus labrax*. *Vet. Immunol. Immunopathol., 117,* 1-16.

Cho, Y. S., Lee, S. Y., Kim, K. H., Kim, S. K., Kim, D. S. & Nam, Y. K. (2009). Gene structure and differential modulation of multiple rockbream (*Oplegnathus fasciatus*) hepcidin isoforms resulting from different biological stimulations. *Dev. Comp. Immunol., 33,* 46-58.

Choi, B. Y., Sohn, Y. S., Choi, Ch.. & Chae, Ch. (2004). Lectin histochemistry for glycoconjugates in the small intestines of piglets naturally infected with *Isospora suis*. *J. Vet. Med. Sci., 65,* 389-392.

Clem, L. W., Bly, J. J., Ellsaesser, C. F., Lobb, C. J. & Miller, N. W. (1990). Channel catfish is an unconventional model for immunological studies. *J. Exp. Zool. Suppl., 4,* 123-125.

Coombes, J. L. & Powrie, F. (2008). Dendritic cells in intestinal immune regulation. *Nature Rev. Immunol., 8,* 435-446.

Creagh, E. M. & O'Neill, L. A. J. (2006). TLRs, NLRs and RLRs: a trinity of pathogen sensors that co-operate in innate immunity. *Trends Immunol., 27,* 352-357.

Cross, P. c. & Mercer, K. L. (1993). Cell and Tissue Ultrastrcuture. A functional perspective. *Freeman and Company.*, New York, 420 pp.

Cuesta, A., Esteban, M. A. & Meseguer, J. (2006b). Cloning, distribution and up-regulation of the teleost fish MHC class II alpha suggest a role for granulocytes as antigen-presenting cells. *Mol. Immunol., 46,* 1275-1285.

Cuesta, Meseguer, J. & Esteban, M. A. (2008). The antimicrobial peptide hepcidin exerts an important role in the innate immunity against bacteria in the bony fish gilthead seabream. *Mol. Immunol., 45,* 2333-2342.

Cuesta, A., Muñoz, P., Rodríguez, A., Salinas, I., Sitjà-Bobadilla, A., Alvarez-Pellitero, P., Esteban, M. A. & Meseguer, J. (2006a). Gilthead seabream (*Sparus aurata* L.) innate defence against the parasite *Enteromyxum leei* (Myxozoa). *Parasitology., 132,* 95-104.

Cuesta, A., Salinas, I., Rodríguez, A., Muñoz, P., Sitjà-Bobadilla, A., Alvarez-Pellitero, P., Meseguer, J. & Esteban, M. A. (2006c). Cell-mediated cytotoxicity is the main innate immune mechanism involved in the cellular defence of gilthead seabream (Teleostei: Sparidae) against *Enteromyxum leei* (Myxozoa). *Parasite Immunol., 28,* 657-665.

Dahan, S., Roth-Walter, F., Arnaboldi, P., Agarwal, S. & Mayer, L. (2007). Epithelia: lymphocyte interactions in the gut. *Immunol. Rev., 215,* 243-253.

Dalmo, R. A., Ingebrigtsen, K. & Bogwald, J. (1997). Non-specific defence mechanisms in fish, with particular reference to the reticuloendothelial system (RES). *J. Fish Dis., 20,* 241-273.

Davies, S. J., Smith, S. J., Lim, K. C., Zhang, H., Purchio, A. F., McKerrow, J. H. & West, D. B. (2005). In vivo imaging of tissue eosinophilia and eosinopoietic responses to schistosome worms and eggs. *Int. J. Parasitol., 35,* 851-859.

Dezfuli, B. S., Arrighi, S., Domeneghini, C. & Bosi, G. (2000). Immunohistochemical detection of neuromodulators in the intestine of *Salmo trutta* L. naturally infected with *Cyatocephalus truncatus* Pallas (Cestoda). *J. Fish Dis., 23,* 265-273.

Dezfuli, B. S., Capuano, S. & Manera, M. (1998). A description of rodlet cells from the alimentary canal of *Anguilla anguilla* and their relationship with parasitic helminthes. *J. Fish Biol., 53,* 1084-1095.

Dezfuli, B. S., Giari, L., Simoni, E., Bosi, G. & Manera, M. (2002a). Histopathology, immunohistochemistry and ultrastructure of the intestine of *Leuciscus cephalus* (L.) naturally infected with *Pomphorhynchus laevis* (Acanthocephala). *J. Fish Dis., 25,* 7-14.

Dezfuli, B. S., Giovinazzo, G., Lui, A. & Giari, L. (2008). Inflammatory response to *Dentitruncus truttae* (Acantocephala) in the intestine of brown trout. *Fish Shellfish Immunol., 24,* 276-733.

Dezfuli, B. S., Manera, M., Onestini, S. & Rossi, R. (1997). Histopathology of the alimentary canal of *Anguilla anguilla* L. associated with digenetic trematodes: a light and electron microscopic study. *J. Fish Dis., 20,* 317-322.

Dezfuli, B. S., Pironi, P., Giari, L., Domeneghini, C. & Bosi, G. (2002b). Effect of *Pomphorhynchus laevis* (Acanthocephala) on putative neuromodulators in the intestine of naturally infected *Salmo trutta*. *Dis. Aquat. Organ., 51,* 27-35.

Dezfuli, B. S., Pironi, F., Shinn, A. P., Manera, M. & Giari, L. (2007b). Histopathology and ultrastructure of *Platichthys flesus* naturally infected with *Anisakis simplex* s.l. larvae (Nematoda: Anisakidae). *J.Parasitol., 93,* 1416-1423.

Dezfuli, B. S., Pironi, F., Simnoni, E., Shinn, A. P. & Giari, L. (2007a). Selected pathological, immunohistochemical and ultrastructural changes associated with an infection by *Diphyllobothrium dendriticum* (Nitzsch, 1824) (Cestoda) plerocercoids in *Coregonus lavaretus* (L.) (Coregonidae). *J. Fish Dis., 30,* 471-482.

Dharmani, P., Srivastava, V., Kissoon-Singh, V. & Chadee, K. (2009). Role of intestinal mucins in innate host defense mechanisms against pathogens. *J. Innate Immun., 1,* 123-135.

Didierlaurent, A., Sirard, J. C. & Kraehenbuhl, J. P. (2002). How the gut senses its content. *Cell. Microbiol., 6,* 61-72.

Dixon, B. & Stet, R. J. M. (2001). The relationship between major histocompatibility receptors and innate immunity in teleost fish. *Dev. Comp. Immunol., 25,* 683-699.

Domeneghini, C., Pannelli Strain, R. & Veggetti, A. (1998). Gut glycoconjugates in *Sparus aurata* L. (Pisces, Teleostei). A comparative histochemical study in larval and adult ages. *Histol. Histopathol., 13,* 359-372.

Domeneghini, C., Arrighi, S., Radaelli, G., Bosi, G. & Veggetti, A. (2005). Histochemical analysis of glycoconjugate secretion in the alimentary canal of *Anguilla anguilla L. Acta Histochem.*, *106,* 477-487.

Dommett, R., Zilbauer, M., George, J. T. & Bajaj-Elliott, M. (2005). Innate immune defence in the human gastrointestinal tract. *Mol. Immunol.*, *42,* 903-912.

Donato, R. (2003). Intracellular and extracellular roles of S100 proteins. *Microsc. Res. Techn.*, *60,* 540-551.

Dorin, D., Martín, P., Sire, M. F., Smal, J. & Vernier, J. M. (1993). Protein uptake by intestinal macrophages and eosinophilic granulocytes in trout – an in-vivo study. *Biol. Cell*, *79,* 37-44.

Du Pasquier, L. (2001). The immune system of invertebrates and vertebrates. *Comp. Biochem. Physiol. B.*, *129,* 1-15.

Ducoroy, P., Lesourd, M., Padros, M. R. & Tournefier, A. (1999). Natural and induced apoptosis during lymphocyte development in the axolotl. *Dev. Comp. Immunol.*, *23,* 241-252.

Eckmann, L. (2003). Mucosal defences against *Giardia. Parasit. Immunol.*, *25,* 259-270.

Eichinger, D. A. (2001). Role for a galactose lectin and its ligands during encystment of *Entamoeba*l. *J. Eukaryot. Microbiol.*, *48,* 17-21.

Eisenbarth, S. C. & Flavell, R. A. (2009). Innate instruction of adaptive immunity revisited: the inflammasome. *EMBO Mol. Med.*, *1,* 92-98.

Else, K. J. (2005). Have gastrointestinal nematodes outwitted the immune system? *Parasite Immunol.*, *27,* 4407-4415.

Estensoro, I., Bermúdez, R, Losada, A. P., Quiroga, M. I., Pérez-Sánchez, J. P. Alvarez-Pellitero' P. & Sitjà-Bobadilla, A. (2009). Effect of E*nteromyxum leei* (Myxozoa) on gastrointestinal neuromodulators and cell apoptosis of gilthead sea bream (S*parus aurata*). *14th EAFP International Conference. Prague*, 14-19 September 2009, *Book of Abstracts*, 264-265.

Ewart, K. V., Johnson, S. C. & Ross, N. W. (1999). Identification of a pathogen-binding lectin in salmon serum. *Comp. Biochem. Physiol.*, C *123,* 9-15.

Farthing, M. J., Pereira, M. E. A. & Keusch, G. T. (1986). Description and characterization of a surface lectin from *Giardia lamblia. Infect. Immun.*, *51,* 661-667.

Fenton, N. B., Arreguin, L. B., Méndez, C. F. & Arreguin, R. A. (2004). Purification and characterization of liver lectins from a lizard, *Sceleropus spinosus. Prep. Biochem. Biotechnol.*, *34,* 153-168.

Ferri, D., Liquori, G. E., Natale, L., Santarelli, G. & Scillitani, G. (2001). Mucin histochemistry of the digestive tract of the red-legged frog *Rana aurora aurora. Acta Histochem. 103,* 225-237.

Fiertak, A. & Kilarski, W. M. (2002). Glycoconjugates of the intestinal goblet cells of four cyprinids. *Cell Mol. Life Sci.*, *59,* 1724-1733.

Fischer, U., Utke, K., Somamoto, T., Kollner, B., Ototake, M. & Nakanishi, T. (2006). Cytotoxic activities of fish leucocytes. *Fish Shellfish Immunol.*, *20,* 209-226.

Flajnik, M. F. & Du Pasquier, L. (2004). Evolution of innate and adaptive immunity: can we draw a line? *Trends Immunol.*, *25,* 640-644.

Flajnik, M. F. & Kasahara, M. (2001). Comparative genomics of the MHC: Glimpses into the evolution of the adaptive immune system. *Immunity.*, *15,* 351-362.

Fleurance, R., Sauvegrain, C., Marques, A., Le Breton, A., Guereaud, C., Cherel, Y. & Wyers, M. (2008). Histopathological changes caused by *Enteromyxum leei* infection in farmed sea bream *Sparus aurata*. *Dis. Aquat. Organ.*, *79,* 219-228.

Foott, J. S., Harmon, R. & Stone, R. (2004). Effect of water temperature on non-specific immune function and ceratomyxosis in juvenile chinook salmon and steelhead from the Klamath River. *Calif. Fish Game.*, *90,* 71-84.

Fournier-Betz, V., Quentel, C., Lamour, F. & Le Ven, A. (2000). Immunocytochemical detection of Ig-positive cells in blood, lymphoid organs and the gut associated lymphoid tissue of the turbot (*Scophthalmus maximus*). *Fish Shellfish Immunol.*, *10,* 187-202.

Frederick, J. R. & Petri W. A. Jr. (2005). Roles of glactose/N-acetylgalactosamine-binding lectin of *Entamoeba* in parasite virulence and differentiation. *Glycobiology.*, *15,* 53R-59R.

Fry, A. E. & Kaltenbach, J. C. (1999). Histology and lectin-binding patterns in the digestive tract of the carnivorous larvae of the anuran *Ceratoprhys ornata*. *J. Morphol.*, *241,* 19-32.

Fujita, T., Matsushita, M. & Endo, Y. (2004). The lectin-complement pathway - its role in innate immunity and evolution. *Immunol. Rev.*, *198,* 185-202.

Galluzzi, L. & Kroemer, G. (2008). Necroptosis: A specialized pathway of programmed necrosis. *Cell.*, *135,* 1161-1163.

Garate, M., Cubillos, I., Marchant, J. & Pajwani, N. (2005). Biochemical characterization and functional studies of *Acanthamoeba* mannose-binding protein. *Infect. Immun.*, *73,* 5775-5781.

Gasque, Ph. (2004). Complement: a unique innate immune sensor for danger signals. *Mol. Immunol.*, *41,* 1089-1098.

Gaucher, D. & Chadee, K. (2002). Construction and immunogenicity of a codon-optimized *Entamoeba histolytica* Gal-lectin-based DNA vaccine. *Vaccine.*, *20,* 3244-3253.

Gay, N. J. & Gangloff, M. (2007). Structure and function of toll receptors and their ligands. *Ann. Rev. Biochem.*, *76,* 141-165.

Gazinelli, R. T. & Denkers, E. Y. (2006). Protozoan encounters with Toll-like receptor signaling pathways: implications for host parasitism. *Nat. Rev. Immunol.*, *6,* 895-906.

Glaberman, S., Du Pasquier, L. & Caccone, A. (2008). Characterization of a nonclassical class I MHC gene in a reptile, the Galápagos marine iguana (*Amblyrhynchus cristatus*). *PLoS One, 3*: e2859. doi:10.1371/journal.pone.0002859.

Glaberman, S., Moreno, M. A. & Caccone, A. (2009). Characterization and evolution of MHC class II B genes in Galápagos marine iguanas (*Amblyrhynchus cristatus*). *Dev. Comp. Immunol., 33*, 939-947.

Glazebrook, J. C. & Campbell, R. S. F. (1987). Diseases of barramundi (*Lates calcarifer*) in Australia: a review. In: Copland J. W., Gery, D. L. (Eds.) Management of wild and cultured sea bass/barramundi (*Lates calcarifer*). *Australian Centre for International Agricultural Research, Canberra.*, 204-206.

Golomazou, E., Athanassopoulou, F., Karagouni, E., Tsagozis, P., Tsantilas, H. & Vagianou, St. 2006. Experimental transmission of *Enteromyxum leei* Diamant, Lom and Dykova, 1994 in sharpsnout sea bream, *Diplodus puntazzo* C. and the effect on some innate immune parameters. *Aquaculture.*, *260,* 44-53.

Goyal, R. K. & Hirano, I. (1996). The enteric nervous system. *New Engl. J. Med.*, *334,* 1106-1115.

Graczyk, T. K., Cranfield, M. R., Mann J. & Strandberg, J. D. (1998). Intestinal *Cryptosporidium* infection in the Egyptian tortoise, *Testudo kleinmanni. Int. J. Parasitol., 28,* 1885-1888.

Grant, J., Mahanty, S., Khadir, A., McLean, D., Kokoskin, E., Yeager, B., Joseph, L., Diaz, J., Gotuzzo, E., Mainville, N. & Ward, B. J. (2001). Wheat germ supplement reduces cyst and trophozoite passage in people with giardiasis. *Am. J. Trop. Med. Hyg., 65,* 705-710.

Green, S. L., Bouley, D. M., Josling, C. A. & Fayer, R. (2003). Cryptosporidiosis associated with emaciation and proliferative gastritis in a laboratory-reared South African clawed frog (*Xenopus laevis*). *Comp. Med., 53,* 81-84.

Grove, S., Johansen, R., Reitan, L. J. & Press, C. M. (2006). Immune- and enzyme histochemical characterisation of leucocyte populations within lymphoid and mucosal tissues of Atlantic halibut (*Hippoglossus hippoglossus*). *Fish Shellfish Immunol., 20,* 693-708.

Gruden-Movsesijan, A. & Milosavljevic, Lj. S. (2006). The involvement of the macrophage mannose receptor in the innate immune response to infection with parasite *Trichinella spiralis. Vet. Immunol. Immunopathol., 109,* 57-67.

Guk, S. M., Yong, T. S. & Chai, J. Y. (2003). Role of murine intestinal intraepithelial lymphocytes and lamina propria lymphocytes against primary and challenge infections with *Cryptosporidium parvum. J.Parasitol., 89,* 270-275.

Gunimaladevi, I., Savan, r. & Sakai, M. (2006). Identification, cloning and characterization of interleukin-17 and its family from zebrafish. *Fish Shellfish Immunol., 21,* 393-403.

Gut, J. & Nelson, R. G. (1999). *Cryptosporidium parvum*: Lectins mediate irreversible inhibition of sporozoite infectivity in vitro. *J. Euk. Microbiol., 46,* 48S-49S.

Hamuro, K., Suetake, H., Saha, R. N., Kikuchi, K. & Suzuki, Y. (2007). A teleost polimeric Ig receptor exhibiting two Ig-like domains transports tetrameric IgM into the skin. *J. Immunol., 178,* 5682-5686.

Hershberg, R. M. & Mayer, L. F. (2000). Antigen processing and presentation by intestinal epithelial cells – polarity and complexity. *Immunol. Today., 21,* 123-128.

Hirabayashi, J., Kusunoki, T. & Kasai, K. (1991). Complete primary structure of a galactose-specific lectin from the venom of the rattlesnake *Crotalus atrox. J. Biol. Chem., 266,* 2320-2326.

Hitomi, J., Christofferson, D. E., Ng, A., Yao, J., Degterev, A. & Xavier, R. J. (2008). Identification of a molecular signaling network that regulates a cellular necrotic cell death pathway. *Cell., 135,* 1311-1323.

Hokke, C. H. & Yazdanbakhsh, M. (2005). Schistosome glycans and innate immunity. *Parasit. Immunol., 27,* 257-264.

Holland, M. C. H. & Lambris, J. D. (2002). The complement system in teleosts. *Fish Shellfish Immunol., 12,* 399-420.

Holmén, J. M., Olson, F. J., Karlsson, H. & Hansson, G. C. (2002). Two glycosylation alterations of mouse intestinal mucins due to infection caused by the parasite *Nippostrongylus brasiliensis. Glyconconj. J., 19,* 67-75.

Holmgren, J. & Czerkinsky, C. (2005). Mucosal immunity and vaccines. *Nat. Med., 11,* S45-S53.

Hölzl, M. A., Hofer, J., Steinberger, P., Pfistershammer, K. & Zlabinger, G. J. (2008). Host antimicrobial proteins as endogenous immunomodulators. *Immunol. Lett., 119,* 4-11.

Honda, S., Kashiwagi, M., Miyamoto, K., Takei, Y. & Hirosi, S. (2000). Multiplicity, structures and endocrine and exocrine natures of eel-fucose-binding lectins. *J. Biol. Chem., 275,* 33151-33157.

Hordvik, I., Torvund, J., Moore, L. & Endresen, C. (2004). Structure and organization of the T cell receptor alpha chain genes in Atlantic salmon. *Mol. Immunol., 41,* 553-559.

Hughes, A. L. & Piontkivska, H. (2008). Functional diversification of the Toll-like receptor gene family. *Immunogenetics., 60,* 249-256.

Huising, M. O., Stet, R. J. M., Savelkoul, H. F. J. & Verburg-Van Kemenade, B. M. L. (2004). The molecular evolution of the interleukin-1 family of cytokines; IL-18 in teleost fish. Dev. *Comp. Immunol., 28,* 395-413.

Igawa, D., Sakai, M. & Savan, R., (2006). An unexpected discovery of two interferon gamma-like genes along with interleukin (IL)-22 and -26 from teleost: IL-22 and -26 genes have been described for the first time outside mammals. *Mol. Immunol., 43,* 999-1009.

Imagawa, T., Hashimoto, Y., Kon, Y. & Sugimura, M. (1990). Lectin histochemistry as special markers for rodlet cells in carp, *Cyprinus carpio* L. *J. Fish Dis., 13,* 537-540.

Imberty, A. & Varrot, A. (2008). Microbial recognition of human cell surface glycoconjugates. *Curr. Op. Struct. Biol., 18,* 567-576.

Ishii, A., Kawasaki, M., Matsumoto, M., Tochinai, S. & Seya, T. (2007b). Phylogenetic and expression analysis of amphibian *Xenopus* Toll-like receptors. *Immunogenetics., 59,* 281-293.

Ishii, A., Matsuo, A., Sawa, H., Tsujita, T., Shida, K., Matsumoto, M. & Seya, T. (2007a). Lamprey TLRs with properties distinct from those of the variable lymphocyte receptors. *J. Immunol., 178,* 397-406.

Ishikawa, H., Naito, T., Iwanaga, T., Takahashi-Iwanaga, H., Suematsu, M., Hibi, T. & Nanno, M. (2007). Curriculum vitae of intestinal intraepithelial T cells: their developmental and behavioural characteristics. *Immunol. Rev., 215,* 154-165.

Ishikawa, N., Horii, Y. & Nawa, Y. (1993). Immune-mediated alteration of the terminal sugars of goblet cell mucins in the small intestine of *Nippostrongylus brasiliensis*-infected rats. *Immunology., 78,* 303-307.

James, E. R. & Green, D. R. (2004). Manipulation of apoptosis in the host-parasite interaction. *Trends Parasitol., 20,* 280-287.

Janeway, C. A. & Medzhitov, R. (2002). Innate immune recognition. *Ann. Rev. Immunol., 20,* 197-216.

Jang, M. H. & Kweon, M. N. & Iwatani, K., Yamamoto, M., Terahara, K., Sasakawa, C., Suzuki, T., Nochi, T., Yokota, Y., Rennert, P. D., Hiroi, T., Tamagawa, H., Lijima, H., Kunisawa, J., Yuki, Y. & Kiyono, H. (2004). Intestinal villous M cells: An antigen entry site in the mucosal epithelium. *Proc. Natl. Acad. Sci., 101,* 6110-6115.

Jendrysek, S., Steinhagen, D., Drommer, W. & Körting, W. (1994). Carp coccidiosis: intestinal histo- and cytopathology under *Goussia carpelli* infection. *Dis. Aquat. Organ., 20,* 171-182.

Jirků, M., Valigurová, A., Koudela, B., Křižek, J., Modrý, D. & Šapeta, J. (2008). New species of *Cryptosporidium* Tyzzer, 1907 (Apicomplexa) from amphibian host: morphology, biology and phylogeny. *Folia Parasitol., 55,* 81-94.

Johansson, C. & Kelsall, B. L. (2005). Phenotype and function of intestinal dendritic cells. *Semin. Immunol., 17,* 284-294.

Jolles, P. & Jolles, J. (1984). What's new in lysozyme research? Always a model system, today as yesterday. *Mol. Cell. Biochem., 63,* 165-189.

Kaattari, S. L., Klemer, J. V. & Evans, D. A. (1999). Teleost antibody structure: simple prototype or elegant alternative? *Bull. Eur. Ass. Fish Pathol., 19,* 245-249.

Kaiko, G. E., Horvat, J. C., Beagley, K. W. & Hansbro, M. (2007). Immunological decision-making: how does the immune system decide to mount a helper T-cell response? *Immunology., 123,* 326-338.

Kamal, M., Wakelin, D., Ouellette, A. J., Smith, A., Podolsky, K. & Mahida, R. (2001). Mucosal T cells regulate Paneth and intermediate cell numbers in the small intestine of *T. spiralis*-infected mice. *Clin. Exp. Immunol., 126,* 117-125.

Karlsson, N. G., Olson, F. J., Jovall, P. A., Andersch, Y., Enerbäck, L. & Hansson, G. C. (2000). Identification of transient glycosylation and sialilated mucin oligosaccharides during infection by the rat intestinal parasite *Nippostrongylus brasiliensis. Biochem. J., 350,* 805-814.

Kawai, T. & Akira, S. (2005). Pathogen recognition with Toll-like receptors. *Curr. Op. Immunol., 17,* 338-344.

Khan, W. I. (2008). Physiological changes in the gastrointestinal tract and host protective immunity: learning from the mouse *Trichinella spiralis* model. *Parasitology., 135,* 671-682.

Klabundc, J., Berger, J., Jensenius, J. C., Klinkert, M. Q., Zelck, U. E., Kremsner, P. G. & Kun, J. F. J. (2000). *Schistosoma mansoni*: adhesion of mannan-binding lectin to surface glycoproteins of cercariae and adult worms. *Exp. Parasitol., 95,* 231-239.

Klein, N. J. (2005). Mannose-binding lectin: do we need it? *Mol. Immunol., 42,* 919-924.

Klion, A. D. & Nutman, T. B. (2004). The role of eosinophils in host defense against helminth parasites. *J. Allergy Clin. Immunol., 113,* 30-37.

Knight, P. A., Brown, J. K. & Pemberton, A. D. (2008). Innate immune response mechanisms in the intestinal epithelium: potential roles for mast cells and goblet cells in the expulsion of adult *Trichinella spiralis. Parasitology, 135,* 655-670.

Kolls, J. K., McCray, P. B. & Chan, Y. R. (2008). Cytokine-mediated regulation of antimicrobial proteins. Nature Rev. *Immunol., 8,* 829-835.

Komatsu, K., Tsutsui, S., Hino, K., Araki, K., Yoshiura, Y., Yamamoto, A., Nakamura, O. & Watanabe, T. (2009). Expression profiles of cytokines releases in intestinal epithelial cells of the rainbow trout, *Oncorhynchus mykiss*, in response to bacterial infection. *Dev. Comp. Immunol., 33,* 499-506.

Komiya, T., Tanigawa, Y. & Hirohashi, S. (1998). Cloning of the novel gene intelectin, which is expressed in intestinal Paneth cells in mice. *Biochem. Byophys. Res. Comm. 251,* 759-762.

Koninkx, J. F., Mirck, M. H., Hendriks, H. G., Mouwen, J. M. & Van Dijk, J. E. (1988). *Nippostrongylus brasiliensis*: histochemical changes in the composition of mucins in goblet cells during infection in rats. *Exp. Parasitol., 65,* 84-90.

Lai, Y. Gallo, R. L. (2009). AMPed up immunity: how antimicrobial peptides have multiple roles in immune defense. *Trends Immunol., 30,* 131-141.

Lee, J., Gonzales-Navajas, J. M. & Raz, E. (2008). The "polarizing-tolerizing" mechanism of intestinal epithelium : its relevance to colonic homeostasis. *Semin. Immunol., 30,* 3-9.

Leknes, I. L. (2007). Eosinophilic granule cells and endocytic cells in intestinal wall of pearl gouramy (Anabantidae: Teleostei). *Fish Shellfish Immunol., 23,* 897-900.

Leulier, F. & Lemaitre, B. (2008). Toll-like receptors – taking and evolutionary approach. *Nat Rev Genet., 9,* 165-178.

Levine, A. D. & Fiocchi, C. (2001). Regulation of life and death in lamina propria T cells. *Semin. Immunol., 13,* 195-199.

Li, J., Barreda, D. R., Zhang, Y. A., Boshra, H., Gelman, A. E., LaPatra, S., Tort, L. & Sunyer, O. (2006). B lymphocytes from early vertebrates have potent phagocytic and microbicidal abilities. *Nature Immunol., 7,* 1116-1124.

Lilleeng, E., Penn, M. H., Haugland, Ø., Xu, Ch., Bakke, A. M., Krogdahl, Å., Landsverk, T. & Frøystad-Saugen, M. K. (2009). Decreased expression of TGF-β, GILT and T-cell markers in the early stages od soybean enteropathy in Atlantic salmon (*Salmo salar* L.). *Fish Shellfish Immunol. 27*, 65-72.

Lin, B., Cao, Z., Su, P., Zhang, H., Li, M., Lin, Y., Zhao, D., Shen, Y., Jing, C., Chen, S. & Xu, A. (2009). Characterization and comparative analyses of zebrafish intelectins: Highly conserved sequences, diversified structures and functions. *Fish Shellfish Immunol., 26,* 396-405.

Linden, S. K., Sutton, P., Karlsson, N. G., Korolik, V. & McGuckin, M. A. (2008). Mucins in the mucosal barrier to infection. *Mucosal Immunol., 1*, 183-197.

Liquori, G., Mastrodonato, M., Zizza, S. & Ferri, D. (2007). Glycoconjugate histochemistry of the digestive tract of *Triturus carnifex* (Amphibia, Caudata). *J. Mol. Hist., 38,* 191-199.

Liu, J., Enomoto, S., Lancto, C. A., Abrahamsen, M. S. & Rutherford, M. S. (2008). Inhibition of apoptosis in *Cryptosporidium parvum*-infected intestinal epithelial cells is dependent on surviving. *Infect. Immun., 76,* 3784-3792.

Lobb, C. J. & Clem, L. W. (1981). Phylogeny of immunoglobulin structure and function. XI. Secretory immunoglobulin in the cutaneous mucus of the sheepshead. *Dev. Comp. Immunol., 5,* 271-282.

Lom, J. & Dyková, I. (1992). Protoozoan parasites of fish. *Developments in Aquaculture and Fisheries Science.*, Vol. *26,* Elsevier, Amsterdam, 315 pp.

López-Castejón, G., Sepulcre, M. P., Mulero, I., Pelegrín, P., Meseguer, J. & Mulero V. (2008). Molecular and functional characterization of gilthead seabream *Sparus aurata* caspase-1: The first identification of an inflammatory caspase in fish. *Mol. Immunol., 45,* 49-57.

Losada, A. P. , Bermúdez, R. Faílde, L. D. Nieto J. M. & Quiroga, M. I. (2009). Caspase-3 mediates apoptosis in the enteromyxosis of turbot (*Psetta maxima* L.). *14th EAFP International Conference, Prague*, 14-19 September 2009, *Book of Abstracts*, 266.

Madrid, J. F., Ballesta, J., Pastor; L. M., Pérez-Tomás, R. & Hernández, F. (1989). Distribution of mucins in the mucosa of the digestive tract of reptiles. A histochemical study. *Acta Histochem., 85,* 117-129.

Maier, V. H., Schmitt, C. N. Z., Gudmunsdottir, S. & Gudmunsson, G. H. (2008). Bacterial DNA indicated as an important inducer of fish cathelicidins. *Mol. Immunol., 45,* 2352-2358.

Maloy, K. J. & Kullberg, M. C. (2008). IL-23 and Th17 cytokines in intestinal homeostasis. *Mucosal Immunol., 1,* 339-349.

Manera, M. & Dezfuli, B. S. (2004). Rodlet cells in teleosts: a new insight into their nature and functions. *J. Fish Biol., 65,* 597-619.

Marchalonis, J. J., Adelman, M. K., Schulter, S. F. & Ramsland, P. A. (2006). The antibody repertoire in evolution: Chance, selection, and continuity. *Dev. Comp. Immunol.*, *30,* 223-247.

Martin-Antonio, B., Jiménez-Cantizano, R. M., Salas-Leiton, E. & Infante, C. (2009). Genomic characterization and gene expression analysis of four hepcidin genes in the redbanded sea bream (*Pagrus auriga*). *Fish Shellfish Immunol.*, *26,* 483-491.

Maruyama, H., Hirabayashi, Y., El-Malky, M., Okamura, S., Aoki, M., Itagaki, T., Nakamura-Uchiyama, F., Nawa, Y., Shimada, S. & Ohta, N. (2002). *Strongyloides venezuelensis*: longitudinal distribution of adult worms in the host intestine is influenced by mucosal sulfated carbohydrates. *Exp. Parasitol.*, *100,* 179-185.

Maruyama, H., Yabu, Y., Yoshida, A., Nawa, Y. & Ohta, N. (2000). A role of mast cell glycosaminoglycans for the immunological expulsion of intestinal nematode, *Strongyloides venezuelensis*. *J. Immunol.*, *164,* 3749-3754.

Mazon, A. F., Huising, M. O., Taverne-Thiele, A. J., Bastiaans, J. & Verburg-Van Kemenade, B. M. L. (2007). The first appearance of rodlet cells in carp (*Cyprinus carpio* L.) ontogeny and their possible roles during stress and parasite infection. *Fish Shellfish Immunol.*, *22,* 27-37.

McCole, D. F., Eckmann, L., Laurent, F. & Kagnoff, M. F. (2000). Intestinal epithelial cell apoptosis following *Cryptosporidium parvum* infection. *Infect. Immun.*, *68,* 1710-1713.

McCraken, V. J. & Lorenz, R. G. (2001). The gastrointestinal ecosystem: a precarious alliance among epithelium, immunity and microbiota. *Cell Microbiol.*, *3,* 1-11.

McDonald, V. (1999). Gut intraepithelial lymphocytes and immunity to coccidia. *Parasitol. Today.*, *15,* 483-487.

McGuinness, D. H., Dehal, P. K. & Pleass, R. J. (2003). Pattern recognition molecules and innate immunity to parasites. *Trends Parasitol.*, *15,* 312-319.

McGreal, E. P., Martinez-Pomares, L. & Gordon, S. (2004). Divergent roles for C-type lectins expressed by cells of the innate immune system. *Mol. Immunol.*, *41,* 1109-1121.

McKay, D. M. & Fairweather, I. (1997). A role for the enteric nervous system in the response to helminth infections. *Parasitology.*, *13,* 63-69.

Medzhitov, J. R. (2007). Recognition of microorganisms and activation of the immune response. *Nature.*, *449,* 819-826.

Medzhitov, J. R. & Janeway, C. A. (2002). Decoding the patterns of self and nonself by the innate immune system. *Science.*, *296,* 298-300.

Meeker, N. D. & Trede N. S. (2008). Immunology and zebrafish: Spawning new models of human disease. *Dev. Comp. Immunol.*, *32,* 745-747.

Miller, H. C., Belov, K. & Daugherty, C. (2006). MHC class I genes in the tatuara (*Sphenodon* spp.): Evolution of the MHC in an ancient reptilian order. *Mol. Biol. Evol.*, *23,* 949-956.

Moncada, D. M. & Chadee, K. (2002). Production, structure, and function of gastrointestinal mucins: In: M. J., Blaser (Ed): *Infections of the Gastrointestinal Tract.* Philadelphia, Lippincott, Williams & Wilkins., 57-79.

Moncada, D. M., Kammanadiminti, S. J. & Chadee, K. (2003). Mucin and toll-like receptors in host defense against intestinal parasites. *Trends Parasitol.*, *19,* 306-311.

Morrison, R. N., Koppang, E. O., Hordvik, I. & Nowak, B. F. (2006). MHC class II+ cells in the gills of Atlantic salmon (*Salmo salar* L.) affected by amoebic gill disease. *Vet. Immunol. Immunpathol.*, *109,* 297-303.

Mulder, I. E., Wadsworth, S. & Secombes, C. J. (2007). Cytokine expression in the intestine of rainbow trout (*Oncorhynchus mykiss*) during infection with *Aeromonas salmonicida. Fish Shellfish Immunol., 23,* 747-759.

Mulero, I., Chaves-Pozo, E., García-Alcázar, A., Meseguer, J., Mulero, V. & García Ayala, A. (2007b). Distribution of the professional phagocytic granulocytes of the bony fish gilthead seabream (*Sparus aurata* L.) during the ontogeny of lymphomyeloid organs and pathogen entry sites. *Dev. Comp. Immunol., 31,* 1024-1033.

Mulero, I., Noga, E. J., Meseguer, J., García-Ayala, A. & Mulero, V. (2008). The antimicrobial peptides piscidins of professional phagocytic granulocytes of fish and are delivered to the bacteria-containing phagosome upon phagocytosis. *Dev. Comp. Immunol., 32,* 1531-1538.

Mulero, I., Sepulcre, M. P., Meseguer, J., García-Ayala, A. & Mulero, V. (2007a). Histamine is stored in mast cells of most evolutionary advanced fish and regulates the fish inflammatory response. *Proc. Natl. Acad. Sci., 104,* 19434-19439.

Muñoz, P., Cuesta, A., Athanassopoulou, F., Golomazou, H., Crespo, S., Padrós, F., Sitjà-Bobadilla, A., Albiñana, G., Esteban, M.A., Alvarez-Pellitero, P. & Meseguer, J. (2007). Sharpsnout sea bream (*Diplodus puntazzo*) humoral immune response against the parasite *Enteromyxum leei* (Myxozoa). *Fish Shellfish Immunol., 23,* 636-645.

Nair, M. G., Guild, K. J., Artis, D. (2006). Novel effector molecules in type 2 inflammation: lessons drawn from helminth infections and allergy. *J. Immunol., 177,* 1393-1399.

Nakamura, O., Inaga, Y., Suzuki, S., Tsutsui, S., Muramoto, K., Kamiya, H. & Watanabe, T. (2007). Possible immune functions of congerin, a mucosal galectin, in the intestinal lumen of Japanese conger eel. *Fish Shellfish Immunol., 23,* 683-692.

Nakamura, O., Watanabe, T., Kamiya, H. & Muramoto, K. (2001). Galectin containing cells in the skin and mucosal tissues in Japanese conger eel, *Conger myriaster*: an immunohistochemical study. *Dev.Comp. Immunol., 25,* 431-437.

Nawa, Y., Ishikawa, N., Tsuchiya, K., Horii, Y., Abe, T., Khan, A., Shi, B., Itoh, H., Ide, H. & Uchiyama, F. (1994). Selective effector mechanisms for the expulsion of intestinal helminths. *Parasite Immunol., 16,* 333-338.

Neuhaus, H., van der Marel, M., Caspari, N., Meyer, W., Enss, M. L. & Steinhagen, D. (2007). Biochemical and histochemical effects of perorally applied endotoxin on intestinal mucin glycoproteins of the common carp *Cyprinus carpio. Dis. Aquat. Organ., 77,* 17-27.

Neumann, N. F., Stafford, J. L. Barreda, D., Ainsworth A. J. & Belosevic, M. (2001). Antimicrobial mechanisms of fish phagocytes and their role in host defense. *Dev. Comp. Immunol., 25,* 807-825.

Nicoletti, C., Temblay, J. N., Regoli, M., Bertelli, E. & Man, A. L. (2006). Differential regulation of dendritic cell-T cell cross talk in the gut-associated lymphoid tissue. *Mol. Immunol., 43,* 542-549.

Nikolakopoulou K. & Zardakis I. K. (2006). Molecular cloning and characterization of two homologues of mannose-binding lectin in rainbow trout. *Fish Shellfish Immunol.,* 21: 305-314.

Noga, E. J., Fan, Z. & Silphaduang, U. (2001). Histone-like proteins from fish are lethal to the parasitic dinoflagellate *Amyloodinium ocellatum. Parasitology., 123,* 57-65.

Noga, E. J., Fan, Z. & Silphaduang, U. (2002). Host site of activity and cytological effects of histone-like proteins on the parasitic dinoflagellate *Amyloodinium ocellatum*. *Dis. Aquat. Organ.*, *52,* 207-215.

Ortega-Barria, E., Ward, H. D., Keusch, G. T. & Pereira, M. E. A. (1994). Growth inhibition of the intestinal parasite *Giardia lamblia* by dietary lectin is associated with arrest of the cell cycle. *Am. Soc. Clin. Invest.*, *94,* 2283-2288.

Ohtsubo, K. & Marth, J. D. (2006). Glycosylation in cellular mechanisms of health and disease. *Cell.*, *126,* 855-867.

Ouellette, A. J. (1999). Mucosal immunity and inflammation. IV. Paneth cell antimicrobial peptides and the biology of the mucosal barrier. *Am. J. Physiol.*, 277 (Gastrointest. Liv. Physiol. 40), G257-G261.

Ouyang, W. & Valdez, P. (2008). IL-22 in mucosal immunity. *Mucosal Immunol. 1,* 335-338.

Pancer, Z. & Cooper, M. D. 2006. The evolution of adaptive immunity. *Ann. Rev. Immunol.*, *24,* 497-518.

Parillo, F., Fagioli, O. & Ceccarelli, P. (2002). Glucidic determinants expressed by the digestive apparatus of *Umbrina cirrosa* (L.) fries as revealed by lectin histochemistry. *Acta Histochem.*, *104,* 209-215.

Pasare, C. & Medzhitov, R. (2004). Toll-like receptors: linking innate and adaptive immunity. *Microb. Infect.*, *6,* 1382-1387.

Patel, N., Kreider, T., Urban, J. F. & Gause, W. C. (2009). Characterisation of effector mechanisms at the host: parasite interface during the immune response to tissue-dwelling intestinal nematode parasites. *Int. J. Parasitol.*, *39,* 13-21.

Paulsen, S. M., Sveinbjornsson, B. & Robertsen, B., (2001). Selective staining and disintegration of intestinal eosinophilic granule cells in Atlantic salmon after intraperitoneal injection of the zinc chelator dithizone. *J. Fish Biol.*, *58,* 768-775.

Pedini, V., Scocco, P., Gargiulo, A. M., Ceccarelli, P. & Lorvik, S., (2002). Glycoconjugate characterization in the intestine of *Umbrina cirrosa* by means of lectin histochemistry. *J. Fish Biol.*, *61,* 1363-1372.

Pemberton, A. D., Knight, P. A., Gamble, J., Colledge, W. H., Leem J. K., Pierce, M. & Miller, H. R. P. (2004). Innate BALB/c enteric epithelial responses to *Trichinella spiralis*: Inducible expression of a novel goblet cell lectin, intelectin-2, and its natural detection in C57BL/10 mice. *J. Immunol.*, *173,* 1894-1901.

Pérez-Tomás, R., Ballesta, J., Madrid, J. F., Pastor, L. M. & Hernández, F. (1990). Histochemical and ultrastructural study of the digestive tract of the tortoise *Testudo graeca* (Testudines). *J. Morphol.*, *204,* 235-245.

Petersen, S. V., Thiel, S. & Jensenius, J. C. 2001. The mannan-binding lectin pathway of complement activation: biology and disease association. *Mol. Immunol.*, *38,* 133-149.

Petri, W. A. Jr., Haque, R. & Mann, B. J. (2002). The bittersweet interface of parasite and host: lectin-carbohydrate interactions during human invasion by the parasite *Entamoeba histolytica*. *Ann. Rev. Microbiol.*, *56,* 39-64.

Pétrilli V., Dostert, C., Muruve, D. A. & Tschopp, J., (2007). The inflammasome: a danger sensing complex triggering innate immunity. *Curr. Opin. Immunol.*, *19,* 615-622.

Pike, K. A. & Ratcliffe, M. J. H. (2002). Cell surface immunoglobulin receptors in B cell development. *Sem. Cell Develop. Biol.*, *14,* 351-358.

Platt, A. M. & Mowat, A. M. (2008). Mucosal macrophages and the regulation of immune responses in the intestine. *Immunol. Lett.*, *119,* 22-31.

Plouffe, D. A., Hanington, P. C., Walsh, J. G., Wilson, E. C. & Belosevic, M. (2005). Comparison of select innate immune mechanisms of fish and mammals. *Xenotransplantation.*, *12,* 266-277.

Purcell, M. K., Smith, K. D., Aderem, A., Hood, L., Winton, J. R. & Roach, J. C. (2006). Conservation of Toll-like receptor signaling pathways in teleost fish. *Comp. Biochem. Physiol., D 1,* 77-88.

Randelli, E., Buonocore, F. & Scapigliati, G. (2008). Cell markers and determinants in fish immunology. *Fish Shellfish Immunol.*, *25,* 326-340.

Redondo, M. J., Cortadellas, N., Palenzuela, O. & Alvarez-Pellitero, P. (2008). Detection of carbohydrate terminals in the enteric parasite *Enteromyxum scophthalmi* (Myxozoa) and possible interactions with its fish host *Psetta maxima. Parasitol. Res.*, *102,* 1257-1267.

Redondo, M. J. & Alvarez-Pellitero, P. (2009a). E*nteromyxum leei* and E. *scophthalmi* (Myxozoa) induce modifications in the intestinal carbohydrate patterns of their hosts, S*parus aurata* L. and P*setta maxima. 14th EAFP International Conference, Prague,* 14-19 September 2009, *Book of Abstracts*, 262-263.

Redondo, M. J. & Alvarez-Pellitero, P. (2009b)Carbohydrate patterns in the digestive tract of *Sparus aurata* L. and *Psetta maxima* (L.) (Teleostei) parasitized by *Enteromyxum leei* and *E. scophthalm*i (Myxozoa). *Parasitol. Int.* (in press).

Redondo, M. J. & Alvarez-Pellitero, P. (2009c). Lectin binding to terminal carbohydrate residues in the enteric myxozoan *Enteromyxum leei* parasitizing gilthead sea bream *Sparus aurata* L. (Pisces:Teleostei). Light and electron microscope studies. *Folia Parasitol.*, *56*, in press.

Reis, M. I. R., do Vale, A., Pinto, C., Nascimento, D. S., Costa-Ramos, C., Silva, D. S.P., Silva, M. T. & dos Santos, N. M. S. (2007a). First molecular cloning and characterisation of caspase-9 gene in fish and its involvement in a Gram negative septicaemia. *Mol. Immunol.*, *44,* 1765-1775.

Reis, M. I. R., Nascimiento, D. S., Vale, A., Silva, M. T. & dos Santos, N. M. S. (2007b). Molecular cloning and characterisation of sea bass (*Dicentrarchus labrax*, L.) caspase-3 gene. *Mol. Immunol.*, *44,* 774-783.

Reite, O. B. (1998). Mast cells/eosinophilic granule cells of teleostean fish: a review focusing on staining properties and functional responses. *Fish Shellfish Immunol.*, *8,* 489-513.

Reite, O. B. & Evensen, O. (2006). Inflammatory cells of teleostean fish: a review focusing on mast cells/eosinophilic granule cells and rodlet cells. *Fish Shellfish Immunol.*, *20,* 192-208.

Renaux, S., Quere, P., Buzoni-Gatel, D., Sewald, B., Vern, Y. L., Coudert, P. & Drouet-Viard, F. (2003). Dynamics and responsiveness of T-lymphocytes in secondary lymphoid organs of rabbits developing immunity to *Eimeria intestinalis*. *Vet. Parasitol.*, *110,* 181-195.

Rinaldi, A. C. (2002). Antimicrobial peptides from amphibian skin: an expanding scenario. *Curr. Opin. Chem. Biol.*, *6,* 799-804.

Roach, J. C., Glusman, G., Rowen, L., Kaur, A., Purcell, M. K. & Smith, K. D. (2005). The evolution of vertebrate toll-like receptors. *Proc. Natl. Acad. Sci.*, *102,* 9577-9582.

Roark, C. L., Simonian, P. L., Fontenot, A. P., Born, W. K. & O'Brien, R. L. (2008). T Cells : an important source of IL-17. *Curr. Opin. Immunol.*, *20,* 353-357.

Roberts, R. J., Young, H. & Milne, J. A., (1971). Studies on the skin of plaice (*Pleuronectes platessa* L.). 1. The structure and ultrastructure of normal plaice skin. *J. Fish Biol., 4,* 87-98.

Rocha, B. (2007). The extrathymic T-cell differentiation in the murine gut. *Immunol. Rev., 215,* 166-177.

Rocha, J. S. & Chiarini-Garcia, H. (2007). Mast cell heterogeneity between two different species of *Hoplias* sp. (Characiformes: Erythrinidae): Response to fixatives, anatomical distribution, histochemical contents and ultrastructural features. *Fish Shellfish Immunol., 22,* 218-229.

Rollins-Smith, L. A., Doersam, J. K., Longcore, J. E., Taylor, S. K., Shamblin, J. C., Carey, C. & Zasloff, M. A. (2002). Antimicrobial peptide defences against pathogens associated with global amphibian declines. *Dev. Comp. Immunol., 26,* 63-72.

Rombout, J. H. M. W., Bot, H. E. & Taverne-Thiele, A. J. (1989). Immunological importance of the second gut segemnt of carp. II. Characterization of mucosal leucocytes. *J. Fish Biol., 35,* 167-168.

Rombout, J. H. M. W., Huttenhuis, H. B. T., Picchietti, S. & Scapigliati, G. (2005). Phylogeny and ontogeny of fish leucocytes. *Fish Shellfish Immunol., 19,* 441-455.

Rombout, J. H. M. W., Taverne-Thiele, A. J. & Villena, M.I. (1993a). The gut-associated lymphoid tissue (GALT) of carp (*Cyprinus carpio* L.): an immunocytochemical analysis. *Dev. Comp. Immunol., 15,* 55-66.

Rombout, J. H. M. W., Taverne, N., van dem Kamp, M. & Taverne-Thiele, A. J. (1993b). Differences in mucus and serum immunoglobulin of carp (*Cyprinus carpio* L.). *Dev. Comp. Immunol., 17,* 309-317.

Rombout, J. H. M. W., van der Tuin, S. J. L., Yang, G., Schopman, N., Mroczek, A., Hermsen, T. & Taverne-Thiele, J. J. (2008). Expression of the polymeric immunoglobulin receptor (pIgR) in mucosal tissues of common carp (*Cyprinus carpio* L.). *Fish Shellfish Immunol., 24,* 620-628.

Ross, M. H. & Pawlina, W. (2006). *Histología. Texto y atlas color con biología celular y molecular.* Editorial Médica Panamericana. Buenos Aires. Argentina., 974 pp.

Russell, S. & Lumsden, J. S. (2005). Function and heterogeneity of fish lectins. *Vet. Immunol. Immunopathol., 108,* 111-120.

Russell, S., Young, K. M., Smith, M. & Hayes, M. A. Lumsden, J. S. (2008). Identification, cloning and tissue localization of a rainbow trout (*Oncorhynchus mykiss*) intelectin-like protein that binds bacteria and chitin. *Fish Shellfish Immunol., 25,* 91-105.

Ryan, U. (2009). *Cryptosporidium* in birds, fish and amphibians. *Exp. Parasitol.* Doi:10.1016/j.exppara., 2009.002.

Salaun, B., Romero, P. & Lebecque, S. (2007). Toll-like receptors' two edge sword: when immunity meets apoptosis. *Eur. J. Immunol., 37,* 3311-3318.

Sanderson, I. R. & Walker, W. A. (2007). TLRs in the gut. I. The role of TLRs/Nods in intestinal development of homeostasis. *Am. J. Physiol. Gastrointest. Liver Physiol., 292,* G6-G10.

Sarasquete, C., Gisbert, E., Ribeiro, L., Vieira, L. & Dinis, M. T. (2001). Glycoconjugates in epidermal, branchial and digestive mucous cells and gastric glands of gilthead sea bream, *Sparus aurata*, sole, *Solea senegalensis* and Siberian sturgeon, *Acipenser baeri* development. *Eur. J. Histochem., 45,* 267-278.

Saurab, S. & Sahoo, P. K. (2008). Lysozyme : an important defence molecule of fish innate immune system. *Aquacult. Res., 39,* 233-239.

Sayed, B. A. & Brown, M. A. (2007). Mast cells as modulators of T-cell responses. *Immunol. Rev., 217,* 53-64.

Schaumburg, F., Hippe, D., Vutova, P. & Luder, C. G. K. (2006). Pro- and anti-apoptotic activities of protozoan parasites. *Parasitology, 132,* S69-S85.

Schroers, V., van der Marel, M., Neuhaus, H. & Steinhagen, D. (2009). Changes in intestinal mucus glycoproteins after peroral application of *Aeromonas hydrophyla* to common carp (*Cyprinus carpio*). *Aquaculture, 288,* 184-189.

Scocco, P., Menghi, G. & Ceccarelli, P. (1997). Histochemical differentiation of glycoconjugates occurring in the tilapine intestine. *J. Fish Biol., 51,* 848-857.

Secombes, C. J., Wang, T., Hong, S., Peddie, S., Crampe, M., Laing, K., Cunningham, C. & Zou, J. (2001). Cytokines and innate immunity of fish. *Dev. Comp. Immunol., 25,* 713-723.

Secombes, C. J., Zou, J., Laing, K., Daniels, G. D. & Cunningham, C. (1999b). Cytokine genes in fish. *Aquaculture, 172,* 93-102.

Sepulcre, M. P., Pelegrin, P., Mulero, V. & Meseguer, J. (2002). Characterisation of gilthead seabream acidophilic granulolcytes by a monoclonal antibody unequivocally points to their involvement in fish phagocytic response. *Cell Tissue Res., 308,* 97-102.

Sepulcre, M. O., Alcaraz-Pérez, F., López-Muñoz, A., Roca, F.J., Meseguer, J., Catuela, M.I. & Mulero, V. (2009). Evolution of lipopolysaccharide (LPS) recognition and signalling: fish TLR4 does not recognize LPS and negatively regulates NF-κB activation. *J. Immunol., 182,* 1836-1845.

Shang., N., Sun, X. F., Hu, W., Wand, Y. P. & Guo, Q. L. (2008). Molecular cloning and characterization of common carp (Cyprinus carpio L.) TCRγ and CD3γ/δ chains. *Fish Shellfish Immunol., 24,* 412-425.

Shao, L., Serrano, D. & Mayer, L. (2001). The role of epithelial cells in immune regulation in the gut. *Sem. Immunol., 13,* 163-175.

Sharon, N. (2006). Carbohydrates as future anti-adhesion drugs for infectious diseases. *Bioch. Biophys.*, Acta, *1760,* 527-537.

Sharon, N. (2008). Lectins: past, present and future. *Biochem. Soc. Trans., 36,* 1457-1460.

Sharon N. & Lis H. (1989). Lectins as cell recognition molecules. *Science*, 246, 227-246.

Shi, M., Huther, S., Burkhardt, E. & Zahner, H. (2001). Lymphocyte subpopulations in the caecum mucosa of rats after infections with *Eimeria separata*: early responses in naive and immune animals to primary and challenge infections. *Int. J. Parasitol., 31,* 49-55.

Silphaduang, U. & Noga, E. J. (2001). Peptide antibiotics in mast cells of fish. *Nature., 414,* 268-269.

Silphaduang, U., Colorni, A. & Noga, E. J. (2006). Evidence for widespread distribution of piscidin antimicrobial peptides in teleost fish. *Dis. Aquat. Organ., 72,* 241-252.

Sire, M. F. & Vernier, J. M. (1995). Partial characterization of eosinophilic granule cells (EGCs) and identification of mast cells of the intestinal lamina propria in rainbow trout (*Oncorhynchus mykiss*). *Biochemical and cytochemical study. Biol., Cell 85,* 35-41.

Sitjà-Bobadilla, A., Calduch-Giner, J., Saera-Vila, A., Palenzuela, O., Alvarez-Pellitero, P. & Pérez-Sánchez, J. (2008). Chronic exposure to the parasite *Enteromyxum leei* (Myxozoa: Myxosporea) modulates the immune response and the expression of growth, redox and immune relevant genes in gilthead sea bream, *Sparus aurata* L. *Fish Shellfish Immunol., 24,* 610-619.

Sitjà-Bobadilla, A., Redondo, M. J., Bermúdez R., Palenzuela, O., Ferreiro, I., Riaza, A., Quiroga, I., Nieto, J.M. & Alvarez-Pellitero, P. (2006). Innate and adaptive immune responses of turbot, *Scophthalmus maximus* (L.), following experimental infection with *Enteromyxum scophthalmi* (Myxosporea: Myxozoa). *Fish Shellfish Immunol., 21,* 485-450.

Smith. L. S. (1989). Digestive functions in teleost fishes. In: Halver, J.E. (Ed.). *Fish nutrition.* Academic Press, San Diego, 331-421.

Soffientino, B., Gómez-Chiarri, M. & Specker, J. (2006). Developmental changes in the stomach, intestine, and skin glycoconjugates in summer flounder (*Paralichthys dentatus*): a lectin histochemical study. *Aquaculture, 253,* 680-687.

Soga, K., Yamauchi, J., Kawai, Y., Yamada, M., Uchikawa, R., Tegoshi, T., Mitsufuji, S., Yoshikawa, T. & Arizono, N. (2008). Alteration of the expression profiles of acidic mucin, sialiltransferase, and sulfotransferases in the intestinal epithelium of rats infected with the nematode *Nippostronglus brasiliensis. Parasitol. Res., 103,* 1427-1434.

Stein, B., Stover, L., Gillem, A., Winters, K., Lee, J. H. & Chauret, C. (2006). The effect of lectins on *Cryptosporidium parvum* oocyst in vitro attachment to host cells. *J. Parasitol., 92,* 1-9.

Steinhagen, D. & Hespe, K. (1997). Carp coccidiosis: activity of phagocytic cells from common carp infected with *Goussia carpelli.* Dis. *Aquat. Organ., 31,* 155-159.

Sterud, E., Poppe, T.T. & Borno, G. (2003). Intracellular infection with *Spironucleus barkhanus* (Diplomonadida: Hexamitidae) in farmed Artic char *Salvelinus alpinus. Dis. Aquat. Organ., 56,* 155-161.

Stroband, H. W. J., Meer, H. V. D. & Timmermans, L. P. M. (1979). Regional functional differentiation in the gut of the grass-carp, *Ctenopharyngodon idella. Histochemistry, 64,* 235-249.

Studnicka, M. & Siwicki, A. (1990). The nonspecific immunological response in carp (*Cyprinus carpio* L.) during natural infection with *Eimeria subepithelialis. Bamidgeh, 42,* 18-21.

Sunyer, J. O. & Lambris, J. D. (1998). Evolution and diversity of the complement system of poikilothermic vertebrates. *Immunol. Rev., 166,* 39-57.

Suzanne, M. & Séller, H. (2009). Letting go : modification of cell adhesion during apoptosis. *J. Biol., 8,* 49.1-49.4.

Suzuki, K. & Fagarasan, S. (2008). How host-bacterial interactions lead to IgA synthesis in the gut. *Trends Immunol., 29,* 523-531.

Sveinbjornsson, B., Olsen, R. & Paulsen, S. (1996). Immunocytochemical localization of lysozyme in intestinal eosinophilic granule cells (EGCs) of Atlantic salmon, *Salmo salar* L. *J. Fish Dis., 19,* 349-355.

Takano, T., Sha, Z., Peatman, E., Terhune, J., Liu, H., Kucuktas, H., Li, P., Edholm, E. S., Wilson, M. & Liu, Z. (2008). The two channel catfish intelectin genes exhibit differential patterns of tissue expression and regulation after infection with *Edwarsiella ictaluri. Dev. Comp. Immunol., 32,* 693-705.

Taylor, I. S., Adam, B., Veverkova, M., Tatner, M. F., Low, C., Secombes, C. & Birkbeck, T. H. (2005). T-cell antigen receptor genes in turbot (*Scophthalmus maximus* L.). *Fish Shellfish Immunol., 18,* 445-448.

Teacher, A. G. F., Garner, T. W. J. & Nichols, R. A. (2009). Evidence for directional selection at a novel major histocompatibility class I marker in wild common frogs (*Rana temporaria*) exposed to a viral pathogen (*Ranavirus*). *Plos One, 4,* e4616. doi:10.1371/ journal.pone.0004616.

Terrell, S. P., Uhl, E. W. & Funk, R. S. (2003). Proliferative enteritis in leopard geckos (*Eublepharis macularius*) associated with *Cryptosporidium* sp. infection. *J. Zoo. Wildl. Med., 34,* 69-75.

Thompson, R. C. A., Olson, M. E., Zhu, G., Enomoto, S., Abrahamsen, M. S. & Hijjawi, N. S. (2005). *Cryptosporidium* and Cryptosporidiosis. *Adv. Parasitol., 59,* 77-162.

Tizard, I. R. (1995). *Immunology. An introduction*. Saunders College Publishing. Orlando, Florida., 544 pp.

Toledo, R., Esteban, J. G. & Fried, B. (2006). Immunology and pathology of intestinal trematodes in their definitive hosts. *Adv. Parasitol., 63,* 285-365.

Trede, N. S., Langenau, D. M., Traver, D., Look, A. T. & Zon, L. I. (2004). The use of zebrafish to understand immunity. *Immunity, 20,* 367-379.

Tse, S. K. & Chadee, K. (1991). The interaction between intestinal mucus glycoproteins and enteric infections. *Parasitol. Today, 7,* 163-172.

Tsujita, T., Ishii, A., Tsukada, H., Matsumoto, M., Che, F. S. & Seya, T. (2006). Fish soluble Toll-like receptor (TLR)5 amplifies human TLR5 response via physical binding to flagellin. *Vaccine, 24,* 2193-2199.

Tsutsui, S., Tasumi, S., Suetake, H. & Suzuki, Y. (2003). Lectins homologous to those of monocotyledoneus plants in the skin mucus and intestine of pufferfish, *Fugu rubripes. J. Biol. Chem., 23,* 20882-20889.

Tsutsui, S., Tasumi, S., Suetake, H., Kikuchi, K. & Suzuki, Y. (2005). Demonstration of the mucosal lectins in the epithelial cells of internal and external body surface tissues in pufferfish (*Fugu rubripes*). *Dev. Comp. Immunol., 29,* 243-253.

Turner, M. W. (2003). The role of mannose-binding lectin in health and disease. *Mol. Immunol., 40,* 423-429.

Uldal, A. & Buchmann, K. (1996). Parasite host relations. *Hexamita salmonis* in rainbow trout *Oncorhynchus mykiss. Dis. Aquat. Organ., 25*, 229-231.

Ullal, A. J., Litaker, R. W. & Noga, E. J. (2008). Antimicrobial peptides derived from hemoglobin are expressed in epithelium of channel catfish (*Ictalarus punctatus*, Rafinesque). *Dev. Comp. Immunol., 32,* 1301-1312.

Urán, P. A., Gonçalves, A. A., Taverne-Thiele, J. J., Schrama, J. W., Verreth, J. A. J. & Rombout, J. H. M. W. (2008). Soybean meal induces intestinal inflammation in common carp (*Cyprinus carpio* L.). *Fish Shellfish Immunol., 25,* 751-760.

do Vale, A., Costa-Ramos, C., Silva, D. S. P., Macedo, P. M., Fernández, R., Sampaio, P., dos Santos, N. M. S. & Silva, M. T. (2007). Cytochemical and ultrastructural study of anoikis and secondary necrosis in enterocytes detached *in vivo*. *Apoptosis., 12,* 1069-1083.

Valentijn, A. J., Zouq, N. & Gilmore, A. P. (2004). Anoikis. *Biochem. Soc. Trans. 22, part 3 Mol. Environ. Integrins.*, 421-425.

Vasta, G. R., Ahmed, H. & Odom, E. W. (2004). Structural and functional diversity of lectin repertoires in invertebrates, protochordates and ectothermic vertebrates. *Curr. Op. Struct. Biol., 14,* 617-630.

Vigliano, F. A., Bermúdez, R., Quiroga, M. I. & Nieto, J. M. (2006). Evidence for melano-macrophage centres of teleost as evolutionary precursors of germinal centres of higher vertebrates: An inmunohistochemical study. *Fish Shellfish Immunol., 21,* 467-471.

Vigliano, F. A., Bermúdez, R., Nieto, J. M. & Quiroga, M. I. (2009). Development of rodlet cells in the gut of turbot (*Psetta maxima* L.): relationship between their morphology and S100 protein immunoreactivity. *Fish Shellfish Immunol., 26,* 146-153.

Vitved L., Kolmskov U., Koch C., Teisner B., Hansen S. & SkjØdt K. (2000). The homologue of mannose-binding lectin in the carp family Cyprinidae is expressed at high level in spleen, and the deduced primary structure predicts affinity for galactose. *Immunogen., 51,* 955-964.

Wang, Q., Wang, Y., Xu, P. & Liu, Z. (2006). NK-lysin of channel catfish: Gene triplication, sequence variation, and expression analysis. *Mol. Immunol., 43,* 1676-1686.

Warr, G. W., Magor, K. E. & Higgins, D. A. (1995). IgY: clues to the origin of modern antibodies. *Immunol. Today, 16,* 392-398.

Wegner, K. M., Reusch, T. B. H. & Kalbe, M. (2003). Multiple parasites are driving major histocompatibility complex polymorphism in the wild. *J. Evol. Biol., 16,* 224-232.

Whyte, S. K. (2007). The innate immune response in finfish. A review of current knowledge. *Fish Shellfish Immunol., 23,* 1127-1151.

Wrackmeyer, U., Hansen, G. H., Seya, T. & Danielsen, E. M. (2006). Intelectin: a novel lipid raft-associated protein in the enterocyte brush border. *Biochemistry, 45,* 9188-9197.

Wynne, J. W., Cook, M. T., Nowak, B. F. & Elliot, N. G. (2007). Major histocompatibility polymorphism associated with resistance towards amoebic gill disease in Atlantic salmon (*Salmo salar* L.). *Fish Shellfish Immunol., 22,* 707-717.

Yarovinsky, F. & Sher, A. 2006. Toll-like receptor recognition of *Toxoplasma gondii*. *Int. J. Parasitol., 36,* 255-259.

Yazawa, R., Cooper, G. A., Beetz-Sargent, M., Robb, A., McKinnel, L., Davidson, W. S. & Koop, B. F. (2008). Functional adaptive diversity of the Atlantic salmon T-cell receptor gamma locus. *Mol. Immunol., 45,* 2150-2157.

Yun, C. H., Lillehoj, H. S. & Lillehoj, E. P. (2000). Intestinal immune responses to coccidiosis. *Dev. Comp. Immunol., 24,* 303-324.

Zapata, A. & Amemiya, C. T. (2000). Phylogeny of lower vertebrates and their immunological structures. Origin and evolution of the vertebrate immune system. *Curr. Top. Microbiol. Immunol., 248,* 67-107.

Zapata, A., Diez, B., Cejalvo, T., Gutierrez-de Frias, C. & Cortes, A. (2006). Ontogeny of the immune system of fish. *Fish Shellfish Immunol., 20,* 126-136.

Zasloff, M. 2002. Antimicrobial peptides of multicellular organisms. *Nature, 415,* 389-395.

Zou, J., Mercier, C., Koussounadis, A. & Secombes, C. (2007). Discovery of multiple beta-defensin like homologues in teleost fish. *Mol. Immunol., 44,* 638-647.

In: Veterinary Immunology and Immunopathology
Editors: L. Nermann, S. Meier
ISBN: 978-1-60876-342-9

Chapter 2

UNRAVELING THE GENETICS OF BOVINE JOHNE'S DISEASE: LESSONS LEARNED FROM HUMAN INFLAMMATORY BOWEL DISEASE

Chris P. Verschoor, Sameer D. Pant and Niel A. Karrow
Centre for Genetic Improvement of Livestock, Department of Animal and Poultry Science, University of Guelph, Guelph, ON Canada

ABSTRACT

Johne's disease is a chronic, granulomatous enteritis of ruminants and some wild-type species that is caused by infection with *Mycobacterium avium* spp. *paratuberculosis* (MAP). This inflammatory bowel disease (IBD) is characterized by a lengthy yet variable asymptomatic subclinical phase in which infected animals can transmit MAP to their progeny and herd mates; this is followed by a clinical phase characterized by chronic diarrhea, wasting, and eventually death. The impact of Johne's disease is particularly evident in the dairy and beef industries, where significant losses due to reduced productivity and fertility, mortality, and premature culling have been reported. Treatment and preventative management strategies have been difficult to implement and are often ineffective for preventing this IBD due to the pathogenic nature of MAP, and the duration of its subclinical phase.

Individual resistance to Johne's disease is a heritable trait; therefore, it may be possible to selectively breed animals with enhanced disease resistance. To date, there have been few attempts to elucidate the genetic mechanisms involved in resistance to Johne's disease. Considering the polygenic nature of the disease, this task is inherently difficult, since there are likely dozens of loci scattered across the entire genome contributing small effects to disease resistance.

Insight into the genetics of Johne's disease may be gained from research on Crohn's disease, a human IBD that parallels Johne's disease in many respects. Extensive epidemiological studies that have been conducted and validated in a number of different human populations have identified numerous susceptibility loci for Crohn's disease. Variants in the corresponding bovine genes may also contribute to resistance to Johne's disease. Therefore, the focus of this review is to examine the bovine immune response during MAP infection, highlight pathological similarities between Johne's disease and Crohn's disease, and lastly, discuss the likelihood of comparable genetic mechanisms and susceptibility loci existing between these diseases.

INTRODUCTION

Bovine Johne's disease (JD), caused by *Mycobacterium avium* subspecies *paratuberculosis* (MAP), is a chronic, infectious, granulomatous inflammatory condition of the intestines with fatal outcomes. MAP is a slow-growing, intracellular, Gram-positive and acid-fast bacterium, belonging to a group of genetically related mycobacteria, collectively referred to as the *Mycobacterium avium* complex. Two principal subtypes of MAP have been identified: type 'C', which primarily infects cattle, and type 'S', which principally infects sheep. These subtypes are not exclusive, and have been shown to infect multiple species [1-4]. The pathology and progression of JD consists of four stages: the infection phase, and the sub-clinical, clinical and advanced clinical phases. Infection with MAP predominantly occurs via the ingestion of milk, colostrum, or feces infected with MAP, though *in utero* transmission has also been reported to occur [5, 6]. There is currently no evidence that vertical transmission can occur from sire to calf, however, mycobacteria can persist in the reproductive organs of bulls, and in fresh [7] and cryopreserved semen intended for artificial insemination [8]. Calves are considered the most susceptible life stage for persistent infection, likely because they have not yet reached full immunocompetence [9]. These infected animals generally do not show clinical symptoms until their second or third lactation [10, 11]. The sub-clinical phase, characterized by the expansion of diffuse granulomatous intestinal lesions and a predominant type 1 (T_h1) cell-mediated immune response driven by cytokines such as interferon (IFN) γ, tumor necrosis factor (TNF) α, and interleukin (IL) 12, is variable in duration and can persist anywhere from two to ten years [12, 13]. Although asymptomatic in many respects, the sub-clinical phase represents the most prevalent health and economic risk, since infected animals are often substandard performers, and actively shed MAP, increasing the relative risk of infection to the rest of the herd [14-16]. The clinical and advanced clinical phases are much more apparent and are characterized by wasting, diarrhea, decreased milk production and eventually death [10]. For every case of clinical JD in an infected herd it has been estimated that at least four other animals in the herd are carriers and in the sub-clinical disease phase [17].

Diagnosis of JD poses various problems, as many of the commercially available diagnostic tests are expensive, and/or suffer from low sensitivity, as well as lengthy turnover. The most common tests and the upper limit of their reported sensitivities include: serum enzyme-linked immunosorbent assays (ELISA) (94%), IFN-γ (85%), direct fecal polymerase chain reaction (PCR) (80%), fecal culture (74%), and milk ELISA (61%) [18-20]. Unfortunately, the extreme variation in sensitivity of these tests indicates that no single test is effective for detecting all cases of JD within herds, thus, screening using multiple tests may be more desirable [21, 22]. The reported prevalence of JD in dairy herds worldwide ranges from 7% in Austria to 60% in New Zealand [14]. In USA, herd level infection rates have been estimated to be approximately 22% overall, and 40% for herds housing more than 300 cows [23]. However, given the low sensitivities of JD diagnostic tests, it is likely that these figures underestimate the true prevalence of the disease. It has been estimated that JD contributes to an annual loss of approximately 1.5 billion dollars to the US cattle industry [24] due to reduced production and reproductive efficiency, premature culling or death from clinical disease, treatment costs, and the implementation of preventative management strategies [15, 25, 26]. Vaccines targeting MAP have been developed and are commercially available, but are only successful for delaying the onset of clinical symptoms and do not eliminate the infection [27].

Human inflammatory bowel disease (IBD) includes a group of related chronic inflammatory conditions of the gastrointestinal tract with either unknown or currently unproven etiology. IBD is prevalent worldwide, but for the most part is considered a disease of the industrialized world [28]. According to estimates from the Crohn's and Colitis Foundation of America, about one million people suffer from IBD in the USA, alone [29]. There are two main types of IBD, namely, Crohn's disease (CD) and ulcerative colitis (UC), both of which differ in location and the nature of inflammation involved. Treatment of these diseases normally includes a long-term regimen of anti-inflammatory as well as immunosuppressive drugs, but surgical removal of damaged intestinal sections is sometimes necessary [30]. Comparative epidemiological studies of the occurrence across and within families, as well as between groups of monozygotic and dizygotic twins, have established that genetic factors contribute to the susceptibility to IBD [31]. Subsequent genetic studies revealed the first susceptibility loci containing the gene encoding nucleotide oligomerization domain 2 (NOD2) in 2001, followed by the identification of at least 30 more genes over the following decade [32, 33].

From a clinical as well as immunophenotypic standpoint, JD and CD parallel each other in many respects. Consequently, many have attempted to establish a zoonotic relationship between humans and cattle, focusing on MAP as the causal organism [reviewed in 34]. The focus of this review will be to examine the bovine immune response during MAP infection, highlight pathological similarities between JD and CD, and lastly, discuss the likelihood of comparable genetic mechanisms and susceptibility loci existing between these diseases.

MAP and the Bovine Immune Response

Host-Pathogen Interactions during Early Infection

Interaction between MAP and the host immune system is similar to most enteric pathogens. Oral ingestion of contaminated materials effectively transports MAP to the gut where it is presumably taken up by intestinal M-cells, a specialized subset of mucosal epithelial cells that actively sample antigens within the intestinal lumen [35]. MAP is then transported to the basolateral surface of the M-cell where it encounters antigen presenting cells (APCs) and lymphocytes, first within a region known as the intraepithelial pocket, and soon after in the underlying lymphoid tissue, the Peyer's patch [36]. This sequence of events, including the active transport of MAP across the intestinal mucosa, and subsequent colonization of Peyer's patches and mesenteric lymph nodes, has been shown to occur as soon one to two hours post-inoculation [37]. Under normal circumstances, these APCs, primarily dendritic cells (DCs) and macrophages, take up invading pathogens and process and present their antigens to surrounding T lymphocytes via major histocompatibility complex (MHC) class II receptors to elicit a cell-mediated or antibody-mediated immune response. MAP and other mycobacteria subspecies, however, have evolved a number of strategies that not only allow them to avoid the bactericidal actions of APCs, but also propagate within these cells.

Within the extracellular environment, the phagocytosis of MAP is initiated by cell-surface receptors such as toll-like receptors (TLRs), CD14, mannose receptors (MR), complement receptors (CR), and β-integrins [38, 39]. Receptor-ligand interaction triggers phagocytosis and normally initiates an anti-microbial response characterized by phagosome acidification and the production of bactericidal molecules such as reactive oxygen species (ROSs). However, as reported for other species of mycobacteria, MAP may have the capacity to alter this response, since it is dependent on the subset of receptors used to mediate phagocytosis [40, 41]. A recent *in vitro* study by Souza and colleagues for example, showed that phagosome acidification could be increased in monocyte-derived macrophages when CR3 (CD11b) -mediated phagocytosis was blocked [42]. A similar study by Kang and colleagues showed that blocking the MR-mediated phagocytosis of beads coated with mannose-capped lipoarabinomannan (LAM) derived from *M. tuberculosis* greatly enhanced subsequent phagosome-lysosome fusion [40]. Interestingly, LAM derived from *M. avium* has been shown to upregulate the expression of suppressors of cytokine signaling (SOCS) proteins in human macrophages *in vitro*, which in turn blocks the expression of IFN-γ and promotes bacterial survival [43]. Studies such as these reveal a potential strategy whereby MAP may preferentially bind to certain APC cell-surface receptors to facilitate phagocytosis without triggering strong anti-microbial responses.

A second complimentary strategy employed by MAP to resist destruction is to minimize phagosome acidification, thereby preventing phagosome maturation that leads to phagosome-lysosome fusion and the subsequent destruction of the microbe [44, 45]. Following phagocytosis, MAP exhibits the ability to remain associated with early endosomes, possibly due to reduced expression of host lysosome associated membrane protein (LAMP) 1, a late marker of phagosome maturation [46, 47]. MAP also has the capacity to limit phagosome acidification, presumably via reduced H^+-ATPase expression [46, 47]. Both of these effects are likely mediated, in part, via the MAPK-p38 signaling pathway [48].

The ability to resist intracellular destruction allows MAP to minimize the potential for MHC-restricted presentation of its antigenic determinants to T lymphocytes, severely limiting the host's ability to mount an effective cell-mediated immune response. This is further achieved by altering the immunophenotype of APCs. Macrophage MHC I and II receptor expression for example, is down-regulated by MAP, even in the presence of the cytokines TNF-α and IFN-γ, which induce macrophage effector function [49]. Reduced IFN-γ receptor expression may also occur, as is the case with *M. tuberculosis* [50]. Some of these effects are likely attributed to increased IL-10 secretion [51, 52], whose regulation may be influenced by MAP [44, 53].

A Role for $CD8^+$ and γδ-T Cells in Early Infection

Considering bacterial degradation products are usually compartmentalized within endosomes of APCs, an appropriate response to MAP infection has traditionally been viewed as being dominated by MHC II-antigen dependant priming of $CD4^+$ T cells. Presentation through MHC I, leading to the activation of $CD8^+$ cytotoxic T cells, usually occurs for antigens that enter the cytosol, such as viral proteins or aberrant proteins expressed by oncogenic cells; such antigens are processed by cytoplasmic proteases [54]. In spite of this, the activation and expansion of $CD8^+$ T cells has been reported to occur during infection with

MAP, as well as other mycobacterial species [38, 55, 56]. In fact, early studies of *M. tuberculosis* infection in mice have shown that modulating the frequency or activity of $CD8^+$ T cells can have a major impact on the susceptibility to infection and the progression of disease [57]. It has been proposed that the cross-presentation of apoptotic bodies released from infected macrophages may be a mechanism facilitating the activation of $CD8^+$ T cells. Apoptotic bodies containing MAP antigens can be taken up by DCs and directed to the cytoplasm, possibly by interactions between the phagosome and endoplasmic reticulum, leading to presentation via the MHC I pathway and subsequent activation of $CD8^+$ T cells [58, 59]. Another hypothesis, offered by Woodworth and colleagues, is that mycobacteria contained within the APC phagosome secrete products that are actively transported into the cytosol [56]. These products would then be available for processing and presentation by the endogenous MHC I pathway. Since the persistence of MAP-infected macrophages is likely a major causal factor in the progression of JD, a strong cytotoxic response to these cells may prove integral in promoting host immunity, ultimately leading to the resolution of infection. A better understanding of the pathways used for presentation of MAP antigens via MHC I, and the subsequent role that $CD8^+$ T cells play during the immune response to MAP definitely warrants further investigation.

γδ-T cells may play a role in the early immune response to MAP infection, since they represent the major circulating T cell population in calves [60, 61], are one of the first leukocyte subsets to encounter pathogens in the gut mucosa [60], and do not require MHC-restricted activation [38]. The expression of innate pattern recognition receptors (PRRs), such as NOD2, TLRs, and CD36, also allows γδ-T cells to respond directly to various pathogen-associated molecular patterns (PAMPs). This interaction can lead to the secretion IFN-γ and TNF-α, as well as chemokines that promote the recruitment and activation of monocytes and neutrophils [60]. In support of this, challenge with MAP antigens promotes the activation of peripheral γδ-T cells in neonatal calves [62], and the frequency of γδ-T cells in the ileal sub-mucosa of cattle diagnosed with clinical JD is increased in comparison to asymptomatic control animals [63]. Interestingly though, bovine γδ-T cells incubated *in vitro* with MAP-infected macrophages are unable to produce significant amounts of IFN-γ, and do not enhance MAP killing by macrophages [64]. A recent human study showed that γδ-T cells can cross-present *M. tuberculosis* antigen to $CD8^+$ T cells, inducing their proliferation more effectively than monocyte-derived DCs [65]. It is therefore possible that the activation of γδ-T cells in the gut of young calves by vaccination may prove to be an excellent strategy to enhance immunity to MAP.

The Host Response during the Sub-Clinical Phase of Infection

During disease progression to the sub-clinical stage, the T_h1 response induced shortly after infection can be detected by measuring increased serum levels of IFN-γ [66]. Low-level shedding of bacteria in feces, and immunoglobulin (Ig) G1-type antibodies in serum may be detected, however, they are generally considered to be unreliable diagnostic markers at this point in time [20, 67].

The infiltration of macrophages and lymphocytes to the site(s) of infection, and the expression of pro-inflammatory cytokines such as TNF-α promote granuloma formation. This organized mass of leukocytes and fibrous tissue has traditionally been thought to be a host barrier designed to confine MAP to the site of infection, thereby slowing its rate of systemic dissemination [67]. However, a recent study using zebrafish and *M. marinum* suggests that mycobacterium may be able to capitalize on this defense strategy to promote the spread of infection to recruited leukocytes during early granuloma formation [68]. Although this model is inherently different in many respects to the formation of granulomas during the bovine response to MAP, it offers another mechanism that MAP might use to ensure its survival within the host, and therefore may impact future treatment strategies for JD.

Although multiple forms of granulomas can be found in the small intestine of MAP infected animals, type II lepromatous granulomas are most abundant [69]. Studies of these lesions, both naturally-occurring and experimentally-induced, report impaired macrophage function and maturation, as indicated by the limited expression of costimulatory molecules, chemokines, and inducible nitric oxide synthase (NOS) [69, 70], as well as decreased or static levels of IFN-γ and TNF-α [69, 71, 72]. This depressed T_h1-associated activity may be due, in part, to the expression of T_h2 cytokines, such as IL-10 and transforming growth factor (TGF) β1, both of which have been shown to be increased at such sites [72, 73]. The expression of IL-10 by macrophages may also be a strategy to ensure that an over-zealous T_h1 response, mediated by activated T cells, does not occur [44]. It has also been proposed that the production of IL-10 can be partially accounted for by a subset of suppressive T cells, known as regulatory T cells (Tregs). These cells are thought to differentiate as a mechanism to limit T_h1 responses, which in turn, limits excessive tissue injury [12, 74]; they will be discussed in greater detail later in this review.

The dominant anti-inflammatory paradigm in MAP-induced granulomas suggests a phenotypic re-programming of local macrophages from M1 to M2 cells. M1 macrophages are generally associated with acute, pro-inflammatory stages of infection, and they secrete cytokines such as IL-12, IL-1β, and TNF, as well as express NOS2. In contrast, M2 macrophages exhibit an anti-inflammatory phenotype, characterized by poor anti-microbial activity, and the secretion of IL-10, TGF-β, and IL-1 receptor agonist [75]. Although the activation phenotype of resident macrophages in MAP-induced granulomas has yet to be revealed, TLR-9 knockout mice challenged with mycobacteria-associated antigens display increased granuloma formation, which coincided with an increase in the number of M2 macrophages [76]. Furthermore, studies involving *in vivo* infection of T_h2 and IL-21R knockout mice with parasitic *Schistosoma mansoni* suggest a prominent role of M2 macrophages in the formation and maintenance of granulomas [77, 78]. Although granulomas increase in size as infection progresses, low concentrations of IFN-γ and TNF-α combined with the potential presence of M2 macrophages suggest that macrophages are likely unable to effectively promote bacterial clearance [67]. Hence, granulomas gradually expand to a level where mucosal tissue function is affected, thereby promoting the emergence of clinical symptoms such as chronic weight loss and diarrhea [10].

In vitro challenge of monocyte-derived macrophages or peripheral blood mononuclear cells (PBMCs) from sub-clinically infected animals with live MAP has been a widely used platform to better understand how newly recruited macrophages respond at sites of infection. A general consensus of such studies is that recruited macrophages phagocytose and process MAP antigen, as well as have the ability to promote T_h1 responses, as indicated by the upregulation of genes such as IFN-γ, H^+-ATPase, and CD14 [79, 80]. However, simultaneous increases in IL-10, SOCS-3, and TGF-β, and decreases in TNF-α, IL-12, LAMP-2, and RANTES/MCP-1 gene expression, suggest that a number of factors are induced to limit overall T_h1 responses, promoting MAP survival and disease progression [52, 80-82]. Within actively infected tissues, namely the ileum and mesenteric lymph nodes, a similar pattern can be found. Pro-inflammatory cytokines IFN-γ, IL-1α, IL-8, and IL-2 are upregulated, and the number of $CD3^+$ and $CD4^+$ T cells is increased. At the same time, IL-16, a stimulator of T cell proliferation, is downregulated, and the number of activated T cells and cells expressing MHC II are decreased [82, 83].

Progression through the sub-clinical phase can be expected to include an expansion of infection sites and granulomas, increased inflammation of the intestinal mucosa, and an increasing likelihood of detecting MAP antibodies in serum and milk or MAP organisms after fecal culture or nested PCR. The host T_h1 response remains high throughout this phase until an eventual drop-off nearing the clinical stage of disease, at which time a T_h2, antibody-mediated response will arise and dominate. It is still uncertain the exact mechanisms behind the decline of the T_h1 response, however a plausible theory offered by Coussens maintains that the extensive sub-clinical period probably leads to the depletion of pro-inflammatory cells [12]. It is likely that IL-10, acting on behalf of a regulatory or suppressive cell population, also plays a prominent role in shifting the host immune response.

Clinical Phase

The host is essentially defenseless against the expansion of MAP infection without the ability to sustain an effective T_h1 response [67]. Therefore, it is not surprising that the switch from T_h1 to T_h2 responses parallels the transition from sub-clinical to clinical disease. During the clinical phase of JD, cattle suffer from excessive intestinal inflammation, and as a result, have difficulty processing feed; this leads to progressive weight loss as well as persistent or intermittent diarrhea. If cattle are not culled by this point, they become extremely lethargic and weak, and eventually die [84]. Since T_h2 responses dominate this phase, the detection of serum IFN-γ may not be consistent, whereas diagnosis using serum or milk ELISA, fecal culture, or direct fecal PCR is more reliable [85].

In infected tissues, the number of T cells is generally deficient, and those that are present are anergic to signals from APCs [61, 67]. Levels of T_h2 cytokines IL-10 and TGF-β remain high, while the expression of IFN-γ is reduced [86]. In comparison to sub-clinically infected animals, clinical cases tend to have a higher frequency of peripheral B cells, which may be of importance given that B cells are highly effective APCs [87, 88]. Furthermore, evidence suggests that the majority of these B cells are of the B-2 phenotype [87]. Also known as "conventional" B cells, this class of B lymphocyte requires additional costimulation from T helper cells via CD40-CD40L interactions in order to proliferate and direct isotype switching [89]. Considering the inherent reduction in number and activation status of T cells at infected tissues, a greater proportion of this B-subset would imply that clinically infected animals have even less protection against MAP than previously thought. This theory is supported by Waters and colleagues, who reported that although the proportion of peripheral B cells was higher in clinically versus sub-clinically infected cattle, the antigen-specific B cell proliferative response was severely hampered [88]. Finally, the balance between IgG1 and IgG2 isotype responses during progression from sub-clinical to clinical disease has been shown to be variable, and largely depends on the antigen [90].

Pathological Similarities between Johne's Disease and Crohn's Disease

In contrast to the well defined stages of JD, CD is generally only classified according to the severity of clinical symptoms; for example, the Crohn's disease activity index (CDAI) [28]. In any case, a number of similarities can be drawn between these two IBDs concerning the presentation of clinical symptoms and immune phenotype.

The clinical signs of JD are first manifested as persistent and intermittent diarrhea and increased weight loss and appetite. Internally, pathological signs are most often localized to the regions of the lower intestinal tract, generally the ileum and sometimes beyond the ileocecal valve, as well as the ileal Peyer's patches and mesenteric lymph nodes [91]. A thickened and edematous intestinal wall, giving the mucosal layer a 'cobblestoned' or 'corrugated' appearance, dilated and thickened serosal and mesenteric lymph vessels, and several lesions consisting of aggregates of lymphoid and epitheloid cells (granulomas) are typically observed [84]. Although not common, the extent of intestinal inflammation can traverse beyond the mucosal layer [91].

Although CD can involve areas of the gastrointestinal tract ranging from the oropharynx to the perianal area, like JD, pathological signs are most commonly restricted to the ileocecal region and the terminal ileum [92]. Considering the location and severity of CD may vary, early symptoms, such as abdominal pain, weight loss, and difficulty in gastric emptying, may also be variable. Pathophysiological signs are also similar to JD, in that a 'cobblestoned' mucosal layer, transmural chronic inflammation, granuloma formation, and dilated lymphatic channels can be found [91]. However, unlike JD, CD is also often accompanied by perianal fistulae, abnormal cavities connecting segments of the gastrointestinal tract, as well as deep anal fissures [91, 93]. Complete diagnosis is usually confirmed by clinical evaluation, and a combination of relevant biochemical, endoscopic, histological, and radiographic findings [93].

MAP has widely been accepted to be the cause of JD since the early 20th century in several types of ruminant livestock and wildlife species [91]. However, there is considerable debate about the causal agents for CD. It is hypothesized that CD is caused by a loss of tolerance to self and non-self antigens, and/or infection by isolated pathological organisms, one of which may include MAP [94].

Immunological tolerance is a state of non-responsiveness to specific antigens induced by prior exposure to a specific concentration of antigen. This typically occurs either in response to food-borne antigens, mucosal microflora, and antigens of self origin that do not induce the threshold of cell activation required to initiate an immune response, or is induced by Tregs [95]. Tregs are immune regulating T lymphocytes characterized by the expression of CD4 or CD8, the IL-2 receptor (CD25), cytotoxic T lymphocyte antigen 4 (CTLA4), glucocorticoid-induced TNF receptor (GITR) and the transcription factor forkhead box p3 (foxp3) [96]. Tregs are divided into two subgroups based on their immunophenotype: natural Tregs ($CD4^+CD25^+$) are derived from the thymus, whereas inducible Tregs ($CD4^+CD25^-$) are derived from peripheral tissues and are activated by IL-2 [97]. Tregs exert their suppressive effects via contact dependant and independent mechanisms. The dependent mechanism is mediated by ligation of CTLA4 to the co-stimulatory molecule B7 (CD80/86) on APCs; this reduces the ability of APCs to present antigen and activate T cells. The independent mechanism is mediated by the secretion of the cytokines IL-10 and TGF-β, which suppress T cell activity and proliferation [98].

A number of murine studies have demonstrated that Tregs can abrogate microflora-induced experimental colitis [99, 100]. Watanabe and colleagues demonstrated that this is mediated by limiting the ability of intestinal DCs to initiate a response, as well as by a reduction in recruitment and activation of antigen-specific effector T cells [101]. Interestingly, Tregs have also been shown to promote a shift toward the M2 macrophage phenotype without exogenous stimulation, attenuating the inflammatory response associated with *in vitro* lipopolysaccharide (LPS) challenge [102]. This may represent a potential strategy to limit unnecessary immune responses to a myriad of harmless antigens passing through the gastrointestinal tract. Although it is yet to be reported in literature, considering the potential link between subclinical JD and the expansion of a regulatory cell population [83], it is not unreasonable to hypothesize that a breakdown of tolerance may contribute to the progression of JD disease [83].

The activation threshold of intestinal APCs likely contributes to the maintenance of tolerance. Hedl and colleagues have shown that chronic stimulation with muramyl dipeptide (MDP), a mycobacterial cell wall component and ligand for the NOD2, renders macrophages hyporesponsive, thereby limiting their ability to initiate a pro-inflammatory response[103]. Furthermore, macrophages from CD patients carrying the leu1007insC NOD2 polymorphism fail to exhibit such tolerance to chronic stimulation, suggesting a response to MAP as a contributing factor in the pathology of CD in at least subset of cases.

Numerous studies have examined the relationship between CD and MAP infection. A meta-analysis by Feller and colleagues showed that there is an increased likelihood of detecting MAP in patients with CD [104]. In contrast, some sub-populations show no evident relationship between CD and MAP infection [105-107]. It has also been suggested that human MAP infection might occur via horizontal transmission from livestock and their byproducts [108], subsequently prompting research into the safety of MAP contaminated dairy products after commercial pasteurization [109, 110]. Nonetheless, it is yet to be proven a definitive link between MAP and CD. Some researchers have hypothesized that a sustained MAP infection, although not the cause, may promote the progression of CD and exacerbate its severity. Chamberlin and colleagues, for example, reported complete remission of a patient with CD following treatment with mycobacterial antibiotics [111]. Another study by Polymeros and colleagues demonstrated that certain MAP antigens could potentially trigger an autoimmune response against self antigens in patients with CD, and this could possibly result in a loss of tolerance as discussed above [112].

The most apparent similarity between CD and JD is the predominance of a pro-inflammatory paradigm, typical of a gut bacterial infection. In the context of JD, MAP is taken up from the intestinal lumen and phagocytosed by APCs such as DCs and macrophages. These cells promote the activation of naïve T cell populations through direct contact and stimulation by secreted IL-1α [61]. Activated T lymphocytes subsequently produce large amounts of IFN-γ, IL-1α, IL-2, and TNF-α during the infection and early-subclinical stage [12, 81, 113-115]. Patients with CD typically have high levels of IFN-γ, TNF-α, IL-12, IL-23 and IL-18 [116-119]. IL-12, IL-23 and IL-18 are all closely associated with the production of IFN-γ by T lymphocytes [120, 121], and TNF-α, produced by APCs when stimulated by pathogens, activates pathways that lead to the secretion of other pro-inflammatory cytokines, chemokines and adhesion molecules [122]. As discussed previously, it is generally accepted that the transition between the sub-clinical and clinical phase of JD parallels a gradual shift from a T_h1 cell-mediated, to a T_h2 antibody-mediated immune response [38]. This "switch", similar to that observed with bovine tuberculosis [123], is marked by a reduction of $CD4^+$ T cells and T_h1 cytokines, such as IFN-γ, and a corresponding predominance of IgG-type antibodies [12, 61, 63, 90]. Recent evidence has shown that such a transition may also occur in patients with CD. Mucosal T cell clones derived from intestinal biopsies of patients suffering from late onset CD exhibited a significant reduction in the expression of IFN-γ and IL-12 receptor β2, and a greater capacity to secrete IL-10 and IL-4, as compared to clones derived from patients with early onset CD [124]. This pattern has also been shown in IL-10 knockout mice following experimentally induced colitis; the T_h1 cytokine, IL-12, was expressed during the early stage of disease, whereas the late stage was dominated by the T_h2 cytokine, IL-4 [125]. Interestingly, IL-4 secretion has also been shown to be induced in blood leukocytes of CD patients infected with MAP, as compared to those without MAP infection, implicating MAP in the transition from T_h1 to T_h2 [126].

The progression from T_h1 to T_h2 immunity during chronic gut inflammation provides an interesting possibly for involvement of $CD4^+CD25^+$ Tregs. As discussed above, these cells have an immunoregulatory role during inflammatory responses, limiting excessive or prolonged T_h1 responses. Although the specific role in which they play is dependent on the origin and magnitude of infection, in the context of gut inflammation Tregs induce immunological tolerance to antigen stimulation in an effort to limit tissue damage [127]. In cattle with sub-clinical JD, the percentage of $CD4^+CD25^+$ T lymphocytes in the gut mucosa is increased more than 3-fold [83]. Studies by Coussens and others have further shown that that the expression levels of IL-10 in PBMCs from sub-clinically infected cattle are distinctly higher than healthy controls, and that the principal source for this cytokine is $CD4^+$ and/or $CD25^+$ T lymphocytes [74]. Furthermore, as reported by Khalifeh and Stabel, IL-10 and TGF-β are consistently higher in cows with clinical versus sub-clinical JD, while IFN-γ is lower [86]. Taken together, these results support the involvement of Tregs in the T_h1 to T_h2 shift exhibited in JD.

A role for Tregs in the pathogenesis of CD has also considered. In patients with active CD, for example, Tregs are believed to migrate from the peripheral blood to the gut mucosa and mesenteric lymph nodes [128]. However, it is debatable whether levels of IL-10 and TGF-β are increased in these patients. A number of older studies have reported increased levels of IL-10 mRNA and IL-10 in the mucosa and serum of patients with CD [129-131], while more recent work indicates that IL-10 levels are similar if not decreased [132-135]. Disagreement is also evident in studies reporting the levels of TGF-β in patients with CD [136-138]. Considering that the infiltration of Tregs in chronic inflammatory conditions is generally synonymous with increased levels of IL-10 and/or TGF-β, as shown above in bovine studies of MAP infection and JD, the reported findings from studies of CD are curious to say the least. Recently, Ebert and colleagues tested an interesting hypothesis in which certain subsets of patients with IBD inherently produce autoantibodies for particular T_h2 cytokines, thereby limiting their ability to mitigate the damaging effects of chronic pro-inflammatory responses [139]. These authors were able to identify an increased number of autoantibodies for IL-10, but not TGF-β, in the serum of patients with CD as compared to healthy controls. Thus, it is possible that the secretion of IL-10 is increased during CD, but local as well as circulating levels of the cytokine are diminished due to the action of specific autoantibodies.

A GENETIC BASIS FOR MAMMALIAN IBDS

JD is a complex disease and its severity and prevalence are determined by the interaction of different factors including host susceptibility, causal agent and the environment [140]. Susceptibility to MAP infection in cattle is known to be hereditary, with heritability estimates ranging from 0.06 to 0.183 [10, 141, 142]. Differences in susceptibility have been observed among different cattle breeds, with Jersey and Shorthorn cattle being reportedly more susceptible to developing JD than Holstein cattle and its crosses [143, 144]. This indicates that the susceptibility to MAP infection is at least partially determined by inherent genetic factors, and thus, selecting for increased JD resistance may be possible.

As described earlier, there are many similarities between JD and CD including disease pathophysiology and the host immune response. It may therefore be possible that both diseases share some common genetic determinants conferring susceptibility or resistance. Therefore, exploring the genetics of CD in humans may provide us with important insights into the genetic basis of JD in cattle. Both genome-wide association (GWA) and candidate gene studies have been used to attempt to unravel the genetics of both JD and CD. The GWA approach uses linkage mapping to establish associations between a disease phenotype and anonymous genetic markers to identify quantitative trait loci (QTL), chromosomal regions harboring markers associated with the given phenotype. This approach requires prior sequence information of the host genome and availability of markers that are densely and evenly spaced throughout the genome. The advantage with this approach is that prior information about the pathogenesis of the disease, and previously established susceptibility loci are not required. Usually, thousands of markers are genotyped and linkage analysis is performed on a genome-wide basis. In contrast, the candidate gene approach focuses to establish associations between the disease phenotype and known polymorphisms in a given gene with an evident or putative role in the development of disease. This approach does not require prior sequence information of the host genome, or dense marker maps, but information about disease pathogenesis or previously established susceptibility loci is beneficial. A candidate gene is selected either on the basis of its putative role in the pathogenesis of disease (functional candidate gene), or on the basis of its location within a chromosomal region having a previously established association with the disease phenotype (positional candidate gene). Thus, whole genome scan studies usually precede the availability of positional candidate genes. The disease phenotype, in the case of both JD and CD is usually binary in nature, although in the case of JD, one GWA study has used the probabilities of infection as a phenotype, based on ELISA sample to positive ratios [145]. Both approaches generally use a case-control design with the underlying hypothesis that different alleles at polymorphic loci confer different susceptibilities to disease by altering either the activity or expression of downstream proteins.

The GWA approach has proved to be very successful in mapping highly penetrant genes in monogenic disorders; genes where the proportion of individuals expressing a particular genotype at the phenotypic level is high [146]. Since susceptibility to complex diseases such as JD and CD, is governed by the interaction of a large number of genes, each having a small effect, this approach is more useful in revealing putative genetic determinants rather than firmly establishing their role in disease predisposition. In most cases, polymorphisms found to be associated with disease susceptibility in a GWA study do not have a functional role themselves, but are in linkage disequilibrium (LD) with polymorphisms of other candidate genes in the surrounding chromosomal region. In the case of JD, additional challenges associated with the identification of disease susceptibility loci using this approach include insufficient sample sizes and an insufficient number of markers. JD is not a notifiable disease in many countries, therefore, the collection and diagnosis of test phenotypes is both difficult and expensive. This ultimately translates into studies being based on smaller sample sizes with insufficient statistical power to identify all disease associated loci. Finally, in the context of the density of available marker maps, in human studies, more than 500,000 single nucleotide polymorphisms (SNPs) are being utilized in genotyping studies of over 10,000 individuals to map susceptibility loci for various diseases [147]. Although, fewer markers will be needed to carry out bovine disease susceptibility studies, because of inbreeding, currently available bovine marker maps contain at most 58,000 SNPs (http://www.ars.usda.gov/).

In the case of CD, numerous studies based on the GWA approach have been published and several important disease susceptibility loci have been revealed [148-160]. Some of the key loci harbor the candidate genes: caspase recruitment domain (CARD) 15, IL-23 receptor (IL-23R), ATG16 autophagy related 16-like 1 (ATG16L1), immunity-related guanosine triphosphatase (IRGM), macrophage stimulating 1 (MST1), NK2 transcription factor-related locus 3 (NKX2-3), and the protein tyrosine phosphatase non-receptor type 2 (PTPN2). Some other genes that have been implicated to play an important role in CD in independent candidate gene studies are nucleotide oligomerization domain 1 (NOD1), TLRs 1,3,4,5 and 9, IL-12β and IL-12 receptor (IL-12R), NRAMP, NF-κβ1 and CD14. Thus far, only one GWA study has been conducted to identify QTLs for bovine JD [145]. This study revealed putative associations between eight chromosomal regions and susceptibility to MAP infection in dairy cattle. Evidence provided by these studies has not only offered numerous positional candidates for follow-up research, but important insights into the pathophysiology of these diseases as well.

In cattle, only a few studies have attempted to use the candidate gene approach to explore the association between candidate host genes and susceptibility to bovine JD [161-164]. Before the first GWA scan for JD in cattle was published, the choice of candidate genes usually depended on either the current understanding of the pathophysiology of the disease, or evidence based on GWA studies mapping susceptibility loci for CD. The following is a brief discussion of some important positional and functional candidate genes.

Potential Susceptibility Loci Conserved across Mammalian IBDs

Card15/nod2

CARD15, also known as NOD2, is a cytosolic PRR that binds to MDP, a minimal bioactive component of peptidoglycan (PGN), formed in phagocytic cells as a consequence of proteolysis of a bacteria's cell wall. Binding of MDP triggers the expression of nuclear factor kappa beta (NF-κβ), a critical transcription factor in the induction of expression of a number of inflammatory mediators and cytokines [165-167]. Interest in CARD15 was first developed when a GWA study for CD revealed the strongest evidence for linkage on a pericentromeric region on human chromosome 16 [154]. Later, fine mapping of this susceptibility locus, and candidate gene studies led to the identification of the gene encoding CARD15 receptor as the first susceptibility gene for CD [32, 168, 169]. The CARD15 receptor in both humans and cattle, consists of two N-terminal CARD domains, a central nucleotide binding domain and ten C - terminal leucine rich repeat (LRR) domains. Although a number of amino acid variants have been reported in humans, three variants Leu1007insC, Arg702Trp and Gly908Arg have been widely reported to have an association with susceptibility to CD [170]. A number of polymorphisms have also been reported in the bovine CARD15 receptor and two SNPs, 2197T>C and 3020A>T, showed associations with susceptibility to paratuberculosis infection [163, 171]. The exact mechanism by which variants in CARD15 have an effect on disease susceptibility has been explored; some have suggested a loss of function of the receptor, while others have suggested a gain in function [172]. However, there is a general consensus that these variants either alter the structure of the LRR domain of the protein or the adjacent region, which ultimately alters the activation of NF-κβ [173].

NOD1

Nucleotide binding oligomerization domain 1 (NOD1) is another cytosolic PRR that binds to g-D-glutamyl-meso-diaminopimelic acid (meso-DAP) and activates the transcription factor NF-κβ [174]. Its ligand, meso-DAP is another structural core motif derived from PGN and is also present in the mycobacterial cell wall. The receptor, itself, is similar in structure to CARD15, and consists of C–terminal LRR domains and a central nucleotide binding domain. However, it consists of a single N–terminal CARD domain as opposed to two N – terminal CARD domains in CARD15. In humans, one variant $ND_1$32656*1, has been reported to have an association to CD susceptibility [175, 176]. The role of NOD1 in bovine paratuberculosis has not yet been investigated.

TLRs

The TLR family consists of a number of structurally conserved membrane PRR that regulate the host immune response upon activation by microbial products [177, 178]. Signaling, mediated by TLR-1-TLR-2/TLR-6 heteromer and TLR-4, helps activated macrophages and DCs to recognize mycobacterial PAMPs [179-181]. Studies also show that polymorphisms in different TLRs can alter their pathogen recognition ability and host immune responses in general [182-185]. At least 13 different TLRs have been reported in mammals and polymorphisms in human TLR 1, 4, 5, and 9 have shown associations with CD in some studies [186-189]. TLR-3 and TLR-4 have been shown to have altered expression in intra-epithelial cells (IECs) of patients with CD [190]. In cattle, there is some evidence of associations between polymorphisms in TLR-2 and TLR-4 genes and susceptibility to MAP infection [162].

IL-12β and IL-12 Receptor

IL-12β encodes the p40 subunit of the pro-inflammatory cytokine IL-12, which is secreted from APCs in response to stimulation by bacterial products. IL-12, mediates critical cell-mediated immune responses by inducing the production of IFN-γ from NK and T cells including γδ-T cells, following infection with intracellular pathogens [191, 192]. It also promotes the generation of $CD8^+$ cytotoxic T cells and the cytotoxic activity of NK cells [193]. The p40 subunit is a shared constituent of both IL-12 and IL-23 cytokines [194]. In humans, this gene is situated on chromosome 5 in a region that has been previously reported to have an association with CD [160]. In cattle, the only GWA study that has been published so far, found a microsatellite marker (BM1853) on bovine autosome 7 (BTA7) to be associated with susceptibility to JD [145]. IL-12β is situated within one megabasepairs of this microsatellite marker on the chromosomal sequence map. This strongly implicates IL-12β as an important positional as well as functional candidate for susceptibility to JD.

The IL-12R, composed of two subunits, IL-12Rβ1 and IL-12Rβ2, is an important cytokine receptor expressed on activated T and NK cells and shares a common subunit with IL-23R [195-198]. Human and murine studies have demonstrated that inactivating mutations within the genes encoding IL-12 or IL-12R can cause increased susceptibility to intracellular infections including those caused by mycobacteria [199-202].

IL-23 Receptor and the IL-23-mediated signaling pathway

IL-23R, composed of IL-23R and IL-12Rβ1 subunits, is an important candidate that has recently been discovered to be associated with susceptibility to CD [151, 203, 204]. Its ligand, IL-23, is secreted by activated macrophages, monocytes and DCs and is known to stimulate a unique subset of $CD4^+$ helper T cells called T_h17 cells that produce IL-6, IL-17, IL-22 and TNF. IL-17 promotes a strong pro-inflammatory response by stimulating monocytes and endothelial cells [205]. Several SNPs within the IL-23R gene have shown significant associations with CD, in particular, SNP Arg381Gln. The rare allele at this locus is under-represented in patients with CD and seems to have a protective effect [151]. As of yet, there have been no attempts to investigate the functional effect of this polymorphism. However, it is conceivable that a nonsynonymous polymorphism in IL-23R could alter its affinity for IL-23 disrupting the IL-23/IL-23R signaling pathway. The key role of IL-23R-mediated signaling in the immunopathogenesis of CD and makes it an important candidate for investigating its possible links with JD in cattle.

GWA studies for CD have also revealed important associations with different components of the IL23/IL23R mediated signaling pathway. IL23 binds to IL23R and activates janus kinase 2 (JAK2), which in turn leads to the activation, homodimerization and nuclear translocation of different STAT (signal transducer and activator of transcription) components including STAT 1, 3, 4 and 5 [206]. Among these, STAT-3 plays an important role in the differentiation of T_h17 cells [121]. Both STAT-3 and JAK2 have been shown to have significant and repeated associations with CD [207]. Functional studies investigating the role of specific polymorphisms within these genes have yet to be undertaken. However, murine studies have demonstrated that lack of STAT-3 expression in macrophages and neutrophils leads to spontaneous enterocolitis [208]. Therefore, it is probable that variation within the genes encoding JAK2 or STAT-3 could contribute to an alteration in IL-23/IL-23R-mediated signaling, and therefore, play a role in the pathogenesis of CD.

ATG16l1

Important associations between autophagy-related genes and CD susceptibility have also been reported in different GWA studies. Autophagy is a catabolic process which involves the lysosomal degradation of a portion of the cell's own cytoplasm. It has been shown that autophagy plays an important role in getting rid of intracellular pathogens such as *M. tuberculosis* [209]. ATG16L1 and IRGM are two such autophagy-related genes that have significant and repeated associations with CD [207]. ATG16L1 is expressed in IECs, macrophages and lymphocytes and its expression by IECs is essential for the autophagy of *Salmonella typhimurium*, an intracellular pathogen of the intestines [158]. One SNP,

Ala281Thr, is highly associated with CD and is of particular interest [153]. Similarly, IRGM encodes a GTP binding protein that induces autophagy and promotes removal of various intracellular pathogens including *M. tuberculosis* [209, 210].

There are a number of other candidates that have been linked to susceptibility to various types of mycobacterial infections and they include the natural resistance associated with macrophage growth (NRAMP), NF-κβ and CD14. NRAMP is a divalent metal transporter that is postulated to restrict microbial access to essential micronutrients like iron and manganese within phagosomes [211]. Murine studies have associated NRAMP with resistance to mycobacterial infections [212] and in humans, polymorphisms in this gene have been associated with pulmonary *M. avium* complex infections [213]. Mammalian NF-κβ group of proteins consists of five different members; NF-κβ1, NF-κβ2, REL, RELA, RELB. NF-κβ1 is a downstream transcription factor shared by both NOD and TLR signaling pathways. A deletion polymorphism '-94delATTG' in NF-κβ1 promoter has been associated with susceptibility to CD and is postulated to alter the activity of the promoter [214]. Similarly, another promoter polymorphism '-159(T/C)' in CD14 has been linked to CD [215, 216]. Lastly, CD14 functions as a co-receptor in the presentation of LPS to TLR-4 and is expressed in monocytes.

Impact of Alternate Mechanisms of Gene Regulation

Alternative splicing

Alternative splicing is a eukaryotic process in which the primary gene transcript (pre-mRNA) is spliced in different ways so as to give rise to different mature mRNAs that may differ in arrangement and sometimes the size of exons. This process contributes to the protein diversity of the cell, generating on average three different proteins from each transcriptional framework [217]. Several studies have shown that alternate transcriptional mechanisms play an important role in innate immune signaling [218-220], and isoforms of immune cell surface receptors are very prevalent [221]. Many candidate genes for CD and JD are known to be alternatively spliced. Human CARD15 for example, widely reported for its association with CD, has 14 isoforms and its transcriptional profile changes after stimulation of the cells with MDP [222]. Human IL-23R is also known to have multiple splice variants that are differentially expressed in NK, $CD3^+$ and $CD4^+$ T cells [223]. Similarly, murine studies have reported the existence of numerous isoforms within the TLR signaling pathways that play an important role in diversification or repression of the inflammatory signaling [224]. It has been reported that expression of CXCR3 splice variants is more prevalent in peripheral blood leukocytes of CD patients compared to controls [225]. CXCR3 is a receptor for the CXC chemokines; it is expressed on the surface of T lymphocytes and plays an important role in their trafficking. Aberrant splicing of key molecules in immune signaling pathways can alter the signaling cascades in which they participate and thus may be associated with disease. Though the specific role that alternative splicing plays in contributing to host susceptibility to CD or JD has not yet been investigated, the fact that many of the key molecules that have been associated with CD or JD are extensively spliced raises the possibility that alternative splicing might play an important role in development of disease.

Epigenetics

Most efforts to understand the genetic basis of CD or JD have focused on trying to establish associations between DNA mutations and the disease phenotype. Given the low heritabilities of most disease phenotypes, it is unlikely that they would be dependent purely on genetic mechanisms. The term epigenetics has been used to explain variations in gene expression that do not entail a change in the DNA sequence itself, but are stably inherited from one generation to the other. Diseases such as Prader-Willi syndrome, Angelman syndrome and Beckwith-Wiedemann syndrome, have been found to have a mixed genetic and epigenetic etiology [226]. This suggests that some patients may exhibit a disease phenotype attributed to genetic etiology, while others may exhibit the same disease phenotype by epigenetic etiology. A new disease model has been constructed in order to explain this phenomenon. The model, called the mixed epigenetic and genetic and mixed de novo and inherited (MEGDI) model, includes epigenetics along with different de novo events to explain the etiology of a phenotype [226]. In recent years there has been increasing evidence that suggests that epigenetic mechanisms play a significant role in controlling gene activity; epigenetic regulation of glucocorticoids is a classic example [227]. Although no direct evidence linking epigenetic mechanisms to the etiology of CD has been found, there are data that emphasize the potential role of epigenetics in IBD. It has been demonstrated for example, that there are differences between the maternal and paternal transmission of CD [228]. Such differences in transmission could be attributed to epigenetic mechanisms like mitochondrial transmission, X linked factors or genomic imprinting. In humans, IBD in general, has been shown to have an association with Turner`s syndrome (X monosomy) [229, 230] and markers on the X chromosome [152]. This reinforces the potential role of X linked factors in disease development. Another human study using monozygotic IBD twins found that the concordance of CD amongst the twins was only about 30% [231, 232]. Traditionally, it has been assumed that monozygotic twins are genetically identical, and any differences in the phenotype of monozygotic twins are due to environmental influences. However, more recent reports provide evidence that disputes this notion and suggests that discordance between monozygotic twins could be due to underlying epigenetic mechanisms [233].

A number of polymorphisms in the promoter region of different functional candidates have also been associated with CD. For example, TNF-α contains a polymorphic G(-308)A site in its promoter. The frequency of one of the alleles was found to be reduced in CD patients as compared to healthy subjects [234]. Although the association was not found to be statistically significant, the study pointed towards the fact that in addition to protein sequence variants, changes in gene expression could also increase the risk for CD. It was also found that the polymorphic G(-308)A site was located within an AP2 transcription factor binding site, and AP2 was sensitive to DNA methylation [235, 236]. This raises the possibility of some degree of interaction between genetic and epigenetic mechanisms. Similar to TNF-α, a C(-511)T polymorphism in the promoter for IL-1β [237, 238], and another G/C polymorphism in the promoter region of kinin B1 receptor gene [239], have been found to be associated with IBD in some studies. Gene expression studies have demonstrated that CD patients carry increased mRNA levels of several cytokines, including IL-1β and TNF-α, compared to healthy subjects [240]. Lastly, it has also been found that NF-kβ is recruited to different promoters with different kinetics, depending on constitutive or inducible H4 acylation modifications in their chromatin structure [241]. Since NF-kβ is an important regulator of inflammatory cytokines and mediators that play an important role in both JD and

CD, this could possibly represent another example of how epigenetic mechanisms can interact with genetic mechanisms, contributing to variation of the expression phenotype.

CONCLUSION

JD is known to be caused by MAP in cattle and while the etiology of CD in human populations has not yet been clearly established, it is evident that MAP plays a role in at least a subset of CD patients. Similarities of the host immune responses and a high degree of structural and functional homology between key molecules involved in the host immune responses indicate a strong possibility that similar host defense mechanisms play a role in contributing to resistance of JD in cattle and CD in humans. Understanding the genetic variation underlying this susceptibility or resistance is primarily geared towards designing novel drug therapies and preventive strategies in humans and devising breeding strategies in livestock populations. Significant advances in livestock productivity have been achieved over the last 50 years by the application of artificial selection to enhance the genetic composition of livestock populations [242]. Establishing genetic variation in the phenotype of interest is a prerequisite to the application of artificial selection for genetic improvement. Heritability is a single, commonly used measure that estimates the fraction of total phenotypic variation that is due to variation in the genotypes between individuals of a population. Studies have estimated low heritabilities for susceptibility to MAP infection in different cattle populations. However, these estimates of MAP infection are imprecise because of the categorical nature of the disease phenotype and low sensitivity of available diagnostic tests. Susceptibility to JD, like any other quantitative trait, can be represented by an underlying liability trait that is normally distributed [243]. However, estimation of heritability for susceptibility to JD is often based on a threshold liability model, where individuals have a phenotype 1 or 0, depending on whether their liability scale exceeds a particular threshold. Such a grouping of phenotype into two distinct categories leads to loss of information and lowers the estimates of heritability [244]. As such, it is conceivable that the actual heritability for susceptibility to MAP infection could be higher. Nevertheless, a lower heritability does not indicate that the genetic variation in the population is small and it is possible for a trait to have high genetic variation in a population despite of a low heritability [244]. Since the response to selection depends on the amount of genetic variation present in a population, it is necessary to explore and understand the genetic basis for susceptibility to JD that would enable livestock breeders to select for increased JD resistance in cattle.

Progress in understanding the genetics of complex diseases has been slower compared to monogenic diseases or disorders. By definition, the genetic basis of monogenic diseases is easier to study as the causal polymorphisms generally reside within a single gene. Such polymorphisms are sufficient in themselves to cause a clear alteration in the function of the associated protein, thereby producing disease irrespective of the state of other polymorphisms existing elsewhere in the genome. In contrast, genetic predisposition to complex diseases like JD and CD usually result from a combination of multiple polymorphisms in more than one gene. Furthermore, the presence of each individual polymorphism neither necessitates, nor is sufficient for disease to arise. Rather, individual polymorphisms may lead to disease only in the presence of other polymorphisms with each individual polymorphism, and as such, contribute very little to an individual's susceptibility phenotype. Favourable or unfavourable alleles at each of these loci add up in their effects to produce substantial alterations to the overall inflammatory response, thus, forming the basis of genetic resistance or susceptibility to inflammatory disorders.

Although a number of genes have been proven to have at least some association with CD susceptibility in humans, a number of candidate genes have yet to be studied; thus, these numbers are only expected to increase. Association studies in cattle are still at a preliminary stage, and many studies, including those from our laboratory, are currently underway. Given that JD draws many parallels from CD in humans, it is highly probable that bovine genes which show homology to those implicated in CD susceptibility, or other components of pathways that have been highlighted to play an important role in CD susceptibility in humans, may also be associated with JD susceptibility. Thus, it would seem logical to focus efforts to try to study the variation underlying the genes and pathways.

There is a high degree of structural and functional homology between the key molecules involved in CD in humans and JD in cattle. This indicates a strong possibility that similar host defense mechanisms play a role in contributing resistance to JD in cattle and CD in humans. It might also very well be that MAP is eventually implicated as the etiological agent of CD in either all, or at least a subset of CD cases, which would further reinforce the necessity to study the genetic basis of JD due to its possible zoonotic potential.

The only genome scan for JD susceptibility loci in cattle has revealed the presence of at least eight putative susceptibility loci. Fine mapping of these loci needs to be undertaken to identify positional candidates. It should be kept in mind that while the genome wide scan approach does provide useful starting points to look for positional candidates, its application to the study of JD might be limited by the fact that susceptibility to JD is likely a complex trait and would be controlled by a number of contributing genes, each with low heritabilities. As such, it is likely that a number of loci that are associated with JD susceptibility might not show up on a whole genome scan. Therefore, putative functional candidates need to be independently studied to identify polymorphisms and to functionally characterize the variants. It has also been demonstrated that that alternate mechanisms of gene regulation, like alternative splicing and epigenetic mechanisms, do operate in CD. Such mechanisms might also be playing an important role in determining the phenotype for susceptibility to JD in cattle. Thus, it is necessary to start shifting our focus from DNA polymorphisms alone to a broader approach that includes investigation of aberrant epigenetic or splicing mechanisms.

References

[1] Collins, DM; Gabric, DM; de Lisle, GW. Identification of two groups of Mycobacterium paratuberculosis strains by restriction endonuclease analysis and DNA hybridization. *J. Clin. Microbiol.*, 1990, 28, 1591-1596.

[2] Florou, M; Leontides, L; Kostoulas, P; Billinis, C; Sofia, M; Kyriazakis, I; Lykotrafitis, F. Isolation of Mycobacterium avium subspecies paratuberculosis from non-ruminant wildlife living in the sheds and on the pastures of Greek sheep and goats. *Epidemiol. Infect.*, 2008, 136, 644-652.

[3] Moloney, BJ; Whittington, RJ. Cross species transmission of ovine Johne's disease from sheep to cattle: an estimate of prevalence in exposed susceptible cattle. *Aust. Vet. J.*, 2008, 86, 117-123.

[4] Ris, DR; Hamel, KL; Ayling, JM. Can sheep become infected by grazing pasture contaminated by cattle with Johne's disease? *N. Z. Vet. J.*, 1987, 35, 137.

[5] Whittington, RJ; Windsor, PA. In utero infection of cattle with Mycobacterium avium subsp. paratuberculosis: a critical review and meta-analysis. *Vet. J.*, 2009, 179, 60-69.

[6] Nielsen, SS; Bjerre, H; Toft, N. Colostrum and milk as risk factors for infection with Mycobacterium avium subspecies paratuberculosis in dairy cattle. *J. Dairy Sci.*, 2008, 91, 4610-4615.

[7] Larsen, AB; Stalheim, OH; Hughes, DE; Appell, LH; Richards, WD; Himes, EM. Mycobacterium paratuberculosis in the semen and genital organs of a semen-donor bull. *J. Am. Vet. Med. Assoc.*, 1981, 179, 169-171.

[8] Jorge, MC; Catena, M; Cabodevila, J; Soto, P. Micobacterias en semen bovino criopreservado (in Spanish). *Veterinaria Argentina*, 1998, 145, 337-340.

[9] Chase, CC; Hurley, DJ; Reber, AJ. Neonatal immune development in the calf and its impact on vaccine response. *Vet. Clin. North Am. Food Anim Pract.*, 2008, 24, 87-104.

[10] Gonda, MG; Chang, YM; Shook, GE; Collins, MT; Kirkpatrick, BW. Genetic variation of Mycobacterium avium ssp. paratuberculosis infection in US Holsteins. *J Dairy Sci*, 2006, 89.

[11] Manning, EJ. Mycobacterium avium subspecies paratuberculosis: a review of current knowledge. *J. Zoo. Wildl. Med.*, 2001, 32, 293-304.

[12] Coussens, PM. Model for immune responses to Mycobacterium avium subspecies paratuberculosis in cattle. *Infect. Immun.*, 2004, 72, 3089-3096.

[13] Pence, M; Baldwin, C; Black, CC, III. The seroprevalence of Johne's disease in Georgia beef and dairy cull cattle. *J. Vet. Diagn. Invest*, 2003, 15, 475-477.

[14] Grant, IR. Zoonotic potential of Mycobacterium avium ssp. paratuberculosis: the current position. *J. Appl. Microbiol.*, 2005, 98, 1282-1293.

[15] Lombard, JE; Garry, FB; McCluskey, BJ; Wagner, BA. Risk of removal and effects on milk production associated with paratuberculosis status in dairy cows. *J. Am. Vet. Med. Assoc.*, 2005, 227, 1975-1981.

[16] Nielsen, SS; Krogh, MA; Enevoldsen, C. Time to the occurrence of a decline in milk production in cows with various paratuberculosis antibody profiles. *J. Dairy Sci.*, 2009, 92, 149-155.

[17] Linnabary, RD; Meerdink, GL; Collins, MT; Stabel, JR; Sweeney, RW; Washington, MK; Wells, SJ. Johne's disease in cattle. Council for Agricultural Science and Technology, 2001.

[18] Lu, Z; Mitchell, RM; Smith, RL; Van Kessel, JS; Chapagain, PP; Schukken, YH; Grohn, YT. The importance of culling in Johne's disease control. *J. Theor. Biol.*, 2008, 254, 135-146.

[19] Mikkelsen, H; Jungersen, G; Nielsen, SS. Association between milk antibody and interferon-gamma responses in cattle from Mycobacterium avium subsp. paratuberculosis infected herds. *Vet. Immunol. Immunopathol.*, 2009, 127, 235-241.

[20] Nielsen, SS & Toft, N. Assessment of management-related risk factors for paratuberculosis in Danish dairy herds using Bayesian mixture models. *Prev. Vet. Med.*, 2007, 81, 306-317.

[21] Clark, DL, Jr.; Koziczkowski, JJ; Radcliff, RP; Carlson, RA; Ellingson, JL. Detection of Mycobacterium avium subspecies paratuberculosis: comparing fecal culture versus serum enzyme-linked immunosorbent assay and direct fecal polymerase chain reaction. *J. Dairy Sci.*, 2008, 91, 2620-2627.

[22] Nielsen, SS. Transitions in diagnostic tests used for detection of Mycobacterium avium subsp. paratuberculosis infections in cattle. *Vet. Microbiol.*, 2008, 132, 274-282.

[23] NAHMS. Johne's disease on U.S. dairy operations. USDA:APHIS:VS, CEAH, *National Animal Health Monitoring System*, 1997.

[24] Stabel, JR. Johne's disease: a hidden threat. *J. Dairy Sci.*, 1998, 81, 283-288.

[25] Gonda, MG; Chang, YM; Shook, GE; Collins, MT; Kirkpatrick, BW. Effect of Mycobacterium paratuberculosis infection on production, reproduction, and health traits in US Holsteins. *Prev. Vet. Med.*, 2007, 80, 103-119.

[26] Ott, SL; Wells, SJ; Wagner, BA. Herd-level economic losses associated with Johne's disease on US dairy operations. *Prev. Vet. Med.*, 1999, 40, 179-192.

[27] Emery, DL & Whittington, RJ. An evaluation of mycophage therapy, chemotherapy and vaccination for control of Mycobacterium avium subsp. paratuberculosis infection. *Vet. Microbiol.*, 2004, 104, 143-155.

[28] Head, K & Jurenka, JS. Inflammatory bowel disease. Part II: Crohn's disease--pathophysiology and conventional and alternative treatment options. *Altern. Med. Rev.*, 2004, 9, 360-401.

[29] Crohn's and colitis foundation of america; How Common is Inflammatory Bowel Disease (IBD)? In. 2009.

[30] Mathew, CG. New links to the pathogenesis of Crohn disease provided by genome-wide association scans. *Nat. Rev. Genet.*, 2008, 9, 9-14.

[31] Binder, V & Orholm, M. Familial occurrence and inheritance studies in inflammatory bowel disease. *Neth. J. Med.*, 1996, 48, 53-56.

[32] Hugot, JP; Chamaillard, M; Zouali, H; Lesage, S; Cezard, JP; Belaiche, J; Almer, S; Tysk, C; O'Morain, CA; Gassull, M; Binder, V; Finkel, Y; Cortot, A; Modigliani, R; Laurent-Puig, P; Gower-Rousseau, C; Macry, J; Colombel, JF; Sahbatou, M; Thomas, G. Association of NOD2 leucine-rich repeat variants with susceptibility to Crohn's disease. *Nature*, 2001, 411.

[33] Zhang, H; Massey, D; Tremelling, M; Parkes, M. Genetics of inflammatory bowel disease: clues to pathogenesis. *Br. Med. Bull.*, 2008, 87, 17-30.

[34] Waddell, LA; Rajic, A; Sargeant, J; Harris, J; Amezcua, R; Downey, L; Read, S; McEwen, SA. The zoonotic potential of Mycobacterium avium spp. paratuberculosis: a systematic review. *Can. J. Public Health*, 2008, 99, 145-155.

[35] Secott, TE; Lin, TL; Wu, CC. Mycobacterium avium subsp. paratuberculosis fibronectin attachment protein facilitates M-cell targeting and invasion through a fibronectin bridge with host integrins. *Infect. Immun.*, 2004, 72, 3724-3732.

[36] Sigurethardottir, OG; Valheim, M; Press, CM. Establishment of Mycobacterium avium subsp. paratuberculosis infection in the intestine of ruminants. *Adv. Drug Deliv. Rev.*, 2004, 56, 819-834.

[37] Wu, CW; Livesey, M; Schmoller, SK; Manning, EJ; Steinberg, H; Davis, WC; Hamilton, MJ; Talaat, AM. Invasion and persistence of Mycobacterium avium subsp. paratuberculosis during early stages of Johne's disease in calves. *Infect. Immun.*, 2007, 75, 2110-2119.

[38] Sohal, JS; Singh, SV; Tyagi, P; Subhodh, S; Singh, PK; Singh, AV; Narayanasamy, K; Sheoran, N; Singh, SK. Immunology of mycobacterial infections: with special reference to Mycobacterium avium subspecies paratuberculosis. *Immunobiology*, 2008, 213, 585-598.

[39] Souza, CD; Evanson, OA; Sreevatsan, S; Weiss, DJ. Cell membrane receptors on bovine mononuclear phagocytes involved in phagocytosis of Mycobacterium avium subsp paratuberculosis. *Am. J. Vet. Res.*, 2007 68, 975-980.

[40] Kang, PB; Azad, AK; Torrelles, JB; Kaufman, TM; Beharka, A; Tibesar, E; DesJardin, LE; Schlesinger, LS. The human macrophage mannose receptor directs Mycobacterium tuberculosis lipoarabinomannan-mediated phagosome biogenesis. *J. Exp. Med.*, 2005, 202, 987-999.

[41] Woo, SR & Czuprynski, CJ. Tactics of Mycobacterium avium subsp. paratuberculosis for intracellular survival in mononuclear phagocytes. *J. Vet. Sci.*, 2008, 9, 1-8.

[42] Souza, CD; Evanson, OA; Weiss, DJ. Role of cell membrane receptors in the suppression of monocyte anti-microbial activity against Mycobacterium avium subsp. paratuberculosis. *Microb. Pathog.*, 2008, 44, 215-223.

[43] Vazquez, N; Greenwell-Wild, T; Rekka, S; Orenstein, JM; Wahl, SM. Mycobacterium avium-induced SOCS contributes to resistance to IFN-gamma-mediated mycobactericidal activity in human macrophages. *J. Leukoc. Biol.*, 2006, 80, 1136-1144.

[44] Weiss, DJ & Souza, CD. Review paper: modulation of mononuclear phagocyte function by Mycobacterium avium subsp. paratuberculosis. *Vet. Pathol.*, 2008, 45, 829-841.

[45] Woo, SR; Heintz, JA; Albrecht, R, Barletta, RG; Czuprynski, CJ. Life and death in bovine monocytes: the fate of Mycobacterium avium subsp. paratuberculosis. *Microb. Pathog.*, 2007, 43, 106-113.

[46] Hostetter, J; Steadham, E; Haynes, J; Bailey, T; Cheville, N. Phagosomal maturation and intracellular survival of Mycobacterium avium subspecies paratuberculosis in J774 cells. *Comp Immunol. Microbiol. Infect. Dis.*, 2003, 26, 269-283.

[47] Rumsey, J; Valentine, JF; Naser, SA. Inhibition of phagosome maturation and survival of Mycobacterium avium subspecies paratuberculosis in polymorphonuclear leukocytes from Crohn's disease patients. *Med. Sci. Monit.*, 2006, 12, BR130-BR139.

[48] Souza, CD; Evanson, OA; Weiss, DJ. Role of the mitogen-activated protein kinase pathway in the differential response of bovine monocytes to Mycobacterium avium subsp. paratuberculosis and Mycobacterium avium subsp. avium. *Microbes. Infect.*, 2007, 9, 1545-1552.

[49] Weiss, DJ; Evanson, OA; McClenahan, DJ; Abrahamsen, MS; Walcheck, BK. Regulation of expression of major histocompatibility antigens by bovine macrophages infected with Mycobacterium avium subsp. paratuberculosis or Mycobacterium avium subsp. avium. *Infect. Immun.*, 2001, 69, 1002-1008.

[50] Singhal, A; Jaiswal, A; Arora, VK; Prasad, HK. Modulation of gamma interferon receptor 1 by Mycobacterium tuberculosis: a potential immune response evasive mechanism. *Infect. Immun.*, 2007, 75, 2500-2510.

[51] Lei, L & Hostetter, JM. Limited phenotypic and functional maturation of bovine monocyte-derived dendritic cells following Mycobacterium avium subspecies paratuberculosis infection in vitro. *Vet. Immunol. Immunopathol.*, 2007, 120, 177-186.

[52] Weiss, DJ; Evanson, OA; Souza, CD. Expression of interleukin-10 and suppressor of cytokine signaling-3 associated with susceptibility of cattle to infection with Mycobacterium avium subsp paratuberculosis. *Am. J. Vet. Res.*, 2005, 66, 1114-1120.

[53] Redpath, S; Ghazal, P; Gascoigne, NR. Hijacking and exploitation of IL-10 by intracellular pathogens. *Trends Microbiol.*, 2001, 9, 86-92.

[54] Kaufmann, SH & Schaible, UE. Antigen presentation and recognition in bacterial infections. *Curr. Opin. Immunol.*, 2005, 17, 79-87.

[55] Villarreal-Ramos, B; McAulay, M; Chance, V; Martin, M; Morgan, J; Howard, CJ. Investigation of the role of CD8+ T cells in bovine tuberculosis in vivo. *Infect. Immun.*, 2003, 71, 4297-4303.

[56] Woodworth, JS; Fortune, SM; Behar, SM. Bacterial protein secretion is required for priming of CD8+ T cells specific for the Mycobacterium tuberculosis antigen CFP10. *Infect. Immun.*, 2008, 76, 4199-4205.

[57] Grotzke, JE & Lewinsohn, DM. Role of CD8+ T lymphocytes in control of Mycobacterium tuberculosis infection. *Microbes. Infect.*, 2005, 7, 776-788.

[58] Giodini, A; Rahner, C; Cresswell, P. Receptor-mediated phagocytosis elicits cross-presentation in nonprofessional antigen-presenting cells. *Proc. Natl. Acad. Sci. U. S. A*, 2009 .

[59] Winau, F; Hegasy, G; Kaufmann, SH; Schaible, UE. No life without death--apoptosis as prerequisite for T cell activation. *Apoptosis.*, 2005, 10, 707-715.

[60] Jutila, MA; Holderness, J; Graff, JC; Hedges, JF. Antigen-independent priming: a transitional response of bovine gammadelta T-cells to infection. *Anim Health Res. Rev.*, 2008, 9, 47-57.

[61] Stabel, JR. Transitions in immune responses to Mycobacterium paratuberculosis. *Vet. Microbiol.*, 2000, 77, 465-473.

[62] Koo, HC; Park, YH; Hamilton, MJ; Barrington, GM; Davies, CJ; Kim, JB; Dahl, JL; Waters, WR; Davis, WC. Analysis of the immune response to Mycobacterium avium subsp. paratuberculosis in experimentally infected calves. *Infect. Immun.*, 2004, 72, 6870-6883.

[63] Koets, A; Rutten, V; Hoek, A; van, MF; Muller, K; Bakker, D; Gruys, E; van, EW. Progressive bovine paratuberculosis is associated with local loss of CD4(+) T cells, increased frequency of gamma delta T cells, and related changes in T-cell function. *Infect. Immun.*, 2002, 70, 3856-3864.

[64] Simutis, FJ; Jones, DE; Hostetter, JM. Failure of antigen-stimulated gammadelta T cells and CD4+ T cells from sensitized cattle to upregulate nitric oxide and mycobactericidal activity of autologous Mycobacterium avium subsp. paratuberculosis-infected macrophages. *Vet. Immunol. Immunopathol.*, 2007, 116, 1-12.

[65] Brandes, M; Willimann, K; Bioley, G; Levy, N; Eberl, M; Luo, M; Tampe, R; Levy, F; Romero, P; Moser, B. Cross-presenting human gammadelta T cells induce robust CD8+ alphabeta T cell responses. *Proc. Natl. Acad. Sci. U. S. A*, 2009, 106, 2307-2312.

[66] Stewart, DJ; Vaughan, JA; Stiles, PL; Noske, PJ; Tizard, ML; Prowse, SJ; Michalski, WP; Butler, KL; Jones, SL. A long-term bacteriological and immunological study in Holstein-Friesian cattle experimentally infected with Mycobacterium avium subsp. paratuberculosis and necropsy culture results for Holstein-Friesian cattle, Merino sheep and Angora goats. *Vet. Microbiol.*, 2007, 122, 83-96.

[67] Coussens, PM. Mycobacterium paratuberculosis and the bovine immune system. *Anim Health Res. Rev.*, 2001, 2, 141-161.

[68] Davis, JM & Ramakrishnan, L. The role of the granuloma in expansion and dissemination of early tuberculous infection. *Cell*, 2009, 136, 37-49.

[69] Lei, L; Plattner, BL; Hostetter, JM. Live Mycobacterium avium subsp. paratuberculosis and a killed-bacterium vaccine induce distinct subcutaneous granulomas, with unique cellular and cytokine profiles. *Clin. Vaccine Immunol.*, 2008, 15, 783-793.

[70] Hostetter, J; Huffman, E; Byl, K; Steadham, E. Inducible nitric oxide synthase immunoreactivity in the granulomatous intestinal lesions of naturally occurring bovine Johne's disease. *Vet. Pathol.*, 2005, 42, 241-249.

[71] Lee, H; Stabel, JR; Kehrli, ME, Jr. Cytokine gene expression in ileal tissues of cattle infected with Mycobacterium paratuberculosis. *Vet. Immunol. Immunopathol.*, 2001, 82, 73-85.

[72] Tanaka, S; Sato, M; Onitsuka, T; Kamata, H; Yokomizo, Y. Inflammatory cytokine gene expression in different types of granulomatous lesions during asymptomatic stages of bovine paratuberculosis. *Vet. Pathol.*, 2005, 42, 579-588.

[73] Munoz, M; Delgado, L; Verna, A; Benavides, J; Garcia-Pariente, C; Fuertes, M; Ferreras, MC; Garcia-Marin, JF; Perez, V. Expression of transforming growth factor-beta 1 (TGF-beta1) in different types of granulomatous lesions in bovine and ovine paratuberculosis. *Comp Immunol. Microbiol. Infect. Dis.*, 2008 .

[74] de Almeida, DE; Colvin, CJ; Coussens, PM. Antigen-specific regulatory T cells in bovine paratuberculosis. *Vet. Immunol. Immunopathol.*, 2008, 125, 234-245.

[75] Benoit, M; Desnues, B; Mege, JL. Macrophage polarization in bacterial infections. *J. Immunol.*, 2008, 181, 3733-3739.

[76] Ito, T; Schaller, M; Hogaboam, CM; Standiford, TJ; Chensue, SW; Kunkel, SL. TLR9 activation is a key event for the maintenance of a mycobacterial antigen-elicited pulmonary granulomatous response. *Eur. J. Immunol.*, 2007, 37, 2847-2855.

[77] Hesse, M; Modolell, M; La Flamme, AC; Schito, M; Fuentes, JM; Cheever, AW; Pearce, EJ; Wynn, TA. Differential regulation of nitric oxide synthase-2 and arginase-1 by type 1/type 2 cytokines in vivo: granulomatous pathology is shaped by the pattern of L-arginine metabolism. *J. Immunol.*, 2001, 167, 6533-6544.

[78] Pesce, J; Kaviratne, M; Ramalingam, TR; Thompson, RW; Urban, JF, Jr.; Cheever, AW; Young, DA; Collins, M; Grusby, MJ; Wynn, TA. The IL-21 receptor augments Th2 effector function and alternative macrophage activation. *J. Clin. Invest*, 2006, 116, 2044-2055.

[79] Coussens, PM; Colvin, CJ; Rosa, GJ; Perez, LJ; Elftman, MD. Evidence for a novel gene expression program in peripheral blood mononuclear cells from Mycobacterium avium subsp. paratuberculosis-infected cattle. *Infect. Immun.*, 2003, 71, 6487-6498.

[80] Weiss, DJ; Evanson, OA; Deng, M; Abrahamsen, MS. Sequential patterns of gene expression by bovine monocyte-derived macrophages associated with ingestion of mycobacterial organisms. *Microb. Pathog.*, 2004, 37, 215-224.

[81] Buza, JJ; Mori, Y; Bari, AM; Hikono, H; Aodon, g; Hirayama, S; Shu, Y; Momotani, E. Mycobacterium avium subsp. paratuberculosis infection causes suppression of RANTES, monocyte chemoattractant protein 1, and tumor necrosis factor alpha expression in peripheral blood of experimentally infected cattle. *Infect. Immun.*, 2003, 71, 7223-7227.

[82] Coussens, PM; Verman, N; Coussens, MA; Elftman, MD; McNulty, AM. Cytokine gene expression in peripheral blood mononuclear cells and tissues of cattle infected with Mycobacterium avium subsp. paratuberculosis: evidence for an inherent proinflammatory gene expression pattern. *Infect. Immun.*, 2004, 72, 1409-1422.

[83] Weiss, DJ; Evanson, OA; Souza, CD. Mucosal immune response in cattle with subclinical Johne's disease. *Vet. Pathol.*, 2006, 43, 127-135.

[84] Tiwari, A; VanLeeuwen, JA; McKenna, SL; Keefe, GP; Barkema, HW. Johne's disease in Canada Part I: clinical symptoms, pathophysiology, diagnosis, and prevalence in dairy herds. *Can. Vet. J.*, 2006, 47, 874-882.

[85] Nielsen, SS & Toft, N. Ante mortem diagnosis of paratuberculosis: a review of accuracies of ELISA, interferon-gamma assay and faecal culture techniques. *Vet. Microbiol.*, 2008, 129, 217-235.

[86] Khalifeh, MS & Stabel, JR. Upregulation of transforming growth factor-beta and interleukin-10 in cows with clinical Johne's disease. *Vet. Immunol. Immunopathol.*, 2004, 99, 39-46.

[87] Stabel, JR & Khalifeh, MS. Differential expression of CD5 on B lymphocytes in cattle infected with Mycobacterium avium subsp. paratuberculosis. *Vet. Immunol. Immunopathol.*, 2008, 126, 211-219.

[88] Waters, WR; Stabel, JR; Sacco, RE; Harp, JA; Pesch, BA; Wannemuehler, MJ. Antigen-specific B-cell unresponsiveness induced by chronic Mycobacterium avium subsp. paratuberculosis infection of cattle. *Infect. Immun.*, 1999, 67, 1593-1598.

[89] Piatelli, MJ; Tanguay, D; Rothstein, TL; Chiles, TC. Cell cycle control mechanisms in B-1 and B-2 lymphoid subsets. *Immunol. Res.*, 2003, 27, 31-52.

[90] Koets, AP; Rutten, VP; de, BM; Bakker, D; Valentin-Weigand, P; van, EW. Differential changes in heat shock protein-, lipoarabinomannan-, and purified protein derivative-specific immunoglobulin G1 and G2 isotype responses during bovine Mycobacterium avium subsp. paratuberculosis infection. *Infect. Immun.*, 2001, 69, 1492-1498.

[91] Chacon, O; Bermudez, LE; Barletta, RG. Johne's disease, inflammatory bowel disease, and Mycobacterium paratuberculosis. *Annu. Rev. Microbiol.*, 2004, 58, 329-363.

[92] Hendrickson, BA; Gokhale, R; Cho, JH. Clinical aspects and pathophysiology of inflammatory bowel disease. *Clin. Microbiol. Rev.*, 2002, 15, 79-94.

[93] Panes, J & Ghosh, S. The classics in perspective. *Gut*, 2007, 56, 1184-1186.

[94] Chamberlin, WM & Naser, SA. Integrating theories of the etiology of Crohn's disease. On the etiology of Crohn's disease: questioning the hypotheses. *Med. Sci. Monit.*, 2006, 12, RA27-RA33.

[95] Strobel, S. Oral tolerance, systemic immunoregulation, and autoimmunity. *Ann. N. Y. Acad. Sci.*, 2002, 958, 47-58.

[96] Joosten, SA & Ottenhoff, TH. Human CD4 and CD8 regulatory T cells in infectious diseases and vaccination. *Hum. Immunol.*, 2008, 69, 760-770.

[97] Le, NT & Chao, N. Regulating regulatory T cells. *Bone Marrow Transplant.*, 2007, 39, 1-9.

[98] Izcue, A; Coombes, JL; Powrie, F. Regulatory T cells suppress systemic and mucosal immune activation to control intestinal inflammation. *Immunol. Rev.*, 2006, 212, 256-271.

[99] Horwitz, BH. The straw that stirs the drink: insight into the pathogenesis of inflammatory bowel disease revealed through the study of microflora-induced inflammation in genetically modified mice. *Inflamm. Bowel. Dis.*, 2007, 13, 490-500.

[100] Izcue, A & Powrie, F. Special regulatory T-cell review: Regulatory T cells and the intestinal tract--patrolling the frontier. *Immunology*, 2008, 123, 6-10.

[101] Watanabe, T; Yamori, M; Kita, T; Chiba, T; Wakatsuki, Y. CD4+CD25+ T cells regulate colonic localization of CD4 T cells reactive to a microbial antigen. *Inflamm. Bowel. Dis.*, 2005, 11, 541-550.

[102] Tiemessen, MM; Jagger, AL; Evans, HG; van Herwijnen, MJ; John, S; Taams, LS. CD4+CD25+Foxp3+ regulatory T cells induce alternative activation of human monocytes/macrophages. *Proc. Natl. Acad. Sci. U. S. A*, 2007, 104, 19446-19451.

[103] Hedl, M; Li, J; Cho, JH; Abraham, C. Chronic stimulation of Nod2 mediates tolerance to bacterial products. *Proc. Natl. Acad. Sci. U. S. A*, 2007, 104, 19440-19445.

[104] Feller, M; Huwiler, K; Stephan, R; Altpeter, E; Shang, A; Furrer, H; Pfyffer, GE; Jemmi, T; Baumgartner, A; Egger, M. Mycobacterium avium subspecies paratuberculosis and Crohn's disease: a systematic review and meta-analysis. *Lancet Infect. Dis.*, 2007, 7, 607-613.

[105] Bernstein, CN; Blanchard, JF; Rawsthorne, P; Collins, MT. Population-based case control study of seroprevalence of Mycobacterium paratuberculosis in patients with Crohn's disease and ulcerative colitis. *J. Clin. Microbiol.*, 2004, 42, 1129-1135.

[106] Ellingson, JL; Cheville, JC; Brees, D; Miller, JM; Cheville, NF. Absence of Mycobacterium avium subspecies paratuberculosis components from Crohn's disease intestinal biopsy tissues. *Clin. Med. Res.*, 2003, 1, 217-226.

[107] Rowbotham, DS; Mapstone, NP; Trejdosiewicz, LK; Howdle, PD; Quirke, P. Mycobacterium paratuberculosis DNA not detected in Crohn's disease tissue by fluorescent polymerase chain reaction. *Gut*, 1995, 37, 660-667.

[108] Scanu, AM; Bull, TJ; Cannas, S; Sanderson, JD; Sechi, LA; Dettori, G; Zanetti, S; Hermon-Taylor, J. Mycobacterium avium subspecies paratuberculosis infection in cases of irritable bowel syndrome and comparison with Crohn's disease and Johne's disease: common neural and immune pathogenicities. *J. Clin. Microbiol.*, 2007, 45, 3883-3890.

[109] Cerf, O; Griffiths, M; Aziza, F. Assessment of the prevalence of Mycobacterium avium subsp. paratuberculosis in commercially pasteurized milk. *Foodborne. Pathog. Dis.*, 2007, 4, 433-447.

[110] Lund, BM; Gould, GW; Rampling, AM. Pasteurization of milk and the heat resistance of Mycobacterium avium subsp. paratuberculosis: a critical review of the data. *Int. J. Food Microbiol.*, 2002, 77, 135-145.

[111] Chamberlin, W; Ghobrial, G; Chehtane, M; Naser, SA. Successful treatment of a Crohn's disease patient infected with bacteremic Mycobacterium paratuberculosis. *Am. J. Gastroenterol.*, 2007, 102, 689-691.

[112] Polymeros, D; Bogdanos, DP; Day, R; Arioli, D; Vergani, D; Forbes, A. Does cross-reactivity between mycobacterium avium paratuberculosis and human intestinal antigens characterize Crohn's disease? *Gastroenterology*, 2006, 131, 85-96.

[113] Burrells, C; Clarke, CJ; Colston, A; Kay, JM; Porter, J; Little, D; Sharp, JM. Interferon-gamma and interleukin-2 release by lymphocytes derived from the blood,

mesenteric lymph nodes and intestines of normal sheep and those affected with paratuberculosis (Johne's disease). *Vet. Immunol. Immunopathol.*, 1999, 68, 139-148.

[114] Shin, SJ; Chang, CF; Chang, CD; McDonough, SP; Thompson, B; Yoo, HS; Chang, YF. In vitro cellular immune responses to recombinant antigens of Mycobacterium avium subsp. paratuberculosis. *Infect. Immun.*, 2005, 73, 5074-5085.

[115] Stabel, JR. Host responses to Mycobacterium avium subsp. paratuberculosis: a complex arsenal. *Anim Health Res. Rev.*, 2006, 7, 61-70.

[116] Coccia, EM; Remoli, ME; Di, GC; Del, ZB; Giacomini, E; Monteleone, G; Boirivant, M. Cholera toxin subunit B inhibits IL-12 and IFN-{gamma} production and signaling in experimental colitis and Crohn's disease. *Gut*, 2005, 54, 1558-1564.

[117] Fuss, IJ; Becker, C; Yang, Z; Groden, C; Hornung, RL; Heller, F; Neurath, MF; Strober, W; Mannon, PJ. Both IL-12p70 and IL-23 are synthesized during active Crohn's disease and are down-regulated by treatment with anti-IL-12 p40 monoclonal antibody. *Inflamm. Bowel. Dis.*, 2006, 12, 9-15.

[118] Kakazu, T; Hara, J; Matsumoto, T; Nakamura, S; Oshitani, N; Arakawa, T; Kitano, A; Nakatani, K; Kinjo, F; Kuroki, T. Type 1 T-helper cell predominance in granulomas of Crohn's disease. *Am. J. Gastroenterol.*, 1999, 94, 2149-2155.

[119] Pages, F; Lazar, V; Berger, A; Danel, C; Lebel-Binay, S; Zinzindohoue, F; Desreumaux, P; Cellier, C; Thiounn, N; Bellet, D; Cugnenc, PH; Fridman, WH. Analysis of interleukin-18, interleukin-1 converting enzyme (ICE) and interleukin-18-related cytokines in Crohn's disease lesions. *Eur. Cytokine Netw.*, 2001, 12, 97-104.

[120] Trinchieri, G. Interleukin-12 and the regulation of innate resistance and adaptive immunity. *Nat. Rev. Immunol.*, 2003, 3, 133-146.

[121] Watford, WT; Hissong, BD; Bream, JH; Kanno, Y; Muul, L; O'Shea, JJ. Signaling by IL-12 and IL-23 and the immunoregulatory roles of STAT4. *Immunol Rev*, 2004 202.

[122] Yoshimura, A. Signal transduction of inflammatory cytokines and tumor development. *Cancer Sci.*, 2006, 97, 439-447.

[123] Welsh, MD; Cunningham, RT; Corbett, DM; Girvin, RM; McNair, J; Skuce, RA; Bryson, DG; Pollock, JM. Influence of pathological progression on the balance between cellular and humoral immune responses in bovine tuberculosis. *Immunology*, 2005, 114, 101-111.

[124] Kugathasan, S; Saubermann, LJ; Smith, L; Kou, D; Itoh, J; Binion, DG; Levine, AD; Blumberg, RS; Fiocchi, C. Mucosal T-cell immunoregulation varies in early and late inflammatory bowel disease. *Gut*, 2007, 56, 1696-1705.

[125] Spencer, DM; Veldman, GM; Banerjee, S; Willis, J; Levine, AD. Distinct inflammatory mechanisms mediate early versus late colitis in mice. *Gastroenterology*, 2002, 122, 94-105.

[126] Ren, Z; Turton, J; Borody, T; Pang, G; Clancy, R. Selective Th2 pattern of cytokine secretion in Mycobacterium avium subsp. paratuberculosis infected Crohn's disease. *J. Gastroenterol. Hepatol.*, 2008, 23, 310-314.

[127] Belkaid, Y. Regulatory T cells and infection: a dangerous necessity. *Nat. Rev. Immunol.*, 2007, 7, 875-888.

[128] Himmel, ME; Hardenberg, G; Piccirillo, CA; Steiner, TS; Levings, MK. The role of T-regulatory cells and Toll-like receptors in the pathogenesis of human inflammatory bowel disease. *Immunology*, 2008, 125, 145-153.

[129] Autschbach, F; Braunstein, J; Helmke, B; Zuna, I; Schurmann, G; Niemir, ZI; Wallich, R; Otto, HF; Meuer, SC. In situ expression of interleukin-10 in noninflamed human gut and in inflammatory bowel disease. *Am. J. Pathol.*, 1998, 153, 121-130.

[130] Kucharzik, T; Stoll, R; Lugering, N; Domschke, W. Circulating antiinflammatory cytokine IL-10 in patients with inflammatory bowel disease (IBD). *Clin. Exp. Immunol.*, 1995, 100, 452-456.

[131] Niessner, M & Volk, BA. Altered Th1/Th2 cytokine profiles in the intestinal mucosa of patients with inflammatory bowel disease as assessed by quantitative reversed transcribed polymerase chain reaction (RT-PCR). *Clin. Exp. Immunol.*, 1995, 101, 428-435.

[132] Bartels, LE; Jorgensen, SP; Agnholt, J; Kelsen, J; Hvas, CL; Dahlerup, JF. 1,25-dihydroxyvitamin D3 and dexamethasone increase interleukin-10 production in CD4+ T cells from patients with Crohn's disease. *Int. Immunopharmacol.*, 2007, 7, 1755-1764.

[133] Hvas, CL; Kelsen, J; Agnholt, J; Hollsberg, P; Tvede, M; Moller, JK; Dahlerup, JF. Crohn's disease intestinal CD4+ T cells have impaired interleukin-10 production which is not restored by probiotic bacteria. *Scand. J. Gastroenterol.*, 2007, 42, 592-601.

[134] Mitsuyama, K; Tomiyasu, N; Takaki, K; Masuda, J; Yamasaki, H; Kuwaki, K; Takeda, T; Kitazaki, S; Tsuruta, O; Sata, M. Interleukin-10 in the pathophysiology of inflammatory bowel disease: increased serum concentrations during the recovery phase. *Mediators. Inflamm.*, 2006, 26875.

[135] Ten, HT; The, OF; Berkhout, M; Bruggeman, JP; Vyth-Dreese, FA; Slors, JF; van Deventer, SJ; Te Velde, AA. Expression of CD45RB functionally distinguishes intestinal T lymphocytes in inflammatory bowel disease. *J. Leukoc. Biol.*, 2004, 75, 1010-1015.

[136] Babyatsky, MW; Rossiter, G; Podolsky, DK. Expression of transforming growth factors alpha and beta in colonic mucosa in inflammatory bowel disease. *Gastroenterology*, 1996 110, 975-984.

[137] Del, ZB; Mumolo, G; Pronio, AM; Montesani, C; Tersigni, R; Boirivant, M. TGF-beta1 production in inflammatory bowel disease: differing production patterns in Crohn's disease and ulcerative colitis. *Clin. Exp. Immunol.*, 2003, 134, 120-126.

[138] Sturm, A; Schulte, C; Schatton, R; Becker, A; Cario, E; Goebell, H; Dignass, AU. Transforming growth factor-beta and hepatocyte growth factor plasma levels in patients with inflammatory bowel disease. *Eur. J. Gastroenterol. Hepatol.*, 2000, 12, 445-450.

[139] Ebert, EC; Panja, A; Das, KM; Praveen, R; Geng, X; Rezac, C; Bajpai, M. Patients with inflammatory bowel disease may have a transforming growth factor-beta-, interleukin (IL)-2- or IL-10-deficient state induced by intrinsic neutralizing antibodies. *Clin. Exp. Immunol.*, 2009, 155, 65-71.

[140] Collins, MT; Wells, SJ; Petrini, KR; Collins, JE; Schultz, RD; Whitlock, RH. Evaluation of five antibody detection tests for diagnosis of bovine paratuberculosis. *Clin Diagn Lab Immunol*, 2005, 12.

[141] Koets, AP; Adugna, G; Janss, LL; van Weering, HJ; Kalis, CH; Wentink, GH; Rutten, VP; Schukken, YH. Genetic variation of susceptibility to Mycobacterium avium subsp. paratuberculosis infection in dairy cattle. *J Dairy Sci*, 2000, 83.

[142] Mortensen, H; Nielsen, SS; Berg, P. Genetic variation and heritability of the antibody response to Mycobacterium avium subspecies paratuberculosis in Danish Holstein cows. *J Dairy Sci*, 2004, 87.

[143] Cetinkaya, B; Erdogan, HM; Morgan, KL. Relationships between the presence of Johne's disease and farm and management factors in dairy cattle in England. *Prev. Vet. Med.*, 1997, 32, 253-266.

[144] Jakobsen, MB; Alban, L; Nielsen, SS. A cross-sectional study of paratuberculosis in 1155 Danish dairy cows. *Prev. Vet. Med.*, 2000, 46, 15-27.

[145] Gonda, MG; Kirkpatrick, BW; Shook, GE; Collins, MT. Identification of a QTL on BTA20 affecting susceptibility to Mycobacterium avium ssp. paratuberculosis infection in US Holsteins. *Anim Genet*, 2007, 38.

[146] Hirschhorn, JN & Daly, MJ. Genome-wide association studies for common diseases and complex traits. *Nat Rev Genet*, 2005, 6.

[147] Goddard, ME. The use of high density genotyping in animal health. *Dev Biol (Basel)*, 2008, 132.

[148] Cho, JH; Nicolae, DL; Gold, LH; Fields, CT; LaBuda, MC; Rohal, PM; Pickles, MR; Qin, L; Fu, Y; Mann, JS; Kirschner, BS; Jabs, EW; Weber, J; Hanauer, SB; Bayless, TM; Brant, SR. Identification of novel susceptibility loci for inflammatory bowel disease on chromosomes 1p, 3q, and 4q: evidence for epistasis between 1p and IBD1. *Proc Natl Acad Sci U S A*, 1998, 95.

[149] Cho, JH; Nicolae, DL; Ramos, R; Fields, CT; Rabenau, K; Corradino, S; Brant, SR; Espinosa, R; LeBeau, M; Hanauer, SB; Bodzin, J; Bonen, DK. Linkage and linkage disequilibrium in chromosome band 1p36 in American Chaldeans with inflammatory bowel disease. *Hum Mol Genet*, 2000, 9.

[150] Duerr, RH; Barmada, MM; Zhang, L; Pfutzer, R; Weeks, DE. High-density genome scan in Crohn disease shows confirmed linkage to chromosome 14q11-12. *Am J Hum Genet*, 2000, 66.

[151] Duerr, RH; Taylor, KD; Brant, SR; Rioux, JD; Silverberg, MS; Daly, MJ; Steinhart, AH; Abraham, C; Regueiro, M; Griffiths, A; Dassopoulos, T; Bitton, A; Yang, H; Targan, S; Datta, LW; Kistner, EO; Schumm, LP; Lee, AT; Gregersen, PK; Barmada, MM; Rotter, JI; Nicolae, DL; Cho, JH. A genome-wide association study identifies IL23R as an inflammatory bowel disease gene. *Science*, 2006, 314.

[152] Hampe, J; Schreiber, S; Shaw, SH; Lau, KF; Bridger, S; MacPherson, AJ; Cardon, LR; Sakul, H; Harris, TJ; Buckler, A; Hall, J; Stokkers, P; van Deventer, SJ; Nurnberg, P; Mirza, MM; Lee, JC; Lennard-Jones, JE; Mathew, CG; Curran, ME. A genomewide analysis provides evidence for novel linkages in inflammatory bowel disease in a large European cohort. *Am J Hum Genet*, 1999, 64.

[153] Hampe, J; Franke, A; Rosenstiel, P; Till, A; Teuber, M; Huse, K; Albrecht, M; Mayr, G; De La Vega, FM; Briggs, J; Gunther, S; Prescott, NJ; Onnie, CM; Hasler, R; Sipos, B; Folsch, UR; Lengauer, T; Platzer, M; Mathew, CG; Krawczak, M; Schreiber, S. A genome-wide association scan of nonsynonymous SNPs identifies a susceptibility variant for Crohn disease in ATG16L1. *Nat Genet*, 2007, 39.

[154] Hugot, JP; Laurent-Puig, P; Gower-Rousseau, C; Olson, JM; Lee, JC; Beaugerie, L; Naom, I; Dupas, JL; Van Gossum, A; Orholm, M; Bonaiti-Pellie, C; Weissenbach, J; Mathew, CG; Lennard-Jones, JE; Cortot, A; Colombel, JF; Thomas, G. Mapping of a susceptibility locus for Crohn's disease on chromosome 16. *Nature*, 1996, 379.

[155] Ma, Y; Ohmen, JD; Li, Z; Bentley, LG; McElree, C; Pressman, S; Targan, SR; Fischel-Ghodsian, N; Rotter, JI; Yang, H. A genome-wide search identifies potential new susceptibility loci for Crohn's disease. *Inflamm Bowel Dis*, 1999, 5.

[156] Massey, DC & Parkes, M. Genome-wide association scanning highlights two autophagy genes, ATG16L1 and IRGM, as being significantly associated with Crohn's disease. *Autophagy*, 2007, 3.

[157] Rioux, JD; Silverberg, MS; Daly, MJ; Steinhart, AH; McLeod, RS; Griffiths, AM; Green, T; Brettin, TS; Stone, V; Bull, SB; Bitton, A; Williams, CN; Greenberg, GR; Cohen, Z; Lander, ES; Hudson, TJ; Siminovitch, KA. Genomewide search in Canadian families with inflammatory bowel disease reveals two novel susceptibility loci. *Am J Hum Genet*, 2000, 66.

[158] Rioux, JD; Xavier, RJ; Taylor, KD; Silverberg, MS; Goyette, P; Huett, A; Green, T; Kuballa, P; Barmada, MM; Datta, LW; Shugart, YY; Griffiths, AM; Targan, SR; Ippoliti, AF; Bernard, EJ; Mei, L; Nicolae, DL; Regueiro, M; Schumm, LP; Steinhart, AH; Rotter, JI; Duerr, RH; Cho, JH; Daly, MJ; Brant, SR. Genome-wide association study identifies new susceptibility loci for Crohn disease and implicates autophagy in disease pathogenesis. *Nat Genet*, 2007, 39.

[159] Satsangi, J; Parkes, M; Louis, E; Hashimoto, L; Kato, N; Welsh, K; Terwilliger, JD; Lathrop, GM; Bell, JI; Jewell, DP. Two stage genome-wide search in inflammatory bowel disease provides evidence for susceptibility loci on chromosomes 3, 7 and 12. *Nat Genet*, 1996, 14.

[160] Wellcome Trust Case Control Consortium. Genome-wide association study of 14,000 cases of seven common diseases and 3,000 shared controls. *Nature*, 2007, 447.

[161] Hinger, M; Brandt, H; Horner, S; Erhardt, G. Short communication: Association analysis of microsatellites and Mycobacterium avium subspecies paratuberculosis antibody response in German Holsteins. *J Dairy Sci*, 2007, 90.

[162] Mucha, R; Bhide, MR; Chakurkar, EB; Novak, M; Mikula, I, Sr. Toll-like receptors TLR1, TLR2 and TLR4 gene mutations and natural resistance to Mycobacterium avium subsp. paratuberculosis infection in cattle. *Vet Immunol Immunopathol*, 2008 .

[163] Pinedo, PJ; Buergelt, CD; Donovan, GA; Melendez, P; Morel, L; Wu, R; Langaee, TY; Rae, DO. Association between CARD15/NOD2 gene polymorphisms and paratuberculosis infection in cattle. *Vet Microbiol*, 2009, 134.

[164] Taylor, KH; Taylor, JF; White, SN; Womack, JE. Identification of genetic variation and putative regulatory regions in bovine CARD15. *Mamm Genome*, 2006, 17.

[165] Baldwin, AS, Jr. The NF-kappa B and I kappa B proteins: new discoveries and insights. *Annu Rev Immunol*, 1996, 14.

[166] Girardin, SE; Boneca, IG; Viala, J; Chamaillard, M; Labigne, A; Thomas, G; Philpott, DJ; Sansonetti, PJ. Nod2 is a general sensor of peptidoglycan through muramyl dipeptide (MDP) detection. *J Biol Chem*, 200,3 278.

[167] Siebenlist, U; Franzoso, G; Brown, K. Structure, regulation and function of NF-kappa B. *Annu Rev Cell Biol*, 1994, 10.

[168] Hampe, J; Cuthbert, A; Croucher, PJ; Mirza, MM; Mascheretti, S; Fisher, S; Frenzel, H; King, K; Hasselmeyer, A; MacPherson, AJ; Bridger, S; van Deventer, S; Forbes, A; Nikolaus, S; Lennard-Jones, JE; Foelsch, UR; Krawczak, M; Lewis, C; Schreiber, S; Mathew, CG. Association between insertion mutation in NOD2 gene and Crohn's disease in German and British populations. *Lancet*, 200,1 357.

[169] Ogura, Y; Bonen, DK; Inohara, N; Nicolae, DL; Chen, FF; Ramos, R; Britton, H; Moran, T; Karaliuskas, R; Duerr, RH; Achkar, JP; Brant, SR; Bayless, TM; Kirschner, BS; Hanauer, SB; Nunez, G; Cho, JH. A frameshift mutation in NOD2 associated with susceptibility to Crohn's disease. *Nature*, 200,1 411.

[170] Economou, M; Trikalinos, TA; Loizou, KT; Tsianos, EV; Ioannidis, JP. Differential effects of NOD2 variants on Crohn's disease risk and phenotype in diverse populations: a metaanalysis. *Am J Gastroenterol*, 200,4 99.

[171] Pinedo, PJ; Wang, C; Li, Y; Rae, DO; Wu, R. Risk haplotype analysis for bovine paratuberculosis. *Mamm Genome*, 2009.

[172] Hugot, JP. CARD15/NOD2 mutations in Crohn's disease. *Ann. N. Y. Acad. Sci.*, 2006, 1072, 9-18.

[173] Quaglietta, L; te Velde, A; Staiano, A; Troncone, R; Hommes, DW. Functional consequences of NOD2/CARD15 mutations in Crohn disease. *J Pediatr Gastroenterol Nutr*, 2007, 44.

[174] Inohara, N; Koseki, T; Lin, J; del Peso, L; Lucas, PC; Chen, FF; Ogura, Y; Nunez, G. An induced proximity model for NF-kappa B activation in the Nod1/RICK and RIP signaling pathways. *J Biol Chem*, 2000, 275.

[175] McGovern, DP; Hysi, P; Ahmad, T; van Heel, DA; Moffatt, MF; Carey, A; Cookson, WO; Jewell, DP. Association between a complex insertion/deletion polymorphism in NOD1 (CARD4) and susceptibility to inflammatory bowel disease. *Hum Mol Genet*, 2005, 14.

[176] Ozen, SC; Dagli, U; Kilic, MY; Toruner, M; Celik, Y; Ozkan, M; Soykan, I; Cetinkaya, H; Ulker, A; Ozden, A; Bozdayi, AM. NOD2/CARD15, NOD1/CARD4, and ICAM-1 gene polymorphisms in Turkish patients with inflammatory bowel disease. *J Gastroenterol*, 2006 41.

[177] Akira, S. Toll-like receptor signaling. *J Biol Chem*, 2003, 278.

[178] Takeda, K; Kaisho, T; Akira, S. Toll-like receptors. *Annu Rev Immunol*, 2003, 21.

[179] Chang, JS; Huggett, JF; Dheda, K; Kim, LU; Zumla, A; Rook, GA. Myobacterium tuberculosis induces selective up-regulation of TLRs in the mononuclear leukocytes of patients with active pulmonary tuberculosis. *J Immunol*, 2006, 176.

[180] Uehori, J; Fukase, K; Akazawa, T; Uematsu, S; Akira, S; Funami, K; Shingai, M; Matsumoto, M; Azuma, I; Toyoshima, K; Kusumoto, S; Seya, T. Dendritic cell maturation induced by muramyl dipeptide (MDP) derivatives: monoacylated MDP confers TLR2/TLR4 activation. *J Immunol*, 2005, 174.

[181] Underhill, DM; Ozinsky, A; Smith, KD; Aderem, A. Toll-like receptor-2 mediates mycobacteria-induced proinflammatory signaling in macrophages. *Proc Natl Acad Sci U S A*, 1999, 96.

[182] Franchimont, D; Vermeire, S; El Housni, H; Pierik, M; Van Steen, K; Gustot, T; Quertinmont, E; Abramowicz, M; Van Gossum, A; Deviere, J; Rutgeerts, P. Deficient host-bacteria interactions in inflammatory bowel disease? The toll-like receptor (TLR)-4 Asp299gly polymorphism is associated with Crohn's disease and ulcerative colitis. *Gut*, 2004, 53.

[183] Henckaerts, L; Pierik, M; Joossens, M; Ferrante, M; Rutgeerts, P; Vermeire, S. Mutations in pattern recognition receptor genes modulate seroreactivity to microbial antigens in patients with inflammatory bowel disease. *Gut*, 2007, 56.

[184] Johnson, CM; Lyle, EA; Omueti, KO; Stepensky, VA; Yegin, O; Alpsoy, E; Hamann, L; Schumann, RR; Tapping, RI. Cutting edge: A common polymorphism impairs cell surface trafficking and functional responses of TLR1 but protects against leprosy. *J Immunol*, 2007, 178.

[185] Schroder, NW & Schumann, RR. Single nucleotide polymorphisms of Toll-like receptors and susceptibility to infectious disease. *Lancet Infect Dis*, 2005, 5.

[186] De Jager, PL; Franchimont, D; Waliszewska, A; Bitton, A; Cohen, A; Langelier, D; Belaiche, J; Vermeire, S; Farwell, L; Goris, A; Libioulle, C; Jani, N; Dassopoulos, T; Bromfield, GP; Dubois, B; Cho, JH; Brant, SR; Duerr, RH; Yang, H; Rotter, JI; Silverberg, MS; Steinhart, AH; Daly, MJ; Podolsky, DK; Louis, E; Hafler, DA; Rioux, JD. The role of the Toll receptor pathway in susceptibility to inflammatory bowel diseases. *Genes Immun*, 2007, 8.

[187] Gewirtz, AT; Vijay-Kumar, M; Brant, SR; Duerr, RH; Nicolae, DL; Cho, JH. Dominant-negative TLR5 polymorphism reduces adaptive immune response to flagellin and negatively associates with Crohn's disease. *Am J Physiol Gastrointest Liver Physiol*, 2006, 290.

[188] Pierik, M; Joossens, S; Van Steen, K; Van Schuerbeek, N; Vlietinck, R; Rutgeerts, P; Vermeire, S. Toll-like receptor-1, -2, and -6 polymorphisms influence disease extension in inflammatory bowel diseases. *Inflamm Bowel Dis*, 2006, 12.

[189] Torok, HP; Glas, J; Tonenchi, L; Bruennler, G; Folwaczny, M; Folwaczny, C. Crohn's disease is associated with a toll-like receptor-9 polymorphism. *Gastroenterology*, 2004, 127.

[190] Cario, E & Podolsky, DK. Differential alteration in intestinal epithelial cell expression of toll-like receptor 3 (TLR3) and TLR4 in inflammatory bowel disease. *Infect Immun*, 2000, 68.

[191] Trinchieri, G. Interleukin-12: a proinflammatory cytokine with immunoregulatory functions that bridge innate resistance and antigen-specific adaptive immunity. *Annu Rev Immunol*, 1995, 13.

[192] Yin, Z; Zhang, DH; Welte, T; Bahtiyar, G; Jung, S; Liu, L; Fu, XY; Ray, A; Craft, J. Dominance of IL-12 over IL-4 in gamma delta T cell differentiation leads to default production of IFN-gamma: failure to down-regulate IL-12 receptor beta 2-chain expression. *J. Immunol.*, 2000, 164, 3056-3064.

[193] Trinchieri, G. Interleukin-12: a cytokine produced by antigen-presenting cells with immunoregulatory functions in the generation of T-helper cells type 1 and cytotoxic lymphocytes. *Blood*, 1994, 84.

[194] Oppmann, B; Lesley, R; Blom, B; Timans, JC; Xu, Y; Hunte, B; Vega, F; Yu, N; Wang, J; Singh, K; Zonin, F; Vaisberg, E; Churakova, T; Liu, M; Gorman, D; Wagner, J; Zurawski, S; Liu, Y; Abrams, JS; Moore, KW; Rennick, D; de Waal-Malefyt, R; Hannum, C; Bazan, JF; Kastelein, RA. Novel p19 protein engages IL-12p40 to form a cytokine, IL-23, with biological activities similar as well as distinct from IL-12. *Immunity*, 2000, 13.

[195] Chua, AO; Chizzonite, R; Desai, BB; Truitt, TP; Nunes, P; Minetti, LJ; Warrier, RR; Presky, DH; Levine, JF; Gately, MK; et al. Expression cloning of a human IL-12 receptor component. A new member of the cytokine receptor superfamily with strong homology to gp130. *J Immunol*, 1994, 153.

[196] Chua, AO; Wilkinson, VL; Presky, DH; Gubler, U. Cloning and characterization of a mouse IL-12 receptor-beta component. *J Immunol*, 1995, 155.

[197] Desai, BB; Quinn, PM; Wolitzky, AG; Mongini, PK; Chizzonite, R; Gately, MK. IL-12 receptor. II. Distribution and regulation of receptor expression. *J Immunol*, 1992, 148.

[198] Presky, DH; Yang, H; Minetti, LJ; Chua, AO; Nabavi, N; Wu, CY; Gately, MK; Gubler, U. A functional interleukin 12 receptor complex is composed of two beta-type cytokine receptor subunits. *Proc Natl Acad Sci U S A*, 1996, 93.

[199] Cooper, AM; Magram, J; Ferrante, J; Orme, IM. Interleukin 12 (IL-12) is crucial to the development of protective immunity in mice intravenously infected with mycobacterium tuberculosis. *J Exp Med*, 1997, 186.

[200] de Jong, R; Altare, F; Haagen, IA; Elferink, DG; Boer, T; van Breda Vriesman, PJ; Kabel, PJ; Draaisma, JM; van Dissel, JT; Kroon, FP; Casanova, JL; Ottenhoff, TH. Severe mycobacterial and Salmonella infections in interleukin-12 receptor-deficient patients. *Science*, 1998, 280.

[201] Wakeham, J; Wang, J; Magram, J; Croitoru, K; Harkness, R; Dunn, P; Zganiacz, A; Xing, Z. Lack of both types 1 and 2 cytokines, tissue inflammatory responses, and immune protection during pulmonary infection by Mycobacterium bovis bacille Calmette-Guerin in IL-12-deficient mice. *J Immunol*, 1998, 160.

[202] Wu, C; Wang, X; Gadina, M; O'Shea, JJ; Presky, DH; Magram, J. IL-12 receptor beta 2 (IL-12R beta 2)-deficient mice are defective in IL-12-mediated signaling despite the presence of high affinity IL-12 binding sites. *J Immunol*, 2000, 165.

[203] Oliver, J; Rueda, B; Lopez-Nevot, MA; Gomez-Garcia, M; Martin, J. Replication of an association between IL23R gene polymorphism with inflammatory bowel disease. *Clin Gastroenterol Hepatol*, 2007, 5.

[204] Tremelling, M; Cummings, F; Fisher, SA; Mansfield, J; Gwilliam, R; Keniry, A; Nimmo, ER; Drummond, H; Onnie, CM; Prescott, NJ; Sanderson, J; Bredin, F; Berzuini, C; Forbes, A; Lewis, CM; Cardon, L; Deloukas, P; Jewell, D; Mathew, CG; Parkes, M; Satsangi, J. IL23R variation determines susceptibility but not disease phenotype in inflammatory bowel disease. *Gastroenterology*, 2007, 132.

[205] McKenzie, BS; Kastelein, RA; Cua, DJ. Understanding the IL-23-IL-17 immune pathway. *Trends Immunol*, 2006, 27.

[206] Parham, C; Chirica, M; Timans, J; Vaisberg, E; Travis, M; Cheung, J; Pflanz, S; Zhang, R; Singh, KP; Vega, F; To, W; Wagner, J; O'Farrell, AM; McClanahan, T; Zurawski, S; Hannum, C; Gorman, D; Rennick, DM; Kastelein, RA; de Waal Malefyt, R; Moore, KW. A receptor for the heterodimeric cytokine IL-23 is composed of IL-12Rbeta1 and a novel cytokine receptor subunit, IL-23R. *J Immunol*, 2002, 168.

[207] Barrett, JC; Hansoul, S; Nicolae, DL; Cho, JH; Duerr, RH; Rioux, JD; Brant, SR; Silverberg, MS; Taylor, KD; Barmada, MM; Bitton, A; Dassopoulos, T; Datta, LW; Green, T; Griffiths, AM; Kistner, EO; Murtha, MT; Regueiro, MD; Rotter, JI; Schumm, LP; Steinhart, AH; Targan, SR; Xavier, RJ; Libioulle, C; Sandor, C; Lathrop, M; Belaiche, J; Dewit, O; Gut, I; Heath, S; Laukens, D; Mni, M; Rutgeerts, P; Van Gossum, A; Zelenika, D; Franchimont, D; Hugot, JP; de Vos, M; Vermeire, S; Louis, E; Cardon, LR; Anderson, CA; Drummond, H; Nimmo, E; Ahmad, T; Prescott, NJ; Onnie, CM; Fisher, SA; Marchini, J; Ghori, J; Bumpstead, S; Gwilliam, R; Tremelling, M; Deloukas, P; Mansfield, J; Jewell, D; Satsangi, J; Mathew, CG; Parkes, M;

Georges, M; Daly, MJ. Genome-wide association defines more than 30 distinct susceptibility loci for Crohn's disease. *Nat Genet*, 2008, 40.

[208] Takeda, K; Clausen, BE; Kaisho, T; Tsujimura, T; Terada, N; Forster, I; Akira, S. Enhanced Th1 activity and development of chronic enterocolitis in mice devoid of Stat3 in macrophages and neutrophils. *Immunity*, 1999, 10.

[209] Gutierrez, MG; Master, SS; Singh, SB; Taylor, GA; Colombo, MI; Deretic, V. Autophagy is a defense mechanism inhibiting BCG and Mycobacterium tuberculosis survival in infected macrophages. *Cell*, 2004, 119.

[210] Collazo, CM; Yap, GS; Sempowski, GD; Lusby, KC; Tessarollo, L; Woude, GF; Sher, A; Taylor, GA. Inactivation of LRG-47 and IRG-47 reveals a family of interferon gamma-inducible genes with essential, pathogen-specific roles in resistance to infection. *J Exp Med*, 2001, 194.

[211] Cellier, MF; Courville, P; Campion, C. Nramp1 phagocyte intracellular metal withdrawal defense. *Microbes Infect*, 2007, 9.

[212] Veazey, RS; Horohov, DW; Krahenbuhl, JL; Taylor, HW; Oliver, JL3; Snider, TG3. Comparison of the resistance of C57BL/6 and C3H/He mice to infection with Mycobacterium paratuberculosis. *Vet Microbiol*, 1995, 47.

[213] Tanaka, G; Shojima, J; Matsushita, I; Nagai, H; Kurashima, A; Nakata, K; Toyota, E; Kobayashi, N; Kudo, K; Keicho, N. Pulmonary Mycobacterium avium complex infection: association with NRAMP1 polymorphisms. *Eur Respir J*, 2007, 30.

[214] Glas, J; Torok, HP; Tonenchi, L; Muller-Myhsok, B; Mussack, T; Wetzke, M; Klein, W; Epplen, JT; Griga, T; Schiemann, U; Lohse, P; Seiderer, J; Schnitzler, F; Brand, S; Ochsenkuhn, T; Folwaczny, M; Folwaczny, C. Role of the NFKB1 -94ins/delATTG promoter polymorphism in IBD and potential interactions with polymorphisms in the CARD15/NOD2, IKBL, and IL-1RN genes. *Inflamm Bowel Dis*, 2006, 12.

[215] Gazouli, M; Mantzaris, G; Kotsinas, A; Zacharatos, P; Papalambros, E; Archimandritis, A; Ikonomopoulos, J; Gorgoulis, VG. Association between polymorphisms in the Toll-like receptor 4, CD14, and CARD15/NOD2 and inflammatory bowel disease in the Greek population. *World J Gastroenterol*, 2005, 11.

[216] Klein, W; Tromm, A; Griga, T; Fricke, H; Folwaczny, C; Hocke, M; Eitner, K; Marx, M; Duerig, N; Epplen, JT. A polymorphism in the CD14 gene is associated with Crohn disease. *Scand J Gastroenterol*, 2002, 37.

[217] Carninci, P; Kasukawa, T; Katayama, S; Gough, J; Frith, MC; Maeda, N; Oyama, R; Ravasi, T; Lenhard, B; Wells, C; Kodzius, R; Shimokawa, K; Bajic, VB; Brenner, SE; Batalov, S; Forrest, AR; Zavolan, M; Davis, MJ; Wilming, LG; Aidinis, V; Allen, JE; Ambesi-Impiombato, A; Apweiler, R; Aturaliya, RN; Bailey, TL; Bansal, M; Baxter, L; Beisel, KW; Bersano, T; Bono, H; Chalk, AM; Chiu, KP; Choudhary, V; Christoffels, A; Clutterbuck, DR; Crowe, ML; Dalla, E; Dalrymple, BP; de Bono, B; Della Gatta, G; di Bernardo, D; Down, T; Engstrom, P; Fagiolini, M; Faulkner, G; Fletcher, CF; Fukushima, T; Furuno, M; Futaki, S; Gariboldi, M; Georgii-Hemming, P; Gingeras, TR; Gojobori, T; Green, RE; Gustincich, S; Harbers, M; Hayashi, Y; Hensch, TK; Hirokawa, N; Hill, D; Huminiecki, L; Iacono, M; Ikeo, K; Iwama, A; Ishikawa, T; Jakt, M; Kanapin, A; Katoh, M; Kawasawa, Y; Kelso, J; Kitamura, H; Kitano, H; Kollias, G; Krishnan, SP; Kruger, A; Kummerfeld, SK; Kurochkin, IV; Lareau, LF; Lazarevic, D; Lipovich, L; Liu, J; Liuni, S; McWilliam, S; Madan Babu, M; Madera, M; Marchionni, L; Matsuda, H; Matsuzawa, S; Miki, H; Mignone, F;

Miyake, S; Morris, K; Mottagui-Tabar, S; Mulder, N; Nakano, N; Nakauchi, H; Ng, P; Nilsson, R; Nishiguchi, S; Nishikawa, S; Nori, F; Ohara, O; Okazaki, Y; Orlando, V; Pang, KC; Pavan, WJ; Pavesi, G; Pesole, G; Petrovsky, N; Piazza, S; Reed, J; Reid, JF; Ring, BZ; Ringwald, M; Rost, B; Ruan, Y; Salzberg, SL; Sandelin, A; Schneider, C; Schonbach, C; Sekiguchi, K; Semple, CA; Seno, S; Sessa, L; Sheng, Y; Shibata, Y; Shimada, H; Shimada, K; Silva, D; Sinclair, B; Sperling, S; Stupka, E; Sugiura, K; Sultana, R; Takenaka, Y; Taki, K; Tammoja, K; Tan, SL; Tang, S; Taylor, MS; Tegner, J; Teichmann, SA; Ueda, HR; van Nimwegen, E; Verardo, R; Wei, CL; Yagi, K; Yamanishi, H; Zabarovsky, E; Zhu, S; Zimmer, A; Hide, W; Bult, C; Grimmond, SM; Teasdale, RD; Liu, ET; Brusic, V; Quackenbush, J; Wahlestedt, C; Mattick, JS; Hume, DA; Kai, C; Sasaki, D; Tomaru, Y; Fukuda, S; Kanamori-Katayama, M; Suzuki, M; Aoki, J; Arakawa, T; Iida, J; Imamura, K; Itoh, M; Kato, T; Kawaji, H; Kawagashira, N; Kawashima, T; Kojima, M; Kondo, S; Konno, H; Nakano, K; Ninomiya, N; Nishio, T; Okada, M; Plessy, C; Shibata, K; Shiraki, T; Suzuki, S; Tagami, M; Waki, K; Watahiki, A; Okamura-Oho, Y; Suzuki, H; Kawai, J; Hayashizaki, Y. The transcriptional landscape of the mammalian genome. *Science*, 2005, 309.

[218] Burns, K; Janssens, S; Brissoni, B; Olivos, N; Beyaert, R; Tschopp, J. Inhibition of interleukin 1 receptor/Toll-like receptor signaling through the alternatively spliced, short form of MyD88 is due to its failure to recruit IRAK-4. *J Exp Med*, 2003, 197.

[219] Hardy, MP & O'Neill, LA. The murine IRAK2 gene encodes four alternatively spliced isoforms, two of which are inhibitory. *J Biol Chem*, 2004, 279.

[220] Pfeffer, LM; Basu, L; Pfeffer, SR; Yang, CH; Murti, A; Russell-Harde, D; Croze, E. The short form of the interferon alpha/beta receptor chain 2 acts as a dominant negative for type I interferon action. *J Biol Chem*, 1997, 272.

[221] Modrek, B; Resch, A; Grasso, C; Lee, C. Genome-wide detection of alternative splicing in expressed sequences of human genes. *Nucleic Acids Res*, 2001, 29.

[222] Leung, E; Hong, J; Fraser, A; Krissansen, GW. Splicing of NOD2 (CARD15) RNA transcripts. *Mol Immunol*, 2007, 44.

[223] Zhang, XY; Zhang, HJ; Zhang, Y; Fu, YJ; He, J; Zhu, LP; Wang, SH; Liu, L. Identification and expression analysis of alternatively spliced isoforms of human interleukin-23 receptor gene in normal lymphoid cells and selected tumor cells. *Immunogenetics*, 2006, 57.

[224] Wells, CA; Chalk, AM; Forrest, A; Taylor, D; Waddell, N; Schroder, K; Himes, SR; Faulkner, G; Lo, S; Kasukawa, T; Kawaji, H; Kai, C; Kawai, J; Katayama, S; Carninci, P; Hayashizaki, Y; Hume, DA; Grimmond, SM. Alternate transcription of the Toll-like receptor signaling cascade. *Genome Biol*, 2006, 7.

[225] Manousou, P; Kolios, G; Drygiannakis, I; Pyrovolaki, K; Bourikas, L; Papadaki, HA; Kouroumalis, E. Expression of a splice variant of CXCR3 in Crohn's disease patients; indication for a lymphocyte--epithelial cell interaction. *J Gastroenterol Hepatol*, 2008, 23.

[226] Jiang, YH; Bressler, J; Beaudet, AL. Epigenetics and human disease. *Annu Rev Genomics Hum Genet*, 2004, 5.

[227] Weaver, IC. Epigenetic effects of glucocorticoids. *Semin. Fetal Neonatal Med.*, 2009, 14, 143-150.

[228] Akolkar, PN; Gulwani-Akolkar, B; Heresbach, D; Lin, XY; Fisher, S; Katz, S; Silver, J. Differences in risk of Crohn's disease in offspring of mothers and fathers with inflammatory bowel disease. *Am J Gastroenterol*, 1997, 92.

[229] Hayward, PA; Satsangi, J; Jewell, DP. Inflammatory bowel disease and the X chromosome. *QJM*, 1996, 89.

[230] Price, WH. A high incidence of chronic inflammatory bowel disease in patients with Turner's syndrome. *J Med Genet*, 1979, 16.

[231] Thompson, NP; Driscoll, R; Pounder, RE; Wakefield, AJ. Genetics versus environment in inflammatory bowel disease: results of a British twin study. *BMJ*, 1996, 312.

[232] Tysk, C; Lindberg, E; Jarnerot, G; Floderus-Myrhed, B. Ulcerative colitis and Crohn's disease in an unselected population of monozygotic and dizygotic twins. A study of heritability and the influence of smoking. *Gut*, 1988, 29.

[233] Singh, SM; Murphy, B; O'Reilly, R. Epigenetic contributors to the discordance of monozygotic twins. *Clin Genet*, 2002, 62.

[234] Louis, E; Satsangi, J; Roussomoustakaki, M; Parkes, M; Fanning, G; Welsh, K; Jewell, D. Cytokine gene polymorphisms in inflammatory bowel disease. *Gut*, 1996, 39.

[235] Ehrlich, M & Ehrlich, KC. Effect of DNA methylation on the binding of vertebrate and plant proteins to DNA. *EXS*, 1993, 64.

[236] Kopp, MU; Winterhalter, KH; Trueb, B. DNA methylation accounts for the inhibition of collagen VI expression in transformed fibroblasts. *Eur J Biochem*, 1997, 249.

[237] Heresbach, D; Alizadeh, M; Dabadie, A; Le Berre, N; Colombel, JF; Yaouanq, J; Bretagne, JF; Semana, G. Significance of interleukin-1beta and interleukin-1 receptor antagonist genetic polymorphism in inflammatory bowel diseases. *Am J Gastroenterol*, 1997, 92.

[238] Nemetz, A; Nosti-Escanilla, MP; Molnar, T; Kope, A; Kovacs, A; Feher, J; Tulassay, Z; Nagy, F; Garcia-Gonzalez, MA; Pena, AS. IL1B gene polymorphisms influence the course and severity of inflammatory bowel disease. *Immunogenetics*, 1999, 49.

[239] Bachvarov, DR; Landry, M; Houle, S; Pare, P; Marceau, F. Altered frequency of a promoter polymorphic allele of the kinin B1 receptor gene in inflammatory bowel disease. *Gastroenterology*, 1998, 115.

[240] Dionne, S; Hiscott, J; D'Agata, I; Duhaime, A; Seidman, EG. Quantitative PCR analysis of TNF-alpha and IL-1 beta mRNA levels in pediatric IBD mucosal biopsies. *Dig Dis Sci*, 1997, 42.

[241] Saccani, S; Pantano, S; Natoli, G. Two waves of nuclear factor kappaB recruitment to target promoters. *J Exp Med*, 2001, 193.

[242] Dekkers, JC & Hospital, F. The use of molecular genetics in the improvement of agricultural populations. *Nat Rev Genet*, 2002, 3.

[243] Dempster, ER & Lerner, IM. Heritability of Threshold Characters. *Genetics*, 1950 35.

[244] Visscher, PM; Hill, WG; Wray, NR. Heritability in the genomics era--concepts and misconceptions. *Nat Rev Genet*, 2008, 9.

In: Veterinary Immunology and Immunopathology
Editors: L. Nermann, S. Meier
ISBN: 978-1-60876-342-9

Chapter 3

CHARACTERIZATION OF FLATFISH MX PROTEINS

[1]E. García-Rosado, [1*] M. C. Alonso,[2] M. A. Fernández-Trujillo,[3] M. Manchado, and J. Béjar[2]

[1]Department of Microbiology, Faculty of Sciences, University of Malaga, 29071 Malaga, Spain

[2]Department of Genetics, Faculty of Sciences, University of Malaga, 29071 Malaga, Spain

[3]IFAPA Centro el Toruño, CICE, Junta de Andalucia, Camino Tiro de Pichon s/n 11500 El Puerto de Santa Maria (Cadiz), Spain

ABSTRACT

Flatfish (*Pleuronectiformes* order) is a broad taxonomical group within teleosts, comprising 11 families and about 500 species worldwide. Farming of several flatfish species, such as Japanese flounder (*Paralichthys olivaceus*), Atlantic halibut (*Hippoglossus hippoglossus*), turbot (*Scophthalmus maximus*) and Senegalese sole (*Solea senegalensis*), has been improved in last years in diverse geographical areas; however, their cultivation can be restricted by the occurrence of viral diseases. Therefore, understanding defence mechanisms of these fish species is a main priority to prevent economic losses.

The innate immune system represents the first defence barrier against microbial pathogens. Interferons (IFNs) are key components of the innate immunity against viral infections. These molecules are cytokines that are divided into two categories: type I, (induced by viruses in most cells) and type II (produced by natural killer cells and T lymphocytes). Type I IFN promotes an antiviral state in neighbouring cells by inducing the expression of several antiviral proteins, such as the Mx proteins, whose antiviral mechanisms remain to be elucidated.

A variable number of Mx genes have been found in mammals, birds, fish and invertebrates. Regarding the *Pleuronectiformes* order, a single Mx gene has been reported in Japanese flounder and Senegalese sole, whereas two genes have been described in

* Corresponding author: Mailing address: Department of Microbiology, Faculty of Sciences, University of Malaga, Campus Teatinos, 29071 Malaga, Spain, Phone: 34-952-137588, Fax: 34-952-136645, E-mail: mdalonso@uma.es

turbot and Atlantic halibut. These genes have been shown to be activated in response to viral factors such as dsRNA, ssRNA or DNA.

The full-length Mx cDNA from these fish species has been cloned, and the resulting sequences have been compared with those obtained from other species, showing the highest identity in the N-terminus. The variability recorded in the C-terminus may be related to the species-specificity of the Mx antiviral activity.

The Mx antiviral activity has been tested for several fish species using cells expressing recombinant Mx. The results derived from this kind of experiments show that Mx reduces the transcription of viral protein, which suggests that Mx may act blocking viral replication.

The understanding of the innate immune system, and particularly the IFN response, may help to develop new strategies and methodologies to control the diseases provoked by microbes and, consequently, to improve the culture conditions in the aquaculture farms. The antiviral activity of the Mx proteins described for several fish species seems to have a key role within the IFN type I response. In addition, Mx can be used as a model to study the pathogen-host interactions.

1. Introduction

The aquaculture industry has been classically focused on the culture of freshwater fish species; however, marine aquaculture is becoming an important activity that is contributing to the diversification of this sector. Newly farmed marine fish species include some flatfish, such as Japanese flounder (*Paralichthys olivaceus*), Atlantic halibut (*Hippoglossus hippoglossus*), turbot (*Scophthalmus maximus*) and Senegalese sole (*Solea senegalensis*) (Naylor & Burke, 2003). In particular, the culture of the Atlantic halibut has been developed in Norway, Scotland, Ireland, Canada, Chile, Iceland and the United States thanks to several advanced hatchery and research programs underway in these countries (Whyte, 2007). The culture of Japanese flounder and Senegalese sole is becoming an important economic activity in Japan and Southern Europe, respectively. Turbot is mainly cultured in Spain, France and Portugal, but also in Denmark, Germany, Iceland, Ireland, Italy, Norway and Wales, although Spain is undoubtedly the world leader of adult turbot production (50% of the total production).

Flatfish constitute the *Pleuronectiformes* order, which is a broad taxonomic group comprising 11 families (*Achiridae, Achiropsettidae, Bothidae, Citharidae, Cynoglossidae, Paralichthydae, Pleuronectidae, Psettodidae, Samaridae, Scophthalmidae* and *Soleidae*) and about 500 species worldwide, mainly with a marine habitat. Main common characteristics of all these species include an asymmetrical body in the adult stage, with both eyes lying on the same side of the head (either the right or the left side, depending on the family), the presence of protrusible eyes, and a bottom-dwelling way of life.

The intensive rearing of flatfish species is resulting in health- and husbandry-related problems, being restricted by the occurrence of diseases with a microbial aetiology. Viral diseases are one of the main threats of the aquaculture industry due to their high mortality, the possibility to establish persistent infections, and the lack of effective treatments. The main viral diseases that have been reported to affect Japanese flounder, Atlantic halibut, turbot and Senegalese sole are caused by viruses included in the *Rhabdoviridae, Iridoviridae, Birnaviridae* and *Nodaviridae* families (Mulcahy, 2002). In particular, the main infections reported to affect Japanese flounder are those caused by Viral Haemorrhagic Septicaemia

Virus (VHSV) and Hirame Rhabdovirus (HIRRV), both belonging to the *Rhabdoviridae* family; marine birnaviruses; Lymphocystis Disease Virus (LCDV), *Iridoviridae* family; and nodaviruses (Isshiki et al., 2001; Jung et al., 2008; Kitamura et al., 2007; Suzuki, 2006; Tanako et al., 2000). Nodaviruses and Infectious Pancreatic Necrosis Virus (IPNV) are the main viruses affecting Atlantic halibut (Bergh et al., 2001; Biering et al., 1994; Danneving et al., 2000; Grotmol et al., 1999; Mortensen et al., 1990). The main viral threat for the turbot culture is VHSV, although IPNV and an iridovirus have also been detected (Cutrin et al., 2005; Johansen et al., 2004; Kalayci et al., 2006; Kim et al., 2005; Mortensen et al., 1990; 1993; Novoa et al., 1993; Ross et al., 1994). Senegalese sole culture could be affected by IPNV, LCDV, and nodaviruses (Alonso et al., 2005; 2007; Rodriguez et al., 1997).

One approach to face this important problem, which may cause vast economic losses, is the study of the immune response of these newly cultured fish species, since the understanding of the protective mechanisms involved in their immunity is essential to successfully manage disease outbreaks and to implement strategies to increase the fish disease resistance, as well as to develop prophylactic measures. The most of the studies on the antiviral immune response are mainly focused on the Interferon (IFN) mediated response, which is thought to be crucial in the primary antiviral defence of fish.

The IFN system is one of the major antiviral mechanisms against viral infections in high vertebrates (Samuel, 2001). It was discovered in chick chorioallantoic membrane cells in response to heat-inactivated influenza virus (Isaacs & Lindenmann, 1957). Since then, a great number of studies have been performed in order to characterize the IFN system with regards to the IFN gene structure, the induction of its expression in response to viral infections, and its mode of action. This kind of studies has led to an early and exhaustive characterization of the IFN system in high vertebrates, which has been considered as a model for the more recent studies on the IFN system in fish. Mx, a long lasting molecule discovered in a strain of influenza-resistant mice (Lindenmann et al., 1963), is one of the most studied IFN-induced antiviral proteins. This protein, in contrast to IFN, may persist for weeks after a single virus infection (Chieux et al., 1998) and, for this reason, it has been frequently used as a model to study the IFN mediated response. Taking into account these considerations, this review will be focused on studies carried out on Mx proteins of economically important flatfish species. Particularly, flatfish Mx proteins will be studied in comparison with those of other fish species, mainly salmonids, which have been deeply studied.

In order to present the available information in a coherent way, this chapter starts with a summary of the current knowledge on the high vertebrate IFN system, which constitutes the base to address the study of the IFN mediated response in fish.

2. Interferon Mediated Response

2.1. Interferon System in High Vertebrates

IFNs constitute a large group of cytokines that are best known for their ability to induce cellular resistance to viral pathogens. Two major families of IFN, types I and II, can be distinguished according to differences in their sequences, protein structures and functional properties.

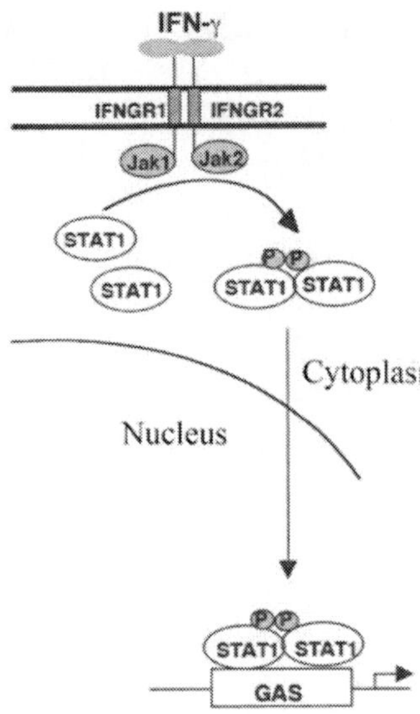

Figure 1. Jak-Stat signalling pathway of IFN-γ.

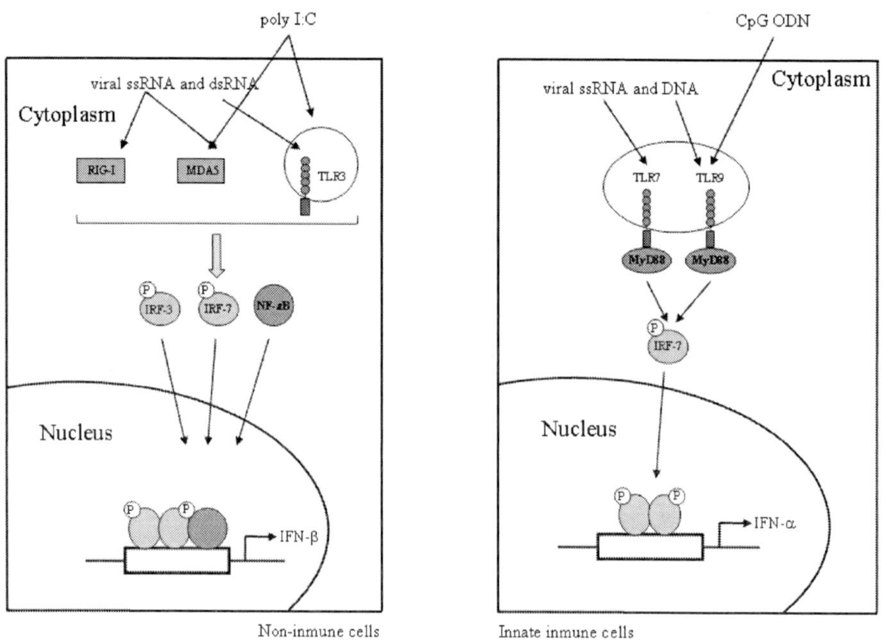

Figure 2. Recognition of viral nucleic acid in non-innate immune cells and innate immune cells (macrophages and pDCs).

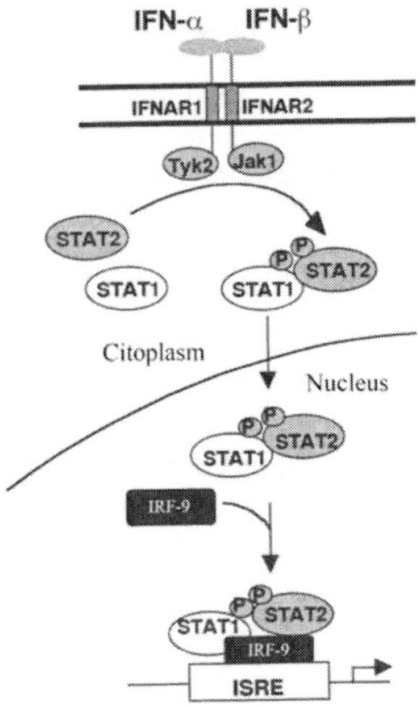

Figure 3. Pathway to stimulate transcription of ISGs by IFN α/β.

Type II IFN (IFN-γ) is encoded by genes interrupted by four introns, and it is exclusively produced by immune cells, such as natural killers (NK), T cell populations, B cells and dendritic cells. This type of IFN binds the extracellular domain of the specific receptors IFNGR1 and IFNGR2, which specifically interact with tyrosin kinase proteins of the Janus Kinase (JAK) family (Levy & Garcia-Sastre, 2001) sited in the cytoplasm (Figure 1). This interaction promotes the phosphorylation and posterior dimerization of the transcription factor STAT-1, which is then translocated to the cell nucleus, where it binds specific DNA sequences of the IFN-γ gene promoter called Gamma Activated Sites (GASs).

Type I IFN constitutes a multigenic family with at least 8 subclasses: IFN-α, β, δ, ε, κ, τ and ω. IFN-α/β is encoded by intronless genes, and it is produced by most cell types in response to viral infections after the recognition of viral nucleic acids by specific ligands. In non-immune cells, the main ligands for viral genome are the RNA helicase (RLH) RIG-I, the melanoma differentiation-associated gene-5 (MDA5), and the Toll-like receptors (TLRs), specifically the TLR3 (Takeuchi & Akira, 2007). RIG-I and MDA5 recognize single-strand (ssRNA) and double-strand (dsRNA) RNA molecules, whereas TLR3 exclusively recognizes dsRNA. Viral RNA recognition by these receptors triggers the activation of the transcription factors IRF3, IRF7 and NF-κB, which induce the transcription of IFN-α/β genes (Robertsen, 2008) (Figure 2). In innate immune cells, such as macrophages and plasmacytoid dendritic cells (pDCs), viral nucleotides are exclusively recognized by TLRs; TLR3 is found in endosomes of macrophages, and TLR7 and TLR9, which recognize ssRNA and non-metilated CpG-motifs (CpG-DNA), respectively, are found in endosomes of pDCs (Takeuchi & Akira, 2007). Binding of viral nucleic acids to TLR7 and TLR9 starts a signalling cascade by the MyD88 pathway, which results in activation of IRF7, and leads to the transcription of type I IFN genes (Birmachu et al., 2007; Haller et al., 2006; Kerkmann et al., 2003; Takeuchi & Akira, 2007) (Figure 2).

IFN-α/β is secreted by infected cells and promotes an antiviral state in neighbouring cells, where it is recognized by specific receptors (IFNAR1 and IFNAR2), whose interaction with the cytokines leads to the phosphorylation of the transcription factors STAT1 and STAT2 thanks to the previous activation of the kinases TYK2 and JAK1 (Levy & Garcia-Sastre, 2001) (Figure 3). STAT1 and STAT2 heterodimerize and bind a DNA binding protein belonging to the IFN Regulatory Factor Family, the IRF9 (p48) protein (Bluyssen & Levy, 1997), resulting in a multimeric transcription factor, called ISGF3 (Levy et al., 1989), which binds specific sequences, called IFN-stimulated response elements (ISREs), that are located in the IFN-α/β-stimulated gene (ISG) promoters (Levy & Garcia-Sastre, 2001).

Table 1. Mx proteins in high vertebrates

Species	Isoforms	Localization	Antiviral specificity	References
Human	MxA	Cytoplasm	FLUAV, THOV, VSV, Measles virus, Hantaan virus, Coxsackie virus B4, Semliki Forest virus	Lee & Vidal, 2002
	MxB	Cytoplasm	Inactive	
Monkey	1	Cytoplasm		
Pig	Mx1	Cytoplasm	VSV	Muller et al., 1992 Asano et al., 2002 Morozumi et al., 2009
	Mx2	Nucleus	FLUAV	
Bovine	Mx1	Cytoplasm	VSV, Rabies virus	Ellinwood et al., 1998 Baise et al., 2004 Gerardin et al., 2004 Leroy et al., 2006
	Mx1a		VSV	
Horse	2	Cytoplasm		Horisberger & Gunst, 1991 Chesters et al., 1997
Cani	M	Cytoplasm	VSV	Naka

ne		x 1			mura et al., 2005
		Mx2			
Cat	1		Cytoplasm		Horisberger & Gunst, 1991
Rab bit	1		Cytoplasm		Horisberger & Gunst, 1991
Rat		M x 1	Nucleus	FLUAV, THOV	Haller & Kochs , 2002
		Mx2	Cytoplasm	VSV, LACV, Rift Valley fever virus	
		Mx3	Cytoplasm	Inactive	
Mou se		M x 1	Nucleus	FLUAV, THOV, Dhori virus	Lee & Vidal, 2002
		Mx2	Cytoplasm	VSV, Hantaan virus	
Chic ken	1		Cytoplasm	FLUAV, VSV	Ko et al., 2002
Duc k	Multiple		Nucleus/Cyto plasm	Inactive	Bazzi gher et al., 1993

IFN-α/β up-regulates the transcription of more than 300 genes in high vertebrates (De Veer et al., 2001; Der et al., 1998), including those coding for antiviral proteins. The most characterized mammalian antiviral proteins induced by type I IFN are the dsRNA-dependent protein kinase (PKR) (Williams, 1999), the 2'-5'-oligoadelinate synthetase/RNaseL (OAS/RNaseL) system (Silverman, 1994) and the Mx proteins (Haller & Kochs, 2002). Nevertheless, Zhou et al. (1999) reported that cells lacking any of these antiviral proteins still exhibit a limited IFN-induced antiviral state, which suggests the existence of additional antiviral pathways. In fact, several authors have reported other antiviral proteins, such as ISG20 (Espert et al., 2003), the promyelocytic leukemia protein (PML) (Regad & Chelbi-Alix, 2001), the guanylate-binding protein 1 (GBP-1) (Anderson et al., 1999; Guo et al., 2000), and the RNA-specific adenosine deaminase 1 (ADAR1) (Samuel, 2001).

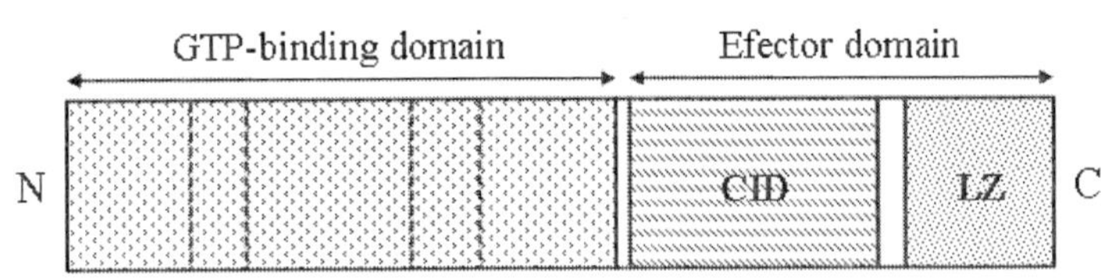

Figure 4. Structure and domains of Mx proteins. Discontinuous vertical lines represent the binding-GTP tripartite elements. CID: Central Interactive Domain; LZ: Leucine Zipper domain.

2.1.1. High vertebrate Mx proteins

Mx proteins have been found in a wide variety of organisms. They are GTPases included into the dynamin family, which are highly conserved in vertebrates (Horisberger & Gunst, 1991; Leong et al., 1998). These proteins contain a conserved GTP-binding domain and a dynamin family signature (LPRGSGIVT) in their amino-terminal half, whereas an efector domain, enclosing the Central Interactive Domain (CID) and the Leucine Zipper (LZ) motif, is present in the C-terminus (Haller et al., 2007) (Figure 4). Mx proteins may form high molecular weight homo-oligomers and self-assemble into ring-like and helical structures, which are critical for GTPase activity, protein stability, and viral recognition (Haller et al., 2007). The homo-oligomerization, and therefore the antiviral activity, seems to be the result of the binding between the LZ region of an Mx molecule and the CID domain of a second neighbouring molecule (Haller & Kochs, 2002).

Regarding the Mx gene structure, there is a large fluctuation in the total length depending on the species considered, ranging from 21 kb for chicken Mx to 26 kb for human MxA. The number of exons is also variable: 17 in human MxA gene (Tazi-Ahnini et al., 2000), 15 in bovine and porcine Mx1 genes (Gerardin et al., 2004; Thomas et al., 2006), 14 in mouse Mx1 gene (Hug et al., 1988), and 13 in chicken Mx gene (Li et al., 2007). In spite of this large variation in genomic sequences, the number of aminoacids of the Mx protein remains close to 662 in all the species. In addition, a variable number of Mx isoforms has also been identified for several species (Table 1). These isoforms may differ in both, antiviral activity and intracellular localization. Mx proteins can localize either in the cytoplasm or in the nucleus, although a cytoplasmic location is more common (Lee & Vidal, 2002). Mx proteins with a nuclear localization contain a Nuclear Localization Signal (NLS). The Mx protein location may account for the specific antiviral activity recorded, since those Mx proteins sited in the nucleus may affect the viral replicative cycle during a different step than those with a cytoplasmic location (Pavlovic et al., 1992). The relation between protein location and antiviral specificity has been supported by translocation studies involving mutant Mx proteins. Johannes et al. (1993) and Pavlovic et al. (1992) reported switches in the antiviral specificity of rat Mx2 and human MxA, respectively, after translocation of these proteins from their natural location in the cytoplasm to the nucleus.

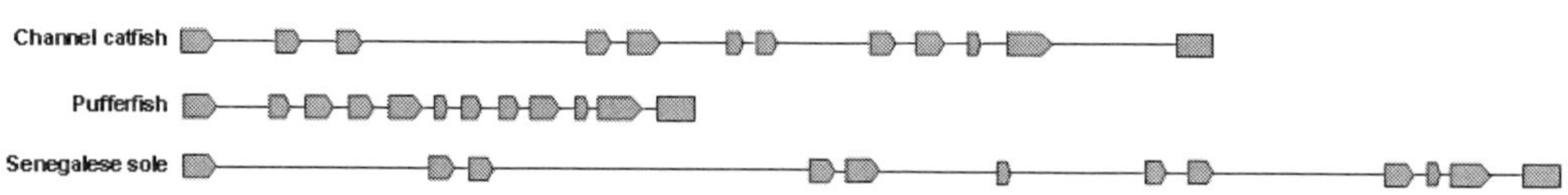

Figure 5. Comparison of the Mx gene structure of pufferfish (accession no. AF525215), channel catfish Mx1 (accession no. EU100454), and Senegalese sole Mx. Exons are represented in grey and introns as black lines.

The Mx proteins inhibit mainly negative sense ssRNA viruses, even though the antiviral activity against some positive sense ssRNA and DNA viruses has also been reported (Chieux et al., 2001; Gordien et al., 2001). Although their antiviral mechanisms remain to be elucidated, it is thought to be mediated by the interaction between Mx and specific viral proteins. This interaction would induce GTP hydrolysis-dependent conformational changes in the protein, and the subsequent oligomerization into spiral structures, which would prevent the virus assembly (Haller & Kochs, 2002). Human MxA has been found to block primary transcription of several viruses (Pavlovic et al., 1990; Schneider-Schaulies et al., 1994; Schnorr et al., 1993). Specifically, this protein inhibits La Crosse virus (LACV) and Thogoto virus (THOV) replication by forming perinuclear complexes, resulting from the Mx-viral nucleoprotein (NP) binding, which interfere the transcriptional activity of the viral ribonucleoprotein (RNPv) (Haller et al., 2007; Kochs et al., 2002). In consequence, Mx prevents incoming viral nucleocapsids from being transported into the nucleus, the assembly of nucleocapsids and the formation of new viral particles (Kochs et al., 2002).

Although the antiviral activity of a great number of Mx proteins from several mammals and birds has been demonstrated (Haller & Kochs, 2002; Haller et al., 1998; Sandrock et al., 2001), there are Mx proteins without a detectable antiviral function, such as human MxB (Pavlovic et al., 1990), rat Mx3 (Meier et al., 1990) or duck Mx. In fact, it is a matter to debate if the antiviral activity of Mx proteins is a luxurious accident of some undefined cellular function or it has evolved to inhibit a set of specific pathogens of every species (Lee & Vidal, 2002). Besides, the similarity between Mx and other high molecular weight GTPases without an antiviral activity suggests that Mx may also have non-immune roles.

2.2. Interferon System in Fish

The innate immune system of fish is poorly described. Although an IFN-like activity, similar to that described for mammals, has been recorded (Ellis, 2001), the genes codifying for type I and II IFNs have just been characterized for a few number of fish species, such as zebrafish (*Danio rerio*), Atlantic salmon (*Salmo salar*), pufferfish (*Takifugu rubripes*), spotted green pufferfish (*Tetraodon nigrovirides*), channel catfish (*Ictalurus punctatus*) and rainbow trout (*Oncorhynchus mykiss*) (Altmann et al., 2003; Long et al., 2004; Lutfalla et al., 2003; Robertsen et al., 2003). Interestingly, the fish IFNs show amino acid sequences similar to those of mammalian IFN-α, although the exon/intron gene structure is similar to that of IFN-λ. In spite of the different evolutionary routes of mammalian and fish IFNs (Lutfalla et al., 2003), studies carried out to date show a great number of similarities between both IFN responses. Fish possess TLRs homologous to those present in mammals (TLR3, TLR7 and TLR9), as well as RIG-I and MDA5. Moreover, fish also show similarities regarding the signalling pathways, showing a JAK-STAT pathway similar to the one described for high vertebrates (Collet & Secombes, 2002; Robertsen, 2006; Robertsen, 2008). These similarities suggest that, on the whole, fish IFN mechanisms do not differ significantly from the IFN system of high vertebrates. As it has been described for mammals, this system may be induced by viral infections, IFN-stimulating agents (poly I:C) or oligodeoxynucleotides CpG (CpGs). Similarly, it has also been demonstrated, both *in vivo* and *in vitro,* that type I IFN induces an antiviral state in some fish species (Jensen & Robertsen, 2002; Jensen et al., 2002; Saint-Jean & Perez-Prieto, 2006; Zhang et al., 2007a).

Type II IFN has just been described in Atlantic salmon (Robertsen, 2006), rainbow trout (Zou et al., 2005), channel catfish (Milev-Milovanovic et al., 2006), pufferfish (Zou et al., 2004), and seabass (*Dicentrarchus labrax*) (Casani et al., 2009).

2.2.1. IFN stimulated genes (ISGs) in fish

Several ISGs have been identified in fish, such as PKR, virus induced-genes (vigs), a ubiquitin-like protein, ISG-15, and Mx proteins (reviewed in Robertsen, 2006).

PKR is a member of the eIF2α kinase family, which possesses two dsRNA-binding domains (dsRBD) at the N-terminus, and a kinase domain at the C-terminus (Garcia et al., 2006). In mammals, dsRNA generated during viral infection binds and activates this protein, which phophorylates the α-subunit of the translation initiation factor eIF2, shutting down the synthesis of viral proteins.

Recently, a different eIF2α-kinase, called Z-DNA binding protein kinase, PKZ, has been cloned from goldfish (*Carassius auratus*) (Hu et al., 2004), zebrafish (Rothenburg et al., 2005), Atlantic salmon (Bergan et al., 2008), and several other fish species (Rothenburg et al., 2008). This protein contains two Zα binding motifs in the N-terminal regulatory domain instead of the dsRBD. Zα motifs specifically bind dsDNA or dsRNA in the left-handed Z conformation (Rich & Zhang, 2003), although there is some diversity depending on the fish species. Zhu et al. (2008) reported the cloning of a PKR containing dsRNA motifs from Japanese flounder, and Rothenburg et al. (2008) described the coexistence of PKR and PKZ in zebrafish. Bergan et al. (2008) demonstrated that Atlantic salmon PKZ also phosphorylates eIF2α *in vitro*, activity that was increased by addition of Z-DNA instead of dsRNA; however, despite this, the results reported suggest a similar mechanism of action for PKZ and PKR.

Several vigs have been identified in fish, 10 of them in rainbow trout (reviewed in Robertsen, 2008). Particularly, vig-1, homologous to the antiviral mammalian protein viperin (Chin & Cresswell, 2001), has been cloned from several experimentally infected fish species, such as Chinook salmon (*Oncorhynchus tshawytscha*), sculpin (*Leptocottus armatus*) (Alonso & Leong, 2002), rainbow trout (O'Farrell et al., 2002), mandarin fish (*Siniperca chuatsi*) (Sun & Nie, 2004) and goldfish (Zhang et al., 2007a). However, its mechanism of action is still unknown.

ISG15-like sequences have recently been identified in several fish species, including goldfish, channel catfish, rainbow trout, and Atlantic salmon (Baerwald et al., 2008; Liu et al., 2002; Røkenes et al., 2007; Zhang et al., 2007b). In mammals, ISG15 is induced in response to IFN and viral infections and it is composed of two ubiquitin-like domains that can be conjugated to target proteins by means of the C-terminal LRLRGG sequence (Haas et al., 1987; Loeb & Haas, 1992). Atlantic salmon ISG-15 is induced by dsRNA and type I IFN, as well as by viral infections, and it seems to conjugate to an Infective Septicaemia Anaemia Virus (ISAV) protein (Røkenes et al., 2007), suggesting an antiviral response of this protein; however, the functionality of this protein is still not understood even in mammals

Table 2. Current knowledge on fish Mx proteins.

Fish species	Isoforms	Intracellular localization	Antiviral specificity		References
			Virus	Family	
Danio rerio	3 possible Mx	Cytoplasm			Altmann et al., 2004
Oncorhynchus mykiss	Mx1	Cytoplasm			Trobridge & Leong, 1995 Trobridge et al., 1997
	Mx2	Nucleus			
	Mx3	Cytoplasm			
Hippoglossus hippoglossus	2 possible Mx	Cytoplasm/nucleus			Jensen & Robertsen, 2000 Bergan & Robertsen, 2004
Paralichthys	1	Cytoplasm	HIRRV	*Rhabdoviridae*	Lee et al., 2000

olivaceus			VHSV		Caipang et al., 2003
Ictalurus punctatus	3 possible Mx				Plant & Thune, 2004
Salmo salar	Mx1	Cytoplasm	IPNV	*Birnaviridae*	Robertsen et al., 1997 Larsen et al., 2004
	Mx2				
	Mx3	Cytoplasm			
Sparus aurata	4				Tafalla et al., 2004 GenBank accession number FJ90555 GenBank accession number FJ652200 GenBank accession number FJ490556
Lates calcarifer	1		VNNV IPNV	*Nodaviridae* *Birnaviridae*	Wu & Chi, 2007
Epinephelus coioides	MxI	Cytoplasm	YGNNV	*Nodaviridae*	Lin et al., 2006
	MxII				
	MxIII				
Solea senegalensis	1	Cytoplasm	IPNV	*Birnaviridae*	Fernandez-Trujillo et al., 2006 Fernandez-Trujillo et al., 2008b
Scophthalmus maximus	Mx type 1				Abollo et al., 2005
	Mx type 2				
Perca fluvialis	1				Staeheli et al., 1989
Takifugu rubripes	1	Cytoplasm			Yap et al., 2003
Carassius auratus	Mx1	Cytoplasm			Zhang et al., 2004
	Mx2	Cytoplasm			

Regarding fish Mx, the high homology between mammalian and fish Mx genes has allowed to clone Mx cDNA sequences from a considerable number of fish species (Table 2), having found the presence of different isoforms in most of them. As in higher vertebrates, fish Mx proteins are typically inducible by type I IFN, poly I:C, and virus infections. They are located either in the cytoplasm or the nucleus, and the antiviral activity has been recently demonstrated for a number of them. The demonstration of antiviral activity of fish Mx proteins is an important step in the understanding of the antiviral mechanisms of IFNs in these animals, in which Mx seems to play a key role.

In the next section of this chapter the current knowledge on flatfish Mx proteins is presented in relation with the information available in other fish species and in higher vertebrates.

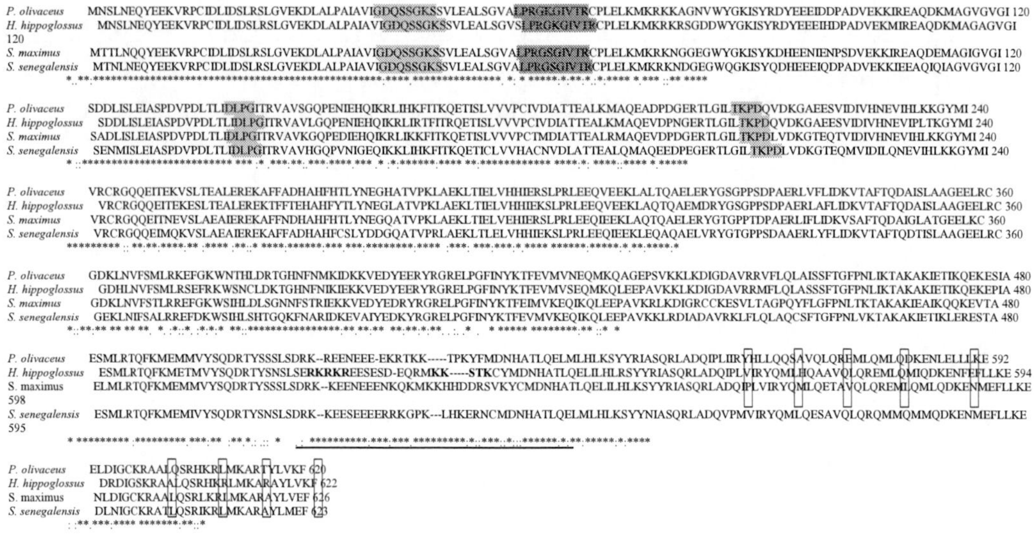

Figure 6. Alignment of flatfish Mx sequences. The tripartite GTP-binding domain appears with light grey background, the dynamin family signature has a dark grey background and the leucine zipper is boxed. The highly variable carboxy-terminal region is underlined.

3. Flatfish Mx Proteins

Mx genes in flatfish were first described in Japanese flounder (Lee et al., 2000). Since then, these genes have been cloned in Atlantic halibut (Jensen & Robertsen, 2000), turbot (Abollo et al., 2005), and Senegalese sole (Fernandez-Trujillo et al., 2006).

Atlantic halibut and turbot possess two Mx isoforms, whereas a single isoform has been detected in Japanese flounder and Senegalese sole. Several isoforms have also been reported for zebrafish, Atlantic salmon, rainbow trout, grouper (*Epinephelus coioides*), channel catfish, sea bream and carp (*Cyprinus carpio*). One approach to study the presence of different isoforms is to determine the number of loci coding for this protein and their genomic arrangement. Using this procedure, a single locus has been reported for the Mx gene in Senegalese sole (Fernandez-Trujillo, 2008) and Japanese flounder (Ooi et al., 2006), whereas two loci have been detected in Atlantic halibut (Jensen & Robertsen, 2000). These data suggest that the genomic situation of the Mx genes is also variable within the flatfish group; however, more information is required to clarify each specific situation, since the entire Mx gene sequence has only been cloned from Senegalese sole (author's unpublished results, GenBank accession number EU717076). Senegalese sole Mx spans 8.2 kb consisting of 12 exons and 11 introns (author's unpublished results). The arrangement of this gene is similar to that of the two fish Mx genes characterized so far, pufferfish and channel catfish, which show the same number of exons and introns (Figure 5). In fact, Senegalese sole Mx gene exons are almost identical in length to the corresponding exons of pufferfish and channel catfish Mx genes, with some of them only differing in one nucleotide. In contrast, large differences between these three species are found regarding the intron length, with Senegalese sole Mx gene showing the longest g-DNA (8273 nucleotides), followed by that of channel catfish

(6182 nucleotides), and pufferfish (3079 nucleotides), which fits its highly condensed genome (Venkatesh, 2003).

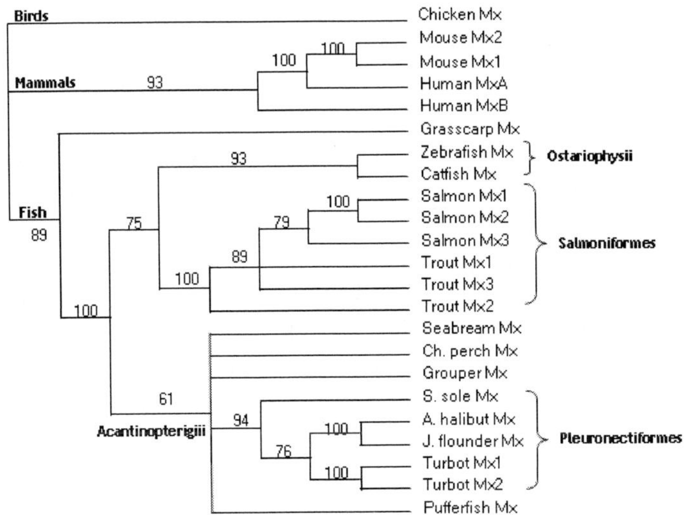

Figure 7. Phylogenetic tree of vertebrate Mx constructed from cDNA sequences. Accession numbers are: chicken: AY695797; mouse Mx2: NM013606; mouse Mx1: NM010846; human MxA: M30817; human MxB: M30818; carp: AY395698; zebrafish: AF533769; channel catfish: AY095349; Atlantic salmon Mx1: U66475; Atlantic salmon Mx2: U66476; Atlantic salmon Mx3: U66477; rainbow trout Mx1: U30253; rainbow trout Mx3: U47946; rainbow trout Mx2: U47945; seabream: AF491302; Chinese perch (*Siniperca chuatsi*): AY392097; grouper: AY574372; Senegalese sole: AY790537; Atlantic halibut: AF245513; Japanese flounder: AB110446; turbot Mx1: AY635932; turbot Mx2: AY635933; pufferfish: AF525215.

In general, the flatfish Mx proteins exhibit all the main domains described previously in this chapter for the mammalian Mx proteins. The alignment represented in Figure 6 shows that flatfish proteins present a highly conserved tripartite GTP-binding domain (positions 40-48; 141-144 and 210-213), the signature of the dynamin family (positions 59-68), and the LZ motif in the carboxyl terminal region. Only the Atlantic halibut Mx sequence contains a nuclear localization signal near the carboxyl end, and its nuclear location has been demonstrated (Bergan & Robertsen, 2004). Studies carried out on Japanese flounder Mx have also shown the presence of a putative bipartite NLS, which is very similar to that described in the Atlantic halibut Mx cDNA (Lin et al., 2005); however, immunofluorescence analyses have demonstrated the cytoplasmic location of this protein. Lin et al. (2005) have suggested several hypotheses to explain these observations, such as the non-functional characteristic of the bipartite NLS, which has been previously reported in the human La (SS-B) autoantigen protein (Simons et al., 1996), or the possibility for this protein to be transported into the nucleus and subsequently exported back to the cytoplasm. The cytoplasmic location of Senegalese sole has been established by immunological techniques in CHSE-214 cells expressing Senegalese sole Mx (Fernandez-Trujillo et al., 2008b). No studies on turbot Mx location have been carried out to date. The existence of a nuclear localization signal in the carboxyl terminus has also been correlated with a nuclear location in rainbow trout Mx2, as it has been demonstrated by the presence of a spotted immunostaining pattern in the cell nucleus (Trobridge et al., 1997).

A comparative analysis of the Mx protein sequences from different flatfish species, as well as from the Atlantic salmon Mx1 and human MxA, is shown in Table 3. The highest identity value corresponds to Japanese flounder and Atlantic halibut Mx sequences (88.9%), followed by the identity between turbot and Japanese flounder sequences (83.2%), turbot and Atlantic halibut sequences (80.2%) and Japanese flounder and Senegalese sole Mx sequences (80.2%). In contrast, all flatfish Mx sequences show a quite similar identity value when compared with the Atlantic salmon Mx1 and human MxA (76.2-78.2%, and about 55%, respectively).

In spite of the high identity values shown, slight variations in cDNA size, which result in small differences in the size of the different flatfish Mx proteins, have been detected. Turbot Mx is the longest protein (626 aminoacids), followed by those of Senegalese sole (623 aminoacids), Atlantic halibut (622 aminoacids), and Japanese flounder (620 aminoacids). This variability is widespread in fish, ranging from 618 to 646 residues, and might be associated with some gaps appearing in the less conserved region of the protein, located near the carboxyl terminus (Figure 6). In fact, when focusing on different domains, it appears that the N-terminal region is the most conserved (80.4-92.6% within flatfish, 83.3-85.9% flatfish *versus* Atlantic salmon, and 70.1-71.2% flatfish *versus* MxA), whereas the carboxyl end, which corresponds to the effector domain, seems to be the most variable (74-85.1% within flatfish, 69.1-70.9% flatfish *versus* Atlantic salmon, and 38.9-40.8% flatfish *versus* human MxA).

Interestingly, the high interspecific variability of the C-terminal region is also reflected at the inter-individual level in human (Zurcher et al., 1992), mouse (Melen & Julkunen, 1994) and chicken (Ko et al., 2002), where a single aminoacidic substitution in the effector domain has been associated with differences in the antiviral specificity of the Mx proteins. These findings prompt for further studies on the inter-individual variability of this region in fish. Actually, divergent sequences in the carboxyl-terminus of other dynamin family proteins are thought to determine their specific localization and functional characteristics, including the interactions with cellular proteins (Floyd & de Camilli, 1998). Taking into account these considerations, it is conceivable that a specialized function of Mx proteins in viral defence had evolved as a result of a direct interaction between the carboxyl-terminus of individual Mx proteins and species-specific viral pathogens (Lee & Vidal, 2002). Hence, it has been demonstrated that the fine-tuning of the mechanisms for host defence against infections, including the non-specific antiviral defences mediated by IFNs, has occurred independently in the different vertebrate lineages (Lutfalla et al., 2003). Therefore, immunogenes provide an interesting perspective on the evolution and diversification during the radiation of vertebrates. For this reason, vertebrate Mx proteins can be grouped according to their sequence similarities into human MxA- and MxB-like subgroups, fish, avian, and rodent subgroups (Haller et al., 2007).

In Figure 7, a phylogenetic tree generated by the Neighbour-Joining method from a ClustalX alignment of vertebrate Mx proteins is drawn. The tree shows that the fish Mx proteins constitute a cluster separated from other vertebrate Mx proteins, which is in agreement with previous studies (Haller et al., 2007; Plant & Thune, 2008; Robertsen et al., 1997; Zhang et al., 2004). Concerning Mx sequences from flatfish species, they appear grouped in a cluster strongly supported by a bootstrapping value of 94%. The topology of the flatfish group basically agrees with that of previous phylogenetic studies based on sequences of different genes (Berendzen et al., 2002; Sotelo et al., 2001; Tinti et al., 1999). This result strongly suggests that Mx proteins show the standard evolutive pattern of flatfish species.

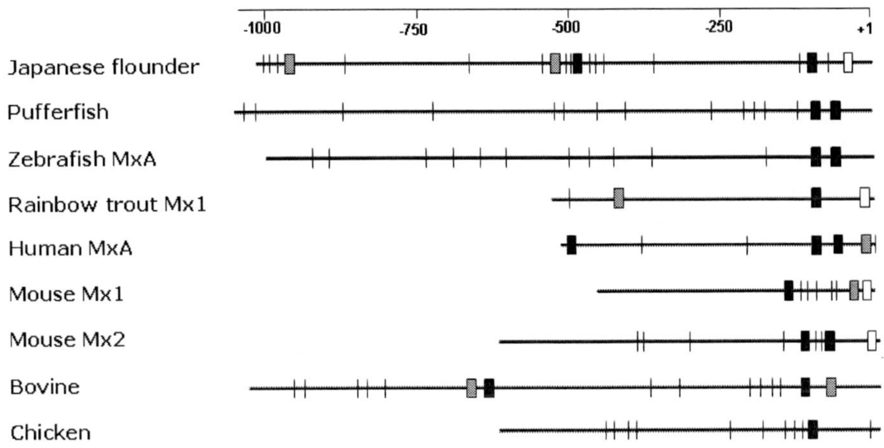

Figure 8. Structure of Mx gene promoters of vertebrate species. A white box represents the TATA box, a black box is the ISRE sequence, a grey box is the Sp1 site and a vertical line is the GAAA/TTTC element. The transcription start site is indicated as +1, except for the zebrafish Mx promoter, where it represents the ATG site. Adapted from Ooi et al. (2006).

3.1. Structure and Regulation of the Flatfish Mx Gene Promoters

The Japanese flounder constitutes the only flatfish species whose Mx gene promoter has been characterized. This promoter is activated by poly I:C and it has been used in functional assays to quantify the response of Mx proteins and to elucidate the regulation pathways of IFN type I in this species. It contains one TATA box (positions -24 to -30 from the transcription start site), two ISRE elements with four GAAA elements between them (ISRE1 positions -69 to -80; ISRE2 positions -508 to -520), fourteen GAAA elements out of the region between the ISRE sequences, which may have a role in IFN-inducibility (Reid et al., 1989), and two Sp1 sites, located at -563 to -572 and -994 to -1003 (Ooi et al., 2006). Both ISRE motifs conform to the (A/G/T)GAAA$(N)_{1\text{-}2}$GAAA(C/G)(A/C/T) ISRE consensus sequence (Collet & Secombes, 2001). Mx promoters with multiple ISREs generally have two ISREs within -150 bp from the transcription start site and in close proximity to each other; however, the two ISRE sequences reported in Japanese flounder Mx gene promoter are more than 400 bp apart from each other.

The structure of the Japanese flounder Mx gene promoter, along with the rainbow trout, pufferfish and zebrafish Mx promoters, is shown in Figure 8. Pufferfish and zebrafish Mx promoters have two closely located proximal ISRE motifs and no TATA boxes or Sp1 sites. In contrast, rainbow trout and Japanese flounder Mx promoters show these elements, although they possess a different number of ISRE sequences.

Table 3. Comparison of flatfish Mx sequences with Atlantic salmon Mx1 and human MxA.

	S. sole	J. flounder	Turbot	Halibut	A. salmon Mx1	Human MxA
S. sole	-------	80.2	79.9	77.2	76.2	55.2
J. flounder		-------	83.2	88.9	78.2	55.5
Turbot			-------	80.2	78.2	55.3
Halibut				-------	76.5	54.5
A. salmon Mx1					-------	55.2
Human MxA						--------

The IFN responsive Mx promoter has also been studied in mammals (Asano et al., 2003; Gerardin et al., 2004; Hug et al., 1988; Ronni et al., 1998) and birds (Schumacher et al., 1994). ISRE motifs are relatively conserved in Mx promoters from vertebrates; however, variations in the number, relative location and contribution to promoter activity regulation have been reported. The Mx promoter of chicken and the murine Mx1 contain a single ISRE, the bovine Mx promoter and the murine Mx2 contain two ISREs, whereas the human MxA shows three ISRE sequences (Figure 8). It is likely that the presence of multiple ISREs enhances the promoter sensitivity to be induced by IFN. In the human MxA gene, a single nucleotide polymorphism in one of the ISRE motifs within its promoter has been reported to affect the gene response to IFN (Hijikata et al., 2000).

ISRE elements have been characterized in the promoters of human MxA (Nakade et al., 1997; Ronni et al., 1998), murine Mx1 (Hug et al., 1988) and chicken Mx (Schumacher et al., 1994). The promoter of the murine Mx1 gene possesses a TATA box and a Sp1-binding site located between positions -43 and -20 from the initiation site (Hug et al., 1988). Spl is a ubiquitous transcription factor, which regulates the tissue specific expression of a variety of genes (Henson, 1994; Zhang et al., 1994). Although TATA boxes determine the initiation of the transcription, they are not consistently found in Mx promoters, being absent in some species of mammals (human and bovine Mx), birds (chicken Mx) and fish.

Ooi et al. (2006) have analyzed the functional activity of the above described Japanese flounder Mx promoter using a green fluorescent protein (GFP) reporter gene, demonstrating a strong GFP fluorescence after poly I:C stimulation. In addition, these authors reported that the deletion of the ISRE1 motif (proximal ISRE) abolished poly I:C inducibility, whereas deletion of ISRE2 (distal ISRE) motif resulted in a reduction of the Mx induction (60%) compared to the full promoter construct assay. This result suggests that IFN-inducibility of the Japanese flounder Mx promoter is due to the presence of the two ISREs, which, despite being relatively distant from each other, work synergistically to regulate the Mx gene expression. The ability of both ISREs to bind transcription factors has been evidenced by a shifted complex using Electrophoresis Mobility Shift Assay (EMSA) experiments. Deletion analyses carried out with other fish Mx promoters containing two ISREs in close proximity, such as zebrafish and pufferfish, have shown a similar pattern, characterized by an important reduction in promoter activity after the deletion of either ISRE motif, which suggests that both ISREs contribute to the inducibility of these Mx promoters. Analyses performed in order to study the regulatory elements of Mx promoters with a single ISRE motif, such as murine Mx1 and chicken Mx, have shown that the deletion of this element results in the incapability to be induced by IFN or virus infection (Hug et al., 1988; Schumacher et al., 1994). No attempt has been performed in order to establish the role of the single ISRE element of the rainbow trout Mx promoter by deletion studies.

3.2. Studies on the Expression of Flatfish Mx Proteins

The characterization of the Mx protein expression includes the study on the regulatory mechanisms of the Mx gene promoter at constitutive levels, as well as on the induction of its expression under several conditions, such as treatment with poly I:C or viral infection. The Mx expression is regulated by type I IFN, and *in vitro* nuclear run-off experiments have shown that this regulation occurs mainly at a gene transcription level (Staeheli et al., 1986). However, the existence of different regulatory mechanisms has been suggested by King et al. (2004), who showed differential expressions of the human MxA and MxB proteins in response to IFN.

Constitutive expression of flatfish Mx has been reported in several tissues of Japanese flounder (Lee et al., 2000), turbot (Abollo et al., 2005) and Senegalese sole (Fernandez-Trujillo et al., 2006). These results agree with those obtained for other fish species, such as Atlantic salmon (Jensen et al., 2002), pufferfish (Yap et al., 2003), zebrafish (Altmann et al., 2004), gilt-head seabream (*Sparus aurata*) (Tafalla et al., 2004), and for mammals (Horisberger & Gunst, 1991), where it has been attributed to cell sensitivity to the low basal levels of circulating endogenous IFN (Staeheli & Haller, 1987). *In vitro*, constitutive Mx expression has also been demonstrated in the Japanese flounder HINAE cell line (Caipang et al., 2003).

Expression of Mx after induction with poly I:C treatment has been described for three out of the four flatfish species studied in this chapter. Fernandez-Trujillo et al. (2008a) reported that the intraperitoneal poly I:C injection of Senegalese sole resulted in a significant increase of Mx expression, reaching maximum levels at 48 h post-injection (p.i.). The induction of Mx expression in liver of poly I:C-injected Atlantic halibut was possible from 2 days to, at least, 14 days p.i., with the maximum intensity of the Mx protein band at 3-4 days p.i. (Bergan & Robertsen, 2004). Yasuike et al. (2007) reported the Mx induction in Japanese flounder kidney samples in response to poly I:C injection. In contrast, Mx expression did not increase after poly I:C injection in turbot (Abollo et al., 2005).

However, *in vitro* expression of Mx in turbot leukocytes was significantly increased by poly I:C 72 h after induction (Abollo et al., 2005). In addition, in the turbot cell line TV-1 a brief expression of Mx was recorded at 12 and 24 h after stimulation (Garcia-Rosado et al., 2008). This expression profile is drastically different from those described for different fish cell lines after poly I:C induction, such as the Atlantic salmon cell lines AS, CHSE-214, SHK-1 and TO, as well as the fibroblastic cell lines RTG-2 (derived for rainbow trout), SAF-1 and HINAE (Collet et al., 2007; Garcia-Rosado et al., 2008; Jensen & Robertsen, 2002; Lin et al., 2005; Nygaard et al., 2000). These cells showed the maximum level of Mx expression 48 h after stimulation, although certain level of expression could be detected at 6 h after poly I:C treatment. All these cell lines showed a long lasting induction of Mx expression after a single treatment with poly I:C. In addition, the stimulation of turbot Mx expression requires a high concentration of poly I:C (10 μg/ml). Therefore, these results may suggest that turbot cells are low-sensitive to this inductor.

Analysis of tissues from other fish species, such as rainbow trout and Atlantic salmon, has also demonstrated the Mx induction after poly I:C treatment (Jensen & Robertsen, 2000; Robertsen et al., 1997; Trobridge & Leong, 1995). Detection of Mx proteins 14 days after a single poly I:C injection has been reported in organs of Atlantic salmon (Jensen et al., 2002). Although previous studies showed a maximum level of Mx mRNA at 2-4 days p.i., an evident decrease was recorded after 7 days (Jensen & Robertsen, 2000). These results suggest that the detection of Mx proteins up to 14 days p.i. is a consequence of the protein stability rather than of the continuous gene transcription. Other analyzed organs, such as kidney, spleen, gills, heart and intestine, also showed the presence of the Mx protein 4 days after poly I:C injection.

Finally, the induction of the Mx expression after a viral infection has also been reported in flatfish species, such as in leukocytes of Japanese flounder (after HIRRV and an aquabirnavirus infection), in HINAE cells (after HIRRV inoculation), in Atlantic halibut, Japanese flounder and Senegalese sole (after an IPNV infection) (Fernandez-Trujillo et al., 2008a; Jensen & Robertsen, 2000; Lee et al., 2000; Pakingking et al., 2004). HIRRV infection induces Mx mRNA expression in Japanese flounder at 48 h p.i., showing a maximum level at 72 h p.i., as was previously demonstrated (Yabu et al., 1998). On the other hand, aquabirnavirus injection of healthy Japanese flounder resulted in the Mx gene expression from 1 day p.i. to, at least, 21 days, showing maximum levels of Mx mRNA at 3-7 days p.i. Inoculation of Atlantic halibut with IPNV stimulated the Mx production in liver from 4 to, at least, 35 days p.i. The time course profile of Mx expression in Senegalese sole after IPNV injection is different from that described after poly I:C induction. The onset of Mx expression was first detected 36 h after IPNV inoculation.

The induction of Mx expression by aquabirnavirus has also been demonstrated *in vivo,* in Atlantic salmon post-smolts immediately after infection (Lockhart et al., 2004), and *in vitro*, in SAF-1 and TV-1 cells, derived from gilt-head seabream and turbot, respectively (Garcia-Rosado et al., 2008). The kinetic of Mx expression in TV-1 cells after IPNV infection is similar to that described after poly I:C treatment.

3.3. Characterization of the Antiviral Activity of the Flatfish Mx Proteins

In flatfish, Mx antiviral activity has been reported for Japanese flounder Mx protein against the rhabdoviruses HIRRV and VHSV (Caipang et al., 2003), and for Senegalese sole Mx, which inhibits IPNV replication (Fernandez-Trujillo et al., 2008b). The antiviral activity of the Atlantic halibut and turbot Mx proteins remains to be investigated.

The inhibition of fish rhabdoviruses (HIRRV and VHSV) by the Japanese flounder Mx has been shown by *in vitro* studies carried out using the homologous HINAE cell line stably expressing this protein, since transfected cells show significantly higher survival rates after infection compared to the control cells (Caipang et al., 2002; 2003). Moreover, the level of viral glycoprotein and nucleoprotein transcripts was very low and constant over time in the Japanese flounder Mx expressing cells, whereas an increasing level of these transcripts was recorded in infected control cells over time (Caipang et al., 2003). According to these authors, Japanese flounder Mx inhibits only partly primary viral protein transcription, blocking rhabdoviral replication by interfering with subsequent steps in transcription of viral subgenomic mRNAs. Japanese flounder Mx may inhibit transcription by wrapping the nucleocapsid transcript, although some of them may escape, which would account for the partial inhibition of the viral nucleocapsid primary transcripts. A similar mechanism has been described for grouper Mx, which inhibits nodavirus replication by interacting with the capsid protein (Chen et al., 2008). An *in vivo* analysis has shown that the Mx expression induced by a previous aquabirnavirus infection inhibits the subsequent VHSV infection (Pakingking et al., 2004), which supports the Japanese flounder Mx antiviral activity against rhabdoviruses previously reported by *in vitro* assays.

The antiviral activity of Senegalese sole Mx has been demonstrated against aquabirnavirus using the salmonid CHSE-214 cell line over-expressing this protein (Fernandez-Trujillo et al., 2008b). These authors described a delay in the appearance of cytopathic effects (CPE) in Senegalese sole Mx expressing cells compared with no-transfected cells, as well as a higher survival rate after the virus inoculation in those cells expressing Senegalese sole Mx. In addition, a significantly low level of the viral VP2 protein was recorded in Senegalese sole Mx expressing cells 48 p.i. All these results indicate that Senegalese sole Mx protects the cells against viral infection. Furthermore, the virus yield 72 h p.i. was dramatically reduced in Senegalese sole Mx-expressing cells. These results are in concordance with those obtained *in vivo* by Fernandez-Trujillo et al. (2008a), who reported a decrease in the viral VP2 protein expression in kidneys of poly I:C induced soles, suggesting a pivotal role of Senegalese sole Mx in the antiviral response of the IFN system after poly I:C induction.

The antiviral activity of fish Mx has also been demonstrated for other fish species, such as Atlantic salmon Mx1 (against IPNV), barramundi (*Lates calcarifer*) (against nodavirus and birnavirus) and grouper (against yellow grouper nervous necrosis virus, YGNNV) (Chen et al., 2008; Larsen et al., 2004; Wu & Chi, 2007). In contrast, only a minimal inhibition of the IHNV by rainbow trout Mx has been recorded after *in vitro* assays conducted in transfected salmon cells (Trobridge et al., 1997), and barramundi Mx failed in the inhibition of fish iridovirus (Wu & Chi, 2007). All these studies used a similar experimental approach consisting of the *in vitro* overexpression of a recombinant Mx and the analysis of its survival after inoculation with a virus. In summary, these results show that cells constitutively expressing these fish Mx proteins were resistant to the viruses tested. However, cells expressing different fish Mx proteins that have been infected with different viruses show diverse effects on viral RNA transcription, indicating that pathogen-host interactions are involved in the antiviral response and/or the molecular mechanisms of this Mx mediated response in fish.

On the other hand, it is necessary to further investigate the complete antiviral spectrum of fish Mx proteins and the molecular mechanisms behind their antiviral properties.

4. Future Studies on Flatfish Mx Proteins

The knowledge on Mx genes may shed light on the fish health status and on the effects of immunomodulators and vaccines. The challenge of future research is to provide information about different functional aspects of this system. Although a significant progress has been made regarding the characterization of the flatfish Mx proteins, many important issues remain to be addressed. Especially urgent are those concerning the organization of Mx genes, the antiviral specificity of Mx proteins, the regulation of their expression, their mechanism of action and their relevance in relation with the rest of interferon stimulated genes.

A key goal to further understand this system should be the study of the pathogen-host interactions, which involves the identification of viral factors contributing to the virus virulence, the identification of viral genes implicated in the antagonist activity mechanisms against the IFN system, or the identification of all the cellular factors that contribute and/or modulate the Mx (and other proteins) antiviral activity.

Additionally, the interplay between viruses and the IFN system reflects an ongoing evolutionary race between the two genetic systems, and, although diversity has been considered the hallmark of adaptive immunity, evidences supporting the importance of diversity in innate immunity have been accumulated in the last few years (Hoffman et al., 1999). Moreover, differences in innate immunity mechanisms have been shown to be critical in host susceptibility to infections (Cooke & Hill, 2001). Currently, major efforts are underway to identify single-nucleotide polymorphisms in humans and several economically important species. The research on Mx genes, along with studies based on the systematic screening of their antiviral properties using *in vitro* systems, seems a promising avenue to further our understanding of the functional diversity found within fish Mx proteins. This kind of studies will help to understand the evolution of the host-virus interactions, and will give us the opportunity to study both, evolutionary and genetic aspects, of the host antiviral functions.

On the other hand, some of the most interesting developments in this field are likely to come from further insights into the broader role of antiviral proteins, particularly in mediating adaptive immunity. Progress in this area could be crucial to achieve the goal of manipulating the immune system in order to enhance the natural fish resistance to viral pathogens.

Finally, although biotechnological developments in the aquaculture industry have not been as fast as those in land-based food production systems, it is evident that the application of this technology to the culture of fish species may contribute enormously to improve this activity (De Santis et al., 2007; Hew & Fletcher, 2001; Melamed et al., 2002). Some of the biotechnological approaches that have been addressed to date in order to control viral infections in fish farms include the use of DNA vaccines, immunostimulants, antisense technology, as well as the identification of markers for disease resistance and transgenesis (Dunham, 2009; Hew & Fletcher, 2001). Now that the antiviral activity of Mx has been established for a variety of fish species, including some flatfish, it would be of great interest to develop a Mx-based technology in order to control viral diseases. Mx may be useful as a transgene, as a blueprint to engineer novel antiviral substances, or as a gene for genetic selection.

In summary, progress on these issues will probably allow the development of new tools to fight virus infections in fish aquaculture. Particularly, flatfish are important and promising commercial fish species that may be interesting candidates to explore these new strategies in order to improve their natural resistance to viral diseases.

References

Abollo, E., Ordas, C., Dios, S., Figueras, A. & Novoa, B. (2005). Molecular characterisation of a turbot Mx cDNA. *Fish Shellfish Immunol., 19*, 185-190.

Alonso, M. C., Cano, I., Garcia-Rosado, E., Castro, D., Lamas, J., Barja, J. L., Borrego, J. J. & Bergmann, S. M. (2005). Isolation of lymphocystis disease virus from sole, *Solea senegalensis* Kaup, and blackspot sea bream, *Pagellus bogaraveo* (Brunnich). *J. Fish Dis., 28,* 221-228.

Alonso, M. C., Cano, I., Garcia-Rosado, E., Dopazo, C. P., Manchado, M., Castro, D. & Borrego, J. J. (2007). Detection and isolation of Viral Nervous Necrosis Virus (VNNV) from non-nervous tissues of cultured Senegalese sole (*Solea senegalensis*, Kaup). *The 7th International Symposium on Viruses of Lower Vertebrates*, Oslo, Norway.

Alonso, M. & Leong, J. (2002). Suppressive subtraction libraries to identify interferon-inducible genes in fish. *Mar. Biotechnol., 4,* 74-80.

Altmann, S. M., Mellon, M. T., Distel, D. L. & Kim, C. H. (2003). Molecular and functional analysis of an interferon gene from the zebrafish, *Danio rerio. J. Virol., 77,* 1992-2002.

Altmann, S. M., Mellon, M. T., Jonson, M. C., Paw, B. H., Trede, N. S., Zon, L. I. & Kim, C. H. (2004). Cloning and characterization of an Mx gene and its corresponding promoter from the zebrafish, *Danio rerio. Dev. Comp. Immunol., 28,* 295-306.

Anderson, S. L., Carton, J. M., Lou, J., Xing, L. & Rubin, B. Y. (1999). Interferon-induced guanylate binding protein-1 (GBP-1) mediates an antiviral effect against vesicular stomatitis virus and encephalomyocarditis virus. *Virology*, *256,* 8-14.

Asano, A., Ko, J. H., Morozumi, T., Hamashima, N. & Watanabe, T. (2002). Polymorphisms and the antiviral property of porcine Mx1 protein. *J. Vet. Med. Sci., 64,* 1085-1089.

Asano, A., Jin, H. K. & Watanabe, T. (2003). Mouse *Mx2* gene: organization, mRNA expression and the role of the interferon-response promoter in its regulation. *Gene, 306,* 105-113.

Baerwald, M. R., Welsh, A. B., Hedrick, R. P. & May, B. (2008). Discovery of genes implicated in whirling disease infection and resistance in rainbow trout using genome-wide expression profiling. *BMC Genomics, 9.*

Baise, E., Pire, G., Leroy, M., Gerardin, J., Goris, N., De Clercq, K., Kerkhofs, P. & Desmecht, D. (2004). Conditional expression of type I interferon-induced bovine MX1 GTPase in a stable transgenic vero cell line interferes with replication of vesicular stomatitis virus. *J. Interf. Cytok. Res, 24,* 513-521.

Bazzigher, L., Schwarz, A. & Staeheli, P. (1993). No enhanced influenza virus resistance of murine and avian cells expressing cloned duck Mx protein. *Virology, 195,* 100-112.

Berendzen, P. B. & Dimmick, W. W. (2002). Phylogenetic relationships of *Pleuronectiformes* based on molecular evidence. *Copeia, 3*, 642-52.

Bergan, V. & Robertsen, B. (2004). Characterization of Atlantic halibut (*Hippoglossus hippoglossus*) Mx protein expression. *Dev. Comp. Immunol., 28,* 1037-1047.

Bergan, V., Jagus, R., Lauksund, S., Kileng, Ø. & Robertsen, B. (2008). The Atlantic salmon Z-DNA binding protein kinase phosphorylates translation initiation factor 2 alpha and constitutes a unique orthologue to the mammalian dsRNA-activated protein kinase R. *FEBS Journal, 275,* 184-197.

Bergh, O., Nilsen, F. & Samuelsen, O. B. (2001). Diseases, prophylaxis and treatment of the Atlantic halibut *Hippoglossus hippoglossus*: a review. *Dis. Aquat. Org., 48,* 57-74.

Biering, E., Nilsen, R., Rodseth, O. M. & Glette, J. (1994). Susceptibility of Atlantic halibut *Hippoglossus hippoglossus* to infectious pancreatic necrosis virus. *Dis. Aquat. Org., 20,* 183-190.

Birmachu, W., Gleason, R. M., Bulbulian, B. J., Riter, C. L., Vasilakos, J. P., Lipson, K. E. & Nikolsky, Y. (2007). Transcriptional networks in plasmacytoid dendritic cells stimulated with synthetic TLR 7 agonists. *BMC Immunol., 8.*

Bluyssen, H. A. R. & Levy, D. E. (1997). Stat2 is a transcriptional activator that requires sequence-specific contacts provided by Stat1 and p48 for stable interaction with DNA. *J. Biol. Chem., 272,* 4600-4605.

Caipang, C. M. A., Hirono, I. & Aoki, T. (2002). Viral resistance of a recombinant Japanese flounder Mx-transfected fish cell. *Fish Sci., 68,* 1217-1218.

Caipang, C. M. A., Hirono, I. & Aoki, T. (2003). *In vitro* inhibition of fish rhabdoviruses by Japanese flounder, *Paralichthys olivaceus* Mx. *Virology, 317,* 373-382.

Casani, D., Randelli, E., Costantini, S., Facchiano, A. M., Zou, J., Martin, S., Secombes, C. J., Scapigliati, G. & Buonocore, F. (2009). Molecular characterisation and structural analysis of an interferon homologue in sea bass (*Dicentrarchus labrax* L.). *Mol. Immunol., 46,* 943-952.

Chen, Y. M., Su, Y. L., Shie, P. S., Huang, S. L., Yang, H. L. & Cen, T. Y. (2008). Grouper Mx confers resistance to nodavirus and interacts with coat protein. *Dev. Comp. Immunol., 32,* 825-836.

Chesters, P. M., Steele, M., Purewal, A. & Edington, N. (1997). Nucleotide sequence of equine MxA cDNA. *DNA Seq.,, 7,* 239-242.

Chieux, V., Hober, D., Harvey, J., Lion, G., Lucidarme, D., Forzy, G., Duhamel, M., Cousin, J., Ducoulombier, H. & Wattre, P. (1998). The MxA protein levels in whole blood lysates of patients with various viral infections. *J. Virol. Methods, 70,* 183-191.

Chieux, V., Chehadeh, W., Harvey, J., Haller, O., Wattre, P. & Hober, D. (2001). Inhibition of coxachievirus B4 replication in stably transfected cells expressing human MxA protein. *Virology, 283,* 84-92.

Chin, K. & Cresswell, P. (2001). Viperin (cig5), an IFN-inducible antiviral protein directly induced by human cytomegalovirus. *Proc. Natl. Acad. Sci.* USA, *98,* 15125-15130.

Collet, B. & Secombes, C. J. (2001). The rainbow trout (*Oncorhynchus mykiss*) Mx1 promoter. Structural and functional characterization. *Eur. J. Biochem., 268,* 1577-1584.

Collet, B. & Secombes, C. J. (2002). Type I-interferon signalling in fish. *Fish Shellfish Immunol., 12,* 389-397.

Collet, B., Munro, E. S., Gahlawat, S., Acosta, F., Garcia, J., Roemelt, C., Zou, J., Secombes, C. J. & Ellis, A. E. (2007). Infectious pancreatic necrosis virus suppresses type I interferon signalling in rainbow trout gonad cell line but not in Atlantic salmon macrophages. *Fish Shellfish Immunol., 22,* 44-56.

Cooke, G. S. & Hill, A. V. (2001). Genetics of susceptibility to human infectious disease. *Nat. Rev. Genet., 2,* 967-977.

Cutrin, J. M., Lopez-Vazquez, C., Olveira, J. G., Castro, S., Dopazo, C. P. & Bandin, I. (2005). Isolation in cell culture and detection by PCR-based technology of IPNV-like virus from leucocytes of carrier turbot, *Scophthalmus maximus* (L.). *J. Fish Dis., 28,* 713-722.

Dannevig, B. H., Nilsen, R., Modahl, I., Jankowsaka, M., Taksdal, T. & Press, C. M. (2000). Isolation in cell culture of nodavirus from farmed Atlantic halibut *Hippoglossus hippoglossus* in Norway. *Dis. Aquat. Org., 43,* 183-189.

De-Santis, C. & Jerry, D. R. (2007). Candidate growth genes in finfish - Where should we be looking?. *Aquaculture, 272,* 22-38.

De Veer, M. J., Holko, M., Frevel, M., Walker, E., Der, S., Paranjape, J. M., Silverman, R. H. & Williams, B. R. G. (2001). Functional classification of interferon-stimulated genes identified using microarrays. *J. Leukocyte Biol., 69*, 912-920.

Der, S. D., Zhou, A., Williams, B. R. G. & Silverman, R. H. (1998). Identification of genes differentially regulated by interferon α, β, or γ using oligonucleotide arrays. *Proc. Natl. Acad. Sci. USA, 95*, 15623-15628.

Dunham, R. A. (2009). Transgenic fish resistant to infectious diseases, their risk and prevention of escape into the environment and future candidate genes for disease transgene manipulation, *Comp. Immunol. Microb., 32,* 139-161.

Ellinwood, N. M., McCue, J. M., Gordy, P. W. & Bowen, R. A. (1998). Cloning and characterization of cDNAs for a bovine (*Bos taurus*) Mx protein. *J. Interf. Cytok. Res., 18,* 745-755.

Ellis, A. E. (2001). Innate host defense mechanisms of fish against viruses and bacteria. *Dev. Coomp. Immunol., 25*, 827-839.

Espert, L., Degols, G., Gongora, C., Blondel, D., Williams, B. R., Silverman, R. H. & Mechti, N. (2003). ISG20, a new interferon-induced RNase specific for single-stranded RNA, defines an alternative antiviral pathway against RNA genomic viruses. *J. Biol. Chem., 278,* 16151-16158.

Fernandez-Trujillo, M. A. (2008). Clonación y valoración funcional del inmunogen Mx de lenguado (*Solea senegalensis*). Doctoral Thesis. University of Málaga.

Fernandez-Trujillo, M. A., Porta, J., Borrego, J. J., Alonso, M. C., Alvarez, M. C. & Bejar, J. (2006). Cloning and expression analysis of Mx cDNA from Senegalese sole (*Solea senegalensis*). *Fish Shellfish Immunol., 21,* 577-582.

Fernandez-Trujillo, M. A., Ferro, P., Garcia-Rosado, E., Infante, C., Alonso, M. C., Bejar, J., Borrego, J. J. & Manchado, M. (2008a). Poly I:C induces Mx transcription and promotes an antiviral state against sole aquabirnavirus in the flatfish Senegalese sole (*Solea senegalensis* Kaup). *Fish Shellfish Immunol., 24,* 279-285.

Fernandez-Trujillo, M. A., Garcia-Rosado, E., Alonso, M. C., Borrego, J .J., Alvarez, M. C. & Bejar, J. (2008b). *In vitro* inhibition of sole aquabirnavirus by Senegalese sole Mx. *Fish Shellfish Immunol., 24,* 187-193.

Floyd, S. & De Camilli, P. (1998). Endocytosis proteins and cancer: a potential link? *Trends Cell Biol., 8,* 299-301. Review.

Garcia, M. A., Gil, J., Ventoso, I., Guerra, S., Domingo, E., Rivas, C. & Esteban, M. (2006). Impact of protein kinase PKR in cell biology: From antiviral to antiproliferative action. *Microbiol. Mol. Biol. R., 70,* 1032-1060.

Garcia-Rosado, E., Alonso, M. C., Bejar, J., Manchado, M., Cano, I. & Borrego, J. J. (2008). Expression analysis of Mx protein and evaluation of its antiviral activity against sole aquabirnavirus in SAF-1 and TV-1 cell lines. *Vet. Immunol. Immunopathol., 121,* 123-129.

Gerardin, J. A., Baise, E. A., Pire, G. A., Leroy, M. P. P. & Desmecht, D. J. M. (2004). Genomic structure, organisation, and promoter analysis of the bovine (*Bos taurus*) Mx1 gene. *Gene, 326,* 67-75.

Gordien, E., Rosmordue, O., Peltekian, C., Garreau, F., Brechot, C. & Kremsdorf, D. (2001). Inhibition of hepatitis B virus replication by the interferon-inducible MxA protein. *J. Virol., 75,* 2684-2691.

Grotmol, S., Bergh, O. & Totland, G. K. (1999). Transmission of viral encephalopathy and retinopathy (VER) to yolk-sac larvae of the Atlantic halibut *Hippoglossus hippoglossus*: occurrence of nodavirus in various organs and a possible route of infection. *Dis. Aquat. Org., 36,* 95-106.

Guo, J., Peters, K. L. & Sen, G. C. (2000). Induction of the human protein P56 by interferon, double-stranded RNA, or virus infection. *Virology, 267,* 209-219.

Haas, A. L., Ahrens, P., Bright, P. M. & Ankel, H. (1987). Interferon induced a 15-kilodalton protein exhibiting marked homology to ubiquitin. *J. Biol. Chem., 262,* 11315-11323.

Haller, O. & Kochs, G. (2002). Interferon-induced Mx proteins: dynamin-like GTPases with antiviral activity. *Traffic, 3,* 710-717.

Haller, O., Frese, M. & Kochs, G. (1998). Mx proteins: mediators of innate resistance to RNA viruses. *Rev. Sci. Tech., 17,* 220-230.

Haller, O., Kochs, G. & Weber, F. (2006). The interferon response circuit: Induction and suppression by pathogenic viruses. *Virology, 344,* 119-130.

Haller, O., Stertz, S. & Kochs, G. (2007). The Mx GTPase family of interferon-induced antiviral proteins. *Microbes Infect., 9,* 1613-1643.

Henson, J. W. (1994). Regulation of the glial-specific JC virus early promoter by the transcription factor Sp1. *J. Biol. Chem., 269,* 1046-1050.

Hew, C. L. & Fletcher, G. L. (2001). The role of aquatic biotechnology in aquaculture. *Aquaculture, 197*, 191-204.

Hijikata, M., Ohta, Y. & Mishiro, S. (2000). Identification of a single nucleotide polymorphism in the MxA gene promoter (G/T at nt -88) correlated with the response of hepatitis C patients to interferon. *Intervirology*, *43,* 124-127.

Hoffmann, J. A., Kafatos, F. C., Janeway, C. A. & Ezekowitz, R. A. (1999). Phylogenetic perspectives in innate immunity. *Science, 284*, 1313-1318.

Horisberger, M. A. & Gunst, M. C. (1991). Interferon-induced proteins: Identification of Mx proteins in various mammalian species. *Virology, 180,* 185-190.

Hu, C., Zhang, Y., Huang, G., Zhang, Q. & Gui, J. (2004). Molecular cloning and characterisation of a fish PKR-like gene from cultured CAB cells induced by UV-inactivated virus. *Fish Shellfish Immunol., 17,* 353-366.

Hug, H., Costas, M., Staeheli, P., Aebi, M. & Weissmann, C. (1988). Organization of the murine Mx gene and characterization of its interferon- and virus-inducible promoter. *Mol. Cell. Biol., 8,* 3065-3070.

Isaacs, A. & Lindenmann, J. (1957). Virus interference. I. The interferon. *Proc. R. Soc. Lond., 147,* 258-267.

Isshiki, T., Nishizawa, T., Kobayashi, T., Nagano, T. & Miyazaki, T. (2001). An outbreak of VHSV (viral hemorrhagic septicemia virus) infection in farmed Japanese flounder *Paralichthys olivaceus* in Japan. *Dis. Aquat. Org., 47,* 87-99.

Jensen, I., Albuquerque, A., Sommer, A. I. & Robertsen, B. (2002). Effect of poly I:C on the expression of Mx proteins and resistance against infection by infectious salmon anemia virus in Atlantic salmon. *Fish Shellfish Immunol., 13,* 311-326.

Jensen, V. & Robertsen, B. (2000). Cloning of an Mx cDNA from Atlantic halibut (*Hippoglossus hippoglossus*) and characterization of Mx mRNA expression in response to double-stranded RNA or infectious pancreatic necrosis virus. *J. Interf. Cytok. Res., 20,* 701-710.

Jensen, V. & Robertsen, B. (2002). Effect of double-stranded RNA and interferon on the antiviral activity of Atlantic salmon cells against infectious salmon anemia virus and infectious pancreatic necrosis virus. *Fish Shellfish Immunol., 13,* 221-241.

Johannes, L., Arnheiter, H. & Meier, E. (1993). Switch in antiviral specificity of a GTPase upon translocation from the cytoplasm to the nucleus. *J. Virol., 67,* 1653-1657.

Johansen, R., Sommerset, I., Torud, B., Korsnes, K., Hjortaas, M. J., Nilsen, F., Nerland, A. H. & Dannevig, B. H. (2004). Characterization of nodavirus and viral encephalopathy and retinopathy in farmed turbot, *Scophthalmus maximus* (L.). *J. Fish Dis., 27,* 591-601.

Jung, S. J., Kim, S. R., Joung, I. Y., Kitamura, S. I., Ceong, H. T. & Oh, M. J. (2008). Distribution of marine birnavirus in cultured olive flounder *Paralichthys olivaceus* in Korea. *J. Microbiol., 46,* 265-273.

Kalayci, G., Incoglu, S. & Ozkan, B. (2006). First isolation of viral haemorrhagic septicaemia (VHS) virus from turbot (*Scophthalmus maximus*) cultured in the Trabzon coastal area of the Black Sea in Turkey. *Bull. Eur. Ass. Fish Pathol., 26,* 157-162.

Kerkmann, M., Rothenfusser, S., Hornung, V., Towarowski, A., Wagner, M., Sarris, A., Giese, T., Endres, S. & Hartmann, G. (2003). Activation with CpG-A and CpG-B oligonucleotides reveals two distinct regulatory pathways of type I IFN synthesis in human plasmacytoid dendritic cells. *J. Immunol., 170,* 4465-4474.

Kim, W. S., Oh, M. J., Jung, S. J., Kim Y. J. & Kitamura, S. I. (2005). Characterization of an iridovirus detected from cultured turbot *Scophthalmus maximus* in Korea. *Dis. Aquat. Org., 64,* 175-180.

King, M. C., Raposo, G. & Lemmon, M. A. (2004). Inhibition of nuclear import cell-cycle progression by mutated forms of the dynamin-like GTPase MxB. *Proc. Natl. Acad. Sci. USA, 101,* 8957-8962.

Kitamura, S. -I., Ko, J. -Y., Lee, W. -L., Kim, S. -R., Song, J. -Y., Kim, D. -Y., Jung, S. -J. & Oh, M. -J. (2007). Seasonal prevalence of lymphocystis disease virus and aquabirnavirus in Japanese flounder, *Paralichthys olivaceus* and blue mussel, *Mytilus galloprovincialis. Aquaculture, 266,* 26-31.

Ko, J. H., Jin, H. K., Asano, A., Takada, A., Ninomiya, A., Kida, H. et al. (2002). Polymorphisms and the differential antiviral activity of the chicken Mx gene. *Genome Res., 12,* 595-601.

Kochs, G. & Haller, O. (1999). Interferon-induced human MxA GTPase blocks nuclear import of Thogoto virus nucleocapsids. *Proc. Natl. Acad. Sci. USA, 96,* 2082-2086.

Kochs, G., Janzen, C., Hohenberg, H. & Haller, O. (2002). Antivirally active MxA protein sequesters La Crosse virus nucleocapsid protein into perinuclear complexes. *Proc. Natl. Acad. Sci. U.S.A., 99,* 3153-3158.

Larsen, R., Røkenes, T. P. & Robertsen, B. (2004). Inhibition of infectious pancreatic necrosis virus replication by Atlantic salmon Mx1 protein. *J. Virol., 78,* 7938-7944.

Lee, J. Y., Hiromo, I. & Aoki, T. (2000). Cloning and analysis of Mx cDNA in Japanese flounder, *Paralichthys olivaceus. Dev. Comp. Immunol., 24,* 407-415.

Lee, S. H. & Vidal, S. M. (2002). Functional diversity of Mx proteins: variations on a theme of host resistance to infection. *Genome Res., 12,* 527-530.

Leong, J. C., Trobridge, G. D., Kim, C. H. Y., Johnson, M. & Simon, B. (1998). Interferon-inducible Mx proteins in fish. *Immunol. Rev., 166,* 349-363.

Leroy, M., Pire, G., Baise, E. & Desmecht, D. (2006). Expression of the interferon-alpha/beta-inducible bovine Mx1 dynamin interferes with replication of rabies virus. *Neurobiol. Dis., 21,* 515-521.

Levy, D. E. & Garcia-Sastre, A. (2001). The virus battles: IFN induction of the antiviral state and mechanisms of viral evasion. *Cytokine Growth F. R., 12,* 143-156.

Levy, D. E., Kessler, D. S., Pine, R. & Darnell Jr., J. E. (1989). Cytoplasmic activation of ISGF3, the positive regulator of interferon-alpha-stimulated transcription, reconstituted *in vitro*. *Gene Dev., 3,* 1362-1371.

Li, X. Y., Qu, L. J., Hou, Z. C., Yao, J. F., Xu, G. Y. & Yang, N. (2007). Genomic structure and diversity of the chicken Mx gene. *Poult. Sci., 86*, 786-789.

Lin, C., Christopher John, J. A., Lin, C. & Chang, C. (2006). Inhibition of nervous necrosis virus propagation by fish Mx proteins. *Biochem. Biophys. Res. Commun., 351,* 534-539.

Lin, O. E., Ohira, T., Hirono, I., Saito-Taki, T. & Takashi, A. (2005). Immunoanalysis of antiviral Mx protein expression in Japanese flounder (*Paralichthys olivaceus*) cells. *Dev. Comp. Immunol., 29,* 443-455.

Lindenmann, J., Lane, C. A. & Hobson, D. (1963). The resistance of A2G mice to myxoviruses. *J. Immunol., 90,* 942-951.

Liu, M., Reimschuessel, R. & Hassel, B. A. (2002). Molecular cloning of the fish interferon stimulated gene, 15 kDa (ISG15) orthologue: A ubiquitin-like gene induced by nephrotoxic damage. *Gene, 298,* 129-139.

Lockhart, K., Gahlawat, S. K., Soto-Mosquera, D., Bowden, T. J. & Ellis, A. E. (2004). IPNV carrier Atlantic salmon growers do not express Mx mRNA and poly I:C induced response does not cure the carrier state. *Fish Shellfish Immunol., 17,* 347-352.

Loeb, K. R. & Haas, A. L. (1992). The interferon-inducible 15-kDa ubiquitin homolog conjugates to intracellular proteins. *J. Biol. Chem., 267,* 7806-7813.

Long, S., Wilson, M., Bengten, E., Bryan, L., Clem, L. W., Miller, N. W. & Chinchar, V. G. (2004). Identification of a cDNA encoding channel catfish interferon. *Dev. Comp. Immunol., 28,* 97-111.

Lutfalla, G., Crollius, H. R., Stange-thomann, N., Jaillon, O., Mogensen, K. & Monneron, D. (2003). Comparative genomic analysis reveals independent expansion of a lineage-specific gene family in vertebrates: The class II cytokine receptors and their ligands in mammals and fish. *BMC Genomics, 4,* 29.

Meier, E., Kunz, G., Haller, O. & Arnheiter, H. (1990). Activity of rat Mx proteins against a rhabdovirus. *J. Virol., 64,* 6263-6269.

Melamed, P., Gong, Z., Fletcher, G. & Hew, C. L. (2002). The potential impact of modern biotechnology on fish aquaculture. *Aquaculture, 204*, 255-69.

Melen, K. & Julkunen, I. (1994). Mutational analysis of murine Mx1 protein: GTP binding core domain is essential for anti-influenza A activity. *Virology, 205*, 269-79.

Milev-Milovanovic, I., Long, S., Wilson, M., Bengten, E., Miller, N. W. & Chinchar, V. G. (2006). Identification and expression analysis of interferon gamma genes in channel catfish. *Immunogenetics, 58,* 70-80.

Morozumi, T., Naito, T., Lan, P. D., Nakajima, E., Mitsuhashi, T., Mikawa, S., Hayashi, T., Awata, T., Uenishi, H., Nagata, K., Watanabe, T. & Hamasima, N. (2009). Molecular cloning and characterization of porcine Mx2 gene. *Mol. Immunol., 46,* 858-865.

Mortensen, S. H., Hjetnes, B., Todseth, O., Krogsond, J. & Christie, R. E. (1990). Infectious pancreatic virus, serotype N1, isolated from Norwegian halibut (*Hippoglossus*

hippoglossus), turbot (*Scophthalmus maximus*) and scallops (*Pecten maximus*). *Bull. Eur. Ass. Fish Pathol., 10,* 42-43.

Mortensen, S. H., Evensen, O., Rodseth, O. M. & Hjetnes, B. K. (1993). The relevance of infectious pancreatic virus (IPNV) in farmed Norwegian turbot (*Scophthalmus maximus*). *Aquaculture, 115,* 243-252.

Mulcahy, M. F. (2002). Diseases of flatfish. *Bull. Eur. Ass. Fish Pathol., 22,* 86-94.

Muller, M., Winnacker, E. & Brem, G. (1992). Molecular cloning of porcine Mx cDNAs: New members of a family of interferon-inducible proteins with homology to GTP-binding proteins. *J. Interferon Res., 12,* 119-129.

Nakade, K., Handa, H. & Nagata, K. (1997). Promoter structure of the MxA gene that confers resistance to influenza virus. *FEBS Lett., 418,* 315-318.

Nakamura, T., Asano, A., Okano, S., Ko, J., Kon, Y., Watanabe, T. & Agui, T. (2005). Intracellular localization and antiviral property of canine Mx proteins. *J. Interf. Cytok. Res., 25,* 169-173.

Naylor, R. & Burke, M. (2003). Aquaculture and ocean resources: raising tigers of the sea. *Ann. Rev. Environ. Res., 30,* 185-218.

Novoa, B., Figueras, A. J., Puentes, C. F., Ledo A. & Toranzo, A. E. (1993). Characterization of birnavirus isolated from diseased turbot cultured in Spain. *Dis. Aquat. Org., 15,* 163-169.

Nygaard, R., Husgard, S., Sommer, A., Leong, J. A. & Robertsen, B. (2000). Induction of Mx protein by interferon and double-stranded RNA in salmonid cells. *Fish Shellfish Immunol., 10,* 435-450.

O'Farrell, C., Vaghefi, N., Cantonnet, M., Buteau, B., Boudinot, P. & Benmansour, A. (2002). Survey of transcript expression in rainbow trout leukocytes reveals a major contribution of interferon-responsive genes in response to a rhabdovirus infection. *J. Virol., 76,* 8040-8049.

Ooi, E. L., Hirono, I. & Aoki, T. (2006). Functional characterisation of the Japanese flounder, *Paralichthys olivaceus*, Mx promoter. *Fish Shellfish Immunol., 21,* 293-304.

Pakingking Jr, R., Okinaka, Y., Mori, K. -I., Arimoto, M., Muroga, K. & Nakai, T. (2004). *In vivo* and *in vitro* analysis of the resistance against viral haemorrhagic septicaemia virus in Japanese flounder (*Paralichthys olivaceus*) precedingly infected with aquabirnavirus. *Fish Shellfish Immunol., 17,* 1-11.

Pavlovic, J., Zurcher, T., Haller, O. & Staeheli, P. (1990). Resistance to influenza virus and vesicular stomatitis virus conferred by expression of human MxA protein. *J. Virol., 64,* 3370-3375.

Pavlovic, J., Haller, O. & Staeheli, P. (1992). Human and mouse Mx proteins inhibit different steps of the influenza virus multiplication cycle. *J. Virol., 66,* 2564-2569.

Plant, K. P. & Thune, R. L. (2008). Genomic organisation of the channel catfish Mx1 gene and characterisation of multiple channel catfish Mx gene promoters. *Fish Shellfish Immunol., 24,* 575-83.

Regad, T. & Chelbi-Alix, M. K. (2001). Role and fate of PML nuclear bodies in response to interferon and viral infections. *Oncogene, 20,* 7274-7286.

Reid, L. E., Brasnett, A. H., Gilbert, C. S., Porter, A. C., Gewert, D. R., Stark G. R., et al. (1989). A single DNA response element can confer inducibility by both alpha- and gamma-interferons. *Proc. Natl. Acad. Sci. USA, 86,* 840-844.

Rich, A. & Zhang, S. (2003). Z-DNA: The long road to biological function. *Nat. Rev. Genet., 4,* 566-572.

Robertsen, B. (2006). The interferon system of teleost fish. *Fish Shellfish Immunol., 20,* 172-191.

Robertsen, B. (2008). Expression of interferon and interferon-induced genes in salmonids in response to virus infection, interferon-inducing compounds and vaccination. *Fish Shellfish Immunol., 25,* 351-357.

Robertsen, B., Trobridge, G. & Leong, J. A. (1997). Molecular cloning of double-stranded RNA inducible Mx genes from Atlantic salmon (*Salmo salar* L.). *Dev. Comp. Immunol., 21,* 397-412.

Robertsen, B., Bergan, V., Røkenes, T., Larsen, R. & Albuquerque, A. (2003). Atlantic Salmon Interferon Genes: Cloning, Sequence Analysis, Expression, and Biological Activity. *J. Interf. Cytok. Res., 23,* 601-612.

Rodriguez, S., Vilas, P., Gutierrez, M. C. & Perez-Prieto, S. (1997). Isolation and preliminary characterization of a birnavirus from the sole *Solea senegalensis* in Southwest Spain. *J. Aquat. Anim. Health, 9,* 295-300.

Røkenes, T. P., Larsen, R. & Robertsen, B. (2007). Atlantic salmon ISG15: Expression and conjugation to cellular proteins in response to interferon, double-stranded RNA and virus infections. *Mol. Immunol., 44,* 950-959.

Ronni, T., Matikainen, S., Lehtonen, A., Palvimo, J., Dellis, J., Van Eylen, F. et al. (1998). The proximal interferon-stimulated response elements are essential for interferon responsiveness: a promoter analysis of the antiviral MxA gene. *J. Interf. Cytok. Res., 18,* 773-781.

Ross, K., Mc Carthy, U., Huntly, P. J., Wood, B. P., Stuart, D., Rough, E. I., Smail, D. A. & Bruno, D. W. (1994). An outbreak of viral haemorrhagic septicaemia (VHS) in turbot (*Scophthalmus maximus*) in Scotland. *Bull. Eur. Ass. Fish Pathol., 14,* 213-214.

Rothenburg, S., Deigendesch, N., Dittmar, K., Koch-Nolte, F., Haag, F., Lowenhaupt, K. & Rich, A. (2005). A PKR-like eukaryotic initiation factor 2α kinase from zebrafish contains Z-DNA binding domains instead of dsRNA binding domains. *Proc. Natl. Acad. Sci. USA, 102,* 1602-1607.

Rothenburg, S., Deigendesch, N., Dey, M., Dever, T. E. & Tazi, L. (2008). Double-stranded RNA-activated protein kinase PKR of fishes and amphibians: varying the number of double-stranded RNA binding domains and lineage-specific duplications. *BMC Biology, 6,* 12.

Saint-Jean, S. R. & Perez-Prieto, S. I. (2006). Interferon mediated antiviral activity against salmonid fish viruses in BF-2 and other cell lines. *Vet. Immunol. Immunopathol., 110,* 1-10.

Samuel, C. E. (2001). Antiviral actions of interferons. *Clin. Microbiol. Rev., 14,* 778-809.

Sandrock, M., Frese, M., Haller, O. & Kochs, G. (2001). Interferon-induced rat Mx proteins confer resistance to Rift Valley fever virus and other arthropod-borne viruses. *J. Interf. Cytok. Res., 21,* 663-668.

Schneider-Schaulies, S., Scheneider-Schaulies, J., Schuster, A., Bayer, M., Pavlovic, J. & Ter Meulen, V. (1994). Cell-type specific MxA-mediated inhibition of measles virus transcription in human brain cells. *J. Virol., 68,* 6910-6917.

Schnorr, J. J., Scheneider-Schaulies, S., Simon-Jodicke, A., Pavlovic, J., Horisberr, M. A. & Ter Muelen, V. (1993). MxA-dependent inhibition of measles virus glycoprotein synthesis in stably human monocytic cell line. *J. Virol., 67,* 4760-4768.

Schumacher, B., Bernasconi, D., Schultz, U. & Staeheli, P. (1994). The chicken Mx promoter contains an ISRE motif and confers interferon inducibility to a reporter gene in chick and monkey cells. *Virology, 203,* 144-148.

Silverman, R. H. (1994). Fascination with 2-5A-dependent RNase: A unique enzyme that functions in interferon action. *J. Interf. Res., 14,* 101-104.

Simons, F. H., Broers, F. J., Van Venrooij, W. J. & Pruijn, G. J. (1996). Characterization of cis-acting signals for nuclear import and retention of the La (SS-B) autoantigen. *Exp. Cell Res., 224,* 224-236.

Sotelo, C. G., Calo-Mata, P., Chapela, M. J., Perez-Martin, R. I., Rehbein, H., Hold, G. L., Russell, V. J., Pryde, S., Quinteiro, J., Izquierdo, M., Rey-Mendez, M., Rosa, C. & Santos, A. T. (2001). Identification of Flatfish (*Pleuronectiformes*) Species Using DNA-Based Techniques. *J. Agri. Food Chem., 49*, 4562-69.

Staeheli, P. & Haller, O. (1987). Interferon-induced Mx protein: a mediator of cellular resistance to influenza virus. *Interferon, 8,* 1-23.

Staeheli, P., Haller, O., Boll, W., Lindenmann, J. & Weissmann, C. (1986). Mx protein: constitutive expression in 3T3 cells transformed with cloned Mx cDNA confers selective resistance to influenza virus. *Cell, 44,* 147-158.

Staeheli, P., Yu, Y., Grob, R. & Haller, O. (1989). A double-stranded RNA-inducible fish gene homologous to the murine influenza virus resistance gene Mx. *Mol. Cell. Biol., 9,* 3117-3121.

Stranden, A. M., Staeheli, P. & Pavlovic, J. (1993). Function of the mouse Mx1 protein is inhibited by overexpressing of the PB2 protein of influenza virus. *Virology, 197,* 642-651.

Sun, B. J. & Nie, P. (2004). Molecular cloning of the viperin gene and its promoter region from the mandarin fish *Siniperca chuatsi*. *Vet. Immunol. Immunopathol., 101,* 161-170.

Suzuki, K. (2006). Detection of DNAs homologous to betanodavirus genome RNAs in barfin flounder *Verasper moseri* and Japanese flounder *Paralichthys olivaceus*. *Dis. Aquat. Org., 72,* 225-239.

Tafalla, C., Aranguren, R., Secombes, C. J., Figueras, A. & Novoa, B. (2004). Cloning and analysis of expression of a gilthead sea bream (*Sparus aurata*) Mx cDNA. *Fish Shellfish Immunol., 16,* 11-24.

Takeuchi, O. & Akira, S. (2007). Recognition of viruses by innate immunity. *Immunol. Rev., 220,* 214-224.

Tanako, R., Nishizawa, M., Arimoto, M. & Muroga, K. (2000). Isolation of viral haemorrhagic septicaemia virus (VHSV) from wild Japanese flounder, *Paralichthys olivaceus*. *Bull. Eur. Ass. Fish Pathol., 20,* 186-192.

Tazi-Ahnini, R., di Giovine, F. S., McDonagh, A. J. G., Messenger, A. G., Amadou, C., Cox, A., Duff, G. W. & Cork, M. J. (2000). Structure and polymorphism of the human gene for the interferon-induced p78 protein (MX1): evidence of association with alopecia areata in the Down syndrome region. *Hum. Genet., 106*, 639-645.

Thomas, V., Palm, M., Broers, A. D., Zezafoun, H. & Desmecht, D. J. M. (2006). Genomic structure, promoter analysis, and expression of the porcine (*Sus scrofa*) Mx1 gene. *Immunogenetics, 58*, 383-389.

Tinti, F., Colombari, A., Vallisneri, M., Piccinetti, C. & Stagni, A. M. (1999). Comparative analysis of a mitochondrial DNA control region fragment amplified from three Adriatic flatfish species and molecular phylogenesis of *Pleuronectiformes*. *Mar. Biotechnol., 1,* 20-24.

Trobridge, G. D. & Leong, J. A. (1995). Characterization of a rainbow trout Mx gene. *J. Interf. Cytok. Res., 15,* 691-702.

Trobridge, G. D., Chiou, P. P. & Leong J. A. C. (1997). Cloning of the rainbow trout (*Oncorhynchus mykiss*) Mx2 and Mx3 cDNAs and characterization of trout Mx protein expression in salmon cells. *J. Virol., 71,* 5304-5311.

Venkatesh, B. (2003). *Fugu: The Pufferfish Model Genome. Nature Encyclopedia of the Human Genome.* Cooper DN (ed.) *Vol 2*, pp 535-539. London: Nature Publishing Group.

Whyte, S. K. (2007). The innate immune response of finfish-A review of current knowledge. *Fish Shellfish Immunol., 23,* 1127-1151.

Williams, B. R. (1999). PKR: A sentinel kinase for cellular stress. *Oncogene, 18,* 6112-6120.

Wu, Y. C. & Chi, S. C. (2007). Cloning and analysis of antiviral activity of a barramundi (*Lates calcarifer*) Mx gene. *Fish Shellfish Immunol., 23,* 97-108.

Yabu, T., Hirose, H., Hirono, I., Katagiri, T., Aoki, Y. & Yamamoto, E. (1998). Molecular cloning of a novel interferon regulatory factor in Japanese flounder *Paralichthys olivaceus. Mol. Mar. Biol. Biotechnol., 7,* 138-144.

Yap, W. H., Tay, A., Brenner, S. & Venkatesh, B. (2003). Molecular cloning of the pufferfish (*Takifugu rubripes*) Mx gene and functional characterization of its promoter. *Immunogenetics, 54,* 705-713.

Yasuike, M., Kondo, H., Hirono, I. & Aoki, T. (2007). Difference in Japanese flounder, *Paralichthys olivaceus* gene expression profile following hirame rhabdovirus (HIRRV) G and N protein DNA vaccination. *Fish Shellfish Immunol., 23,* 531-541.

Zhang, D. E., Hetherington, C. J., Tan, S., Dziennis, S. E., Gonzalez, D. A., Chen, H. M. & Tenen, D. G. (1994). Sp1 is a critical factor for the monocytic specific expression of human CD14. *J. Biol. Chem., 269,* 11425-11434.

Zhang, Y., Li, Q. & Gui, J. (2004). Differential expression of two *Carassius auratus* Mx genes in cultured CAB cells induced by grass carp hemorrhage virus and interferon. *Immunogenetics, 56,* 68-75.

Zhang, Y., Jiang, J., Chen, Y., Zhu, R., Shi, Y., Zhang, Q. & Gui, J. (2007a). The innate immune response to grass carp hemorrhagic virus (GCHV) in cultured *Carassius auratus* blastulae (CAB) cells. *Dev. Comp. Immunol., 31,* 232-243.

Zhang, Y., Wang, Y. & Gui, J. (2007b). Identification and characterization of two homologues of interferon-stimulated gene ISG15 in crucian carp. *Fish Shellfish Immunol., 23,* 52-61.

Zhou, A., Paranjape, J. M., Der, S. D., Williams, B. R. G. & Silverman, R. H. (1999). Interferon action in triply deficient mice reveals the existence of alternative antiviral pathways. *Virology, 258,* 435-440.

Zhu, R., Zhang, Y., Zhang, Q. & Gui, J. (2008). Functional domains and the antiviral effect of the double-stranded RNA-dependent protein kinase PKR from *Paralichthys olivaceus. J. Virol., 82,* 6889-6901.

Zou, J., Yoshiura, Y., Dijkstra, J. M., Sakai, M., Ototake, M. & Secombes, C. (2004). Identification of an interferon gamma homologue in Fugu, *Takifugu rubripes. Fish Shellfish Immunol., 17,* 403-409.

Zou, J., Carrington, A., Collet, B., Dijkstra, J. M., Yoshiura, Y., Bols, N. & Secombes, C. (2005). Identification and bioactivities of IFN-γ in rainbow trout *Oncorhynchus mykiss*: The first Th1-type cytokine characterized functionally in fish. *J. Immunol., 175,* 2484-2494.

Zurcher, T., Pavlovic, J. & Staeheli, P. (1992). Mechanism of human MxA protein action: Variants with changed antiviral properties. *EMBO J., 11*, 1657-61.

In: Veterinary Immunology and Immunopathology
Editors: L. Nermann, S. Meier
ISBN: 978-1-60876-342-9

Chapter 4

ADVANCES IN DNA VACCINATION: DESIGN, IMMUNOLOGY AND MANUFACTURING

Gabriel Amaro Monteiro[1*], Marcelo Sousa Silva[2], Ana Margarida Henriques[3], Joana Abreu Carvalho[1], Jorge Atouguia[2], Miguel Fevereiro[3] and Duarte Miguel F. Prazeres[1]

[1]IBB-Institute for Biotechnology and Bioengineering, Centre for Biological and Chemical Engineering, Instituto Superior Técnico. Lisboa, Portugal
[2]Centre for Malaria and Tropical Disease, Instituto de Higiene e Medicina Tropical, Lisboa, Portugal
[3]Laboratório Nacional de Investigação Veterinária, Lisboa, Portugal

ABSTRACT

DNA vaccination is a promising immunology and biotechnology tool that effectively triggers the activation of a diversity of immune responses. DNA immunization has been shown to be an efficient way of generating antibodies, activating T helper cells and cytotoxic T lymphocytes, conferring protection against a wide range of infectious and non-infectious diseases. When compared with traditional vaccines, DNA vaccine technology is associated with simple and cost effective production protocols. The increase of the immunogenicity and potency of DNA vaccines is a necessary step for their successful application in larger animals and humans. A new generation of vaccines is being tested in ongoing clinical trials, using more efficient delivery devices and formulations. These efforts have resulted in the licensing of four DNA products for veterinary diseases which constitutes a huge step forward for the establishment of this immunization concept. In this chapter the molecular and cellular mechanisms of DNA vaccines will be addressed, as well as the optimization strategies that are currently pursued to harness the therapeutic and commercial advantages of DNA vaccination.

[*] Corresponding author: Instituto Superior Técnico. Av. Rovisco Pais. 1049-001 Lisboa, Portugal, Phone: +351 - 218 419 195, Fax: +351 - 218 419 062, gabmonteiro@ist.utl.pt

Introduction

DNA vaccination is an efficient immunization technique which is based on the *in vivo* expression of antigens. Successful induction of both humoral and cellular responses, as well as protective immunity against numerous infectious and non-infectious diseases is obtained following DNA immunization. In 1990 Wolff and colleagues described for the first time the detection of protein expression in mice muscle cells following the injection of plasmid DNA (pDNA), without the use of any delivery system or formulation [1]. Subsequently, Tang and colleagues reported a new technique for the *in vivo* delivery of pDNA [2] using a device, named gene gun, that propels pDNA-coated gold beads directly into the skin, which results in the production of antibodies against the encoded reporter gene. Shortly after these first experiments demonstrating the ability of plasmid DNA to elicit the action of the immune system, Robinson and colleagues [3] reported that the use of a DNA vector encoding a replication incompetent influenza hemagglutinin protein conferred protection against a lethal challenge of influenza virus. This study was the first to use an antigen-expressing plasmid as a vaccine.

Much progress has been accomplished since the discovery that plasmid DNA could be used as a way to elicit an immune response. Even though DNA vaccination proved to be a safe platform, it failed to elicit strong immune responses when tested in nonhuman primates and humans. Therefore one of the major challenges has been to increase its immunogenicity and potency in order to establish DNA vaccination as a valid approach to immunize larger animals and humans. Several optimization strategies have been investigated including the optimization of the vector (promoter and poly-A sequence optimization and addition of transcriptional enhancer elements), improvement of protein expression (e.g. codon optimization), use of adjuvants (e.g. liposomes, cytokines) and more effective delivery methods and devices. More efficient immunization protocols such as prime-boost strategies have also been tested.

The concept of DNA vaccination has been tested in an increasing number of clinical trials and successfully applied to the prevention and treatment of several diseases. Four DNA products have recently been licensed for veterinary use, underscoring the potential medical and commercial benefits of this therapeutic approach. An overview of preclinical studies and clinical trials using DNA vaccination will be made in next section. The basic mechanisms of DNA vaccination and the guidelines for DNA production and safety concerns will also be addressed. Finally, the advantages of this approach when compared to traditional vaccines will be reviewed, as well as the optimization strategies that are being pursued to design a new generation of improved DNA vaccines.

Proof of Concept: Preclinical Studies

Several preclinical studies using pDNA expressing antigens associated with a variety of infectious and non-infectious diseases have demonstrated that DNA vaccination is capable of inducing strong cellular and humoral immune responses (Table 1). Additionally, protective immunity has been observed following plasmid immunization in several animal models, including non-human primates. Attempts have been made to use the DNA vaccine approach in the management of diseases like cancer, allergies, autoimmune disorders and several infectious diseases.

Table 1. DNA vaccine prototypes developed against different diseases models

Disease Model	Antigenic candidates	Immune Response Stimulated	Refs.
Alzheimer	Aβ	Reduction of amyloid burden (anti- Aβ antibodies)	[4]
Allergic diseases	Der p 2	Suppression of allergic airway inflammation (reduction of IgE levels and inhibition of the Th2 response)	[5]
Autoimmune diseases	MOG	Cellular response (increase of IFN-β and down regulation of Il-17) Suppression of EAE	[6]
Cancer	HPV18 E6 and E7	Cellular response (CD8 T cells)	[7]
Coronavirus (SARS)	3a gene	Humoral and cellular responses (increase IFN-γ and IL2)	[8]
HIV	CON-S	Cellular response	[9]
Influenza	Has	Humoral response (neutralising antibodies) Broad protection against heterologous strains	[10]
Leishmaniosis	Antigen cocktail (LACKp24, TSA, LmSTI1 and CPa)	Protective immunity Cellular response (Th1 cells)	[11]
Malaria	LSA-3	Protective immunity Cellular response (increase of IFN-γ regulated by CD4 T cells)	[12]
Tuberculosis	Mycobacterial Ag85 antigen	Treatment of MDR-TB Humoral and cellular responses	[13]
Trypanosoma cruzi	ASP-2 and TS	Protective immunity Cellular response (increase of IFN-γ)	[14]
Maedi-Visna	p16 and p25	Humoral response	[15]

Abbreviations: Aβ (Amyloid β); Der p 2 (Dermatophagoides pteronyssinus group 2); MOG (Myelin oligodendrocyte glycoprotein); EAE (Experimental Autoimmune Encephalomyelitis); HPV (Human papilloma-virus); SARS (Severe Acute Respiratory Syndrome); 3a gene (Open-reading frame found in the SARS coronavirus); HIV (Human Immunodeficient Virus); CON-S HIV (Consensus envelope sequence); HAs (influenza A hemagglutinin); LACKp24, TSA, LmSTI1 and CPa (*Leishmania major* antigens); LSA-3 (liver stage antigen 3); MDR-TB (multi-drug resistant tuberculosis) ASP-2 (Amastigote Surface Protein-2); TS (Trans-sialidase).

From the Bench to the Field

In order for DNA vaccines to undergo clinical trials using human volunteers, consistent preclinical data using appropriate animal models must first be collected. Strong scientific data on the vaccine safety and efficacy (i.e. immunogenicity) must therefore be presented. The number of clinical trials that currently use plasmid-based pharmaceuticals represents 18% of the total number of trials which use gene delivery vectors (Figure 1). This figure is demonstrative of the potential and interest in the use of plasmid-based medicines, and namely of DNA vaccines.

DNA vaccines for the treatment of cancer [16-18] and autoimmune diseases (e.g. multiple sclerosis [19]), and for the prophylaxis of malaria [20] and of several viral diseases, including HIV [21, 22], hepatitis B [23], hepatitis C [24], West Nile [25], SARS [26], Ebola [27], Herpes [28], have been evaluated at the different phases of clinical trials.

The first licensed DNA vaccine was developed for West Nile virus prevention in horses [29], while the therapeutic programme using a DNA vaccine against canine malignant melanoma resulted in the reduction of the tumour and prolongation of survival [30]. This was the first licensed cellular immune therapy for cancer. Protective immunity associated with the production of neutralising antibodies against an infectious hematopoietic necrosis virus (IHNV) was observed following DNA vaccination [31]. This vaccine was approved for the prevention of this disease in farm-raised salmons. Finally, the growth hormone-releasing hormone (GHRH) for the prevention of foetal loss in swine was the first approved gene therapy product to be administered by electroporation [32]. The licensing of these four products, all dedicated to veterinary medicine, constitutes an important breakthrough in the field of DNA vaccination that opens promising perspectives for their use in humans.

Figure 1 was re-introduced as the borders were missing

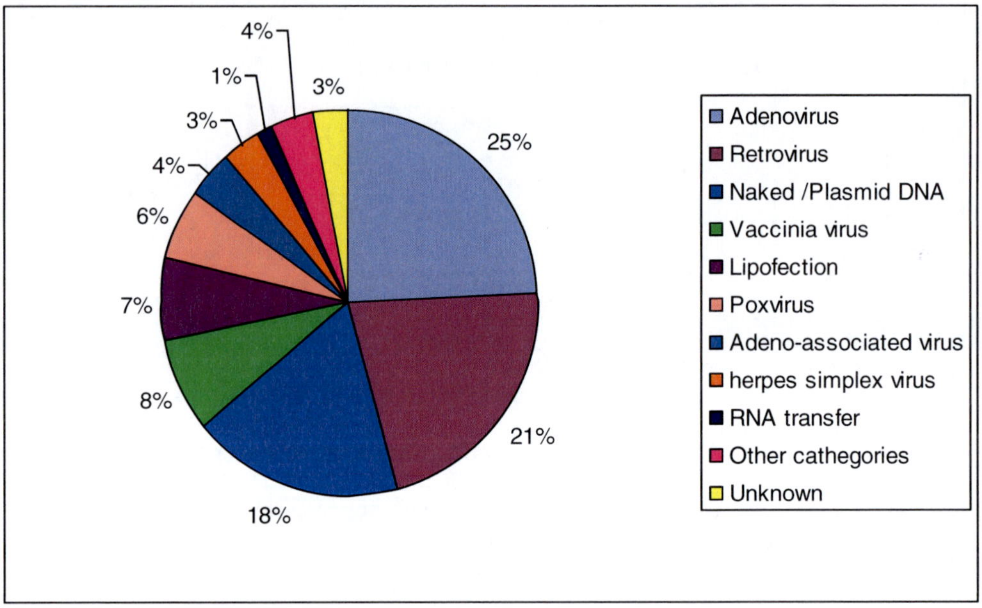

Figure 1. Percentage of clinical trials organized by vector used in the immunization protocol (last update made in September 2008; data available at http://www.wiley.co.uk/genetherapy/clinical/)

Mechanism of Action

Although the concept underlying DNA vaccination is simple, the exact mechanisms by which the plasmid-encoded antigen is delivered to the immune system, as well as the nature of the cells that activate the immune system, are not yet fully understood.

Early reports showed that protein expression is achieved following immunization with pDNA encoding a diversity of reporter genes (i.e. luciferase or beta-galactosidase genes) [1] and other proteins such as hepatitis B surface antigen [33], human-alpha-1-antitrypsin [34], and murine erythropoietin [35].

Following intramuscular injection, most of the transfected cells are non-professional APCs (myocytes and keratinocytes) [1]. Ulmer and colleagues [36] demonstrated that direct transfection of professional APCs is not necessary to trigger the induction of a CTL response. Transplanted myoblasts previously transfected with nucleoprotein (NP) into mice led to the production of anti-NP antibodies and to the development of a CTL response, as well as protective immunity against a viral challenge with influenza virus. This result could mean that myocytes or other non-professional presenting cells could directly prime the immune system. However, there are evidences that point to the fact that these cells lack the necessary co-stimulatory molecules required to prime $CD8^+$ T cells, found in APC [37]. Accordingly with this assumption, several reports provided evidences that bone marrow-derived APC cells are the ones responsible for antigen presentation to the immune system [36, 38-40]. Antigen presentation is likely to occur following protein transfer from transfected muscle cells to APCs, through a mechanism called cross-priming. The transfer mechanisms are still not clear, but antigen uptake by APCs could follow necrosis or apoptosis of transfected myocytes [41]. These exogenous antigens can then be presented by the APC to the immune system in the context of the major histocompatibility complex (MHC) class I molecules [42].

On the other hand, antigen processing via MHC II route is better understood. Exogenous proteins, phagocytised by APCs, are degraded in lysosomal/endosomal compartments and assembled with the MHC II molecules in MHC II vesicles. Subsequently they are transported to the cell surface and presented to the immune system for $CD4^+$ T cells activation (reviewed in [43]).

Two subsets of $CD4^+$ T, characterized by different immune functions, can be differentiated. Th1 cells are involved in macrophage activation, while the Th2 population is responsible for the proliferation of B cells and memory B cells. The induction of a Th1 profile is generally detected following DNA vaccination, due to the presence of CpG motifs inserted into the vector backbone. However, this profile can be altered by using different immunization routes. Prevalence of the Th1 response associated with the production of IgG2a antibodies is observed after intramuscular injection of plasmid DNA, while the administration of a secreted antigen by gene gun immunization elicits a Th2 response characterized by the production of IgG1 antibodies [44].

Both arms of the immune response are therefore activated following plasmid immunization. DNA vaccination is also capable of inducing memory T cells, such as conventional vaccines. A schematic representation of these mechanisms is outlined in figure 2.

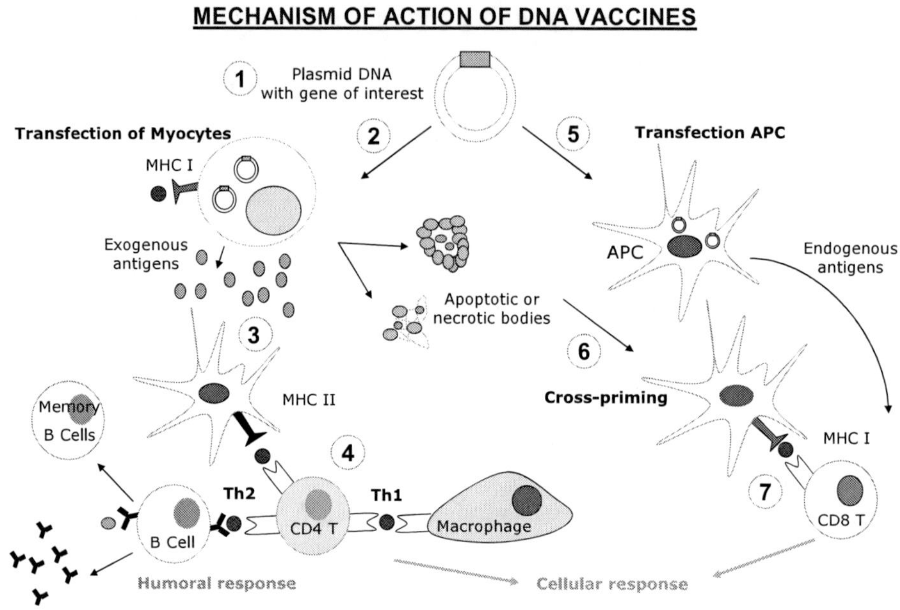

Figure 2. Mechanism of action of DNA vaccines. Following DNA immunization into the muscle, plasmid can be taken up by myocytes or directly transfect antigen–presenting cells. Exogenous antigens, secreted by transfect myocytes can be phagocytised by APC and processed in the context of MHC II pathway and presented to $CD4^+$ T cells. In turn, these cells can activate macrophages or B cells to produce antibodies and differentiate into memory B cells. Endogenous antigens, produced in transfected APCs, are loaded to the MHC I pathway and presented to $CD8^+$ T cells. These cells can also be activated by APC following phagocytosis of necrotic or apoptotic of transfected myocytes, through a process named cross-priming.

Optimized Approaches to Vaccine Design

Optimized designs have been considered to improve the potency of DNA vaccines, including changes in the vector in order to obtain higher yields of protein expression, the use of adjuvants or more efficient delivery methods. Plasmid design can be improved by the insertion of leader sequences that promote antigen targeting to the immune system, in order to improve antigen presentation and processing. DNA vaccination could be used to prime the immune system and combined with the subsequent administration of the recombinant antigen or of a second vector encoding the same antigen as a boost of the immune response (prime-boost strategy). Immunomodulators such as cytokines or chemokines allow the manipulation of the immune response, namely the balance between the Th1 and Th2 profile. These strategies, summarized in figure 3, will be discussed in detail.

Figure 3 was re-introduced as a mistake was detected (lysosome instead of liposome)

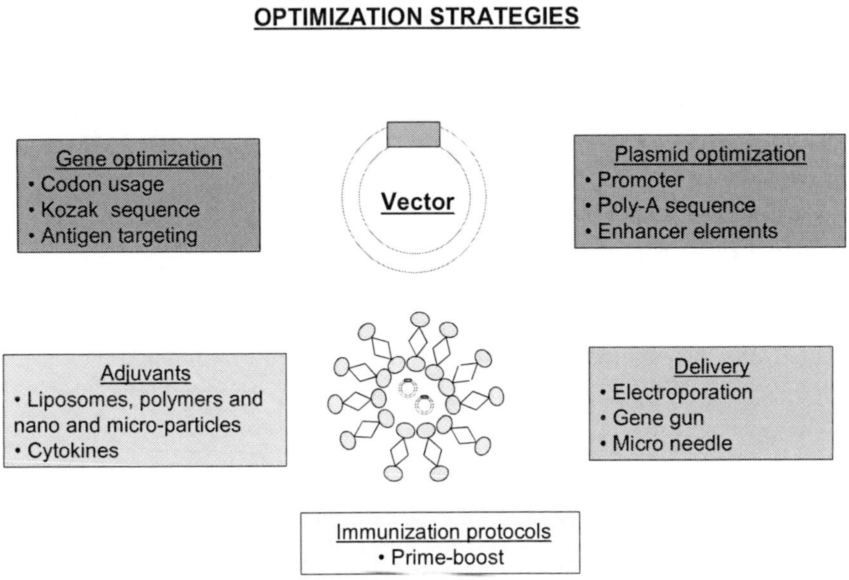

Figure 3. Approaches for the optimization of DNA vaccines

Plasmid Optimization

As previously referred, a DNA vaccine encodes the protein of interest whose gene is cloned into a pDNA vector. The characteristics of this vector may strongly influence the efficiency of the end product. Some features are typically part of a pDNA vector, including: (1) a bacterial origin of replication for growth in prokaryotes, (2) a prokaryotic selectable marker for selection of transformed bacteria, (3) a eukaryotic promoter for expression of the antigen encoding gene in mammalian cells, (4) a multiple cloning site for gene cloning, (5) a transcription termination/polyadenylation element for efficient termination of transcription and polyadenylation and finally (6) immunostimulatory sequences for stimulation of the immune system. These features can be grouped into the two major units that are part of a typical DNA vaccine. The first unit controls the plasmid production and purification processes and includes the origin of replication and the gene used as the selection marker. The second unit controls transcription and includes the elements involved in promoting antigen synthesis, namely the eukaryotic promoter, the antigen-encoding gene and the termination/polyadenylation sequence. The vector backbone also includes immuno-stimulatory sequences (ISS) for induction of cytokines that can modulate protective immunity [45]. The most used sequences consist of unmethylated cytosine-phosphate-guanosine (CpG) dinucleotides flanked, typically, by two 5'purine (optimally a GpA) and two 3' pyrimidine residues (optimally a TpC or TpT) [46]. CpG motifs elicit innate immune responses characterized by the production of IL-6, IL-12, TNF-α, IFN-γ and IFN-α and directly activate B cells to proliferate or secrete antibodies. Natural killer (NK) cells and T cells are indirectly activated by cytokines induced [36]. These motifs therefore act as an endogenous adjuvant that is antigen independent.

The vector design can be improved by the optimization of the transcriptional elements to increase gene transcription and expression, including the choice of a strong promoter that provide high levels of gene expression in a wide range of host cells, such as the SV40 [47] or the human cytomegalovirus (CMV) [48]. More recently the choice of a tissue-specific promoter has been considered to avoid constitutive expression in tissues that are not the desired therapeutic target [49]. The inclusion of a termination site, also named poly (A), is necessary for an efficient transcription and transport of the mRNA from the nucleus. The bovine growth hormone sequence [50] is one of the most used in vaccine development. The impact of changes in the polyadenylation sequence in transgene expression has been investigated to obtain improved vectors [51]. Finally, additional transcriptional transactivators from viral [52] or non viral origin [53] can be inserted into the vector to increase promoter activity.

Gene optimization

The most obvious way of improving gene expression is to alter the sequence of the gene of interest [54-56]. Each organism uses preferential codons assuring an efficient protein synthesis, a phenomenon called codon bias. Codon usage seems to be an evolutionary feature, since taxonomically-related organisms use similar codons, whereas taxonomically-distant organisms are likely to use a different system. Due to limitations in the amount of available tRNA, differences between the codons used in the sequence of the transgene and the host's codon usage may affect its expression. Therefore, translation efficiency may be enhanced by changing the transgene sequence and replacing the wild-type codons by the ones preferentially used by the host.

Another way of achieving higher gene expression yields is through the addition of Kozak sequences. These are regions flanking the AUG initiator codon that have a fundamental role in mRNA recognition by eukaryotic ribosomes, and are the starting site where translation is initiated. Since prokaryotic genes and some eukaryotic genes do not have these sequences, its inclusion might contribute to higher protein expression levels in mammalian cells [57].

Once the antigen is expressed, it has to be presented to the immune system. Several DNA vaccines developed so far failed to perform effectively this fundamental step. To overcome this limitation, several targeting sequences are being used to promote the transport of the expressed antigen to cellular compartments that are a part of the MCH I and MHC II pathways (Figures 4 and 5).

The MHC II antigen processing and presentation pathway conventionally operates in professional APCs. Exogenous antigen are engulfed by APCs and degraded in the lysosomal/endosomal vesicles. The resulting antigenic peptides are then transported to the ER and complexed with the MHC II molecules and taken to the cell surface for presentation to $CD4^+$ T cells (Figure 4) [58]. Since DNA vaccines encode proteins which are expressed intracellularly, the resulting epitopes will not have access to this important pathway. The targeting of antigens to the lysosomes [58, 59] or endosomes [60] might help to overcome this limitation and increase the cellular response and antibody production following DNA immunization.

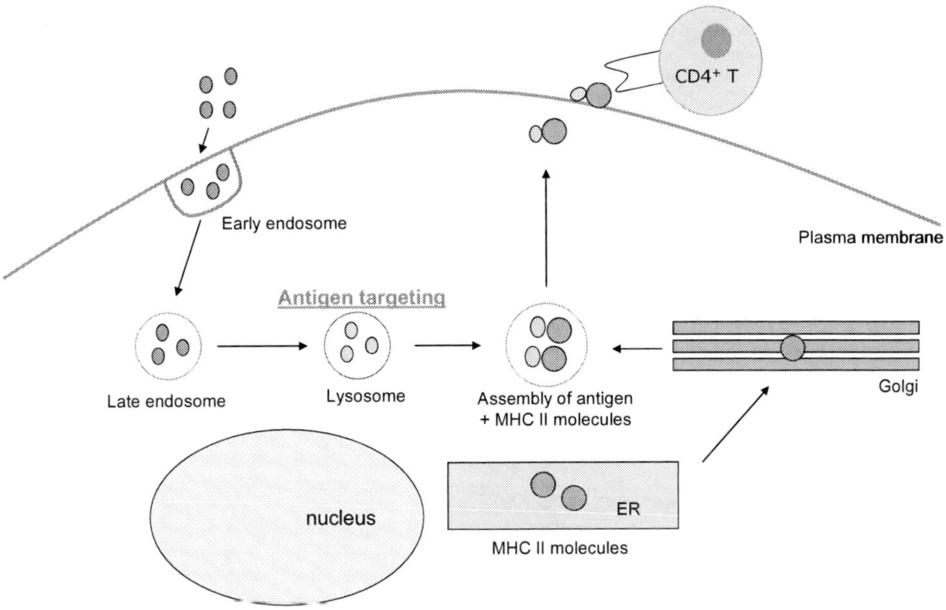

Figure 4. Schematic representation of the MHC II antigen processing pathway. Exogenous antigens are phagocytised by APC and degraded in endosomal/lysosomal compartments. Once complexed with MHC II molecules in MHC II vesicles the antigens are transported to the cell surface for $CD4^+$ T cell presentation. Targeting the antigens to the lysosomes would allow the proteins encoded by the plasmid DNA to gain access to this pathway.

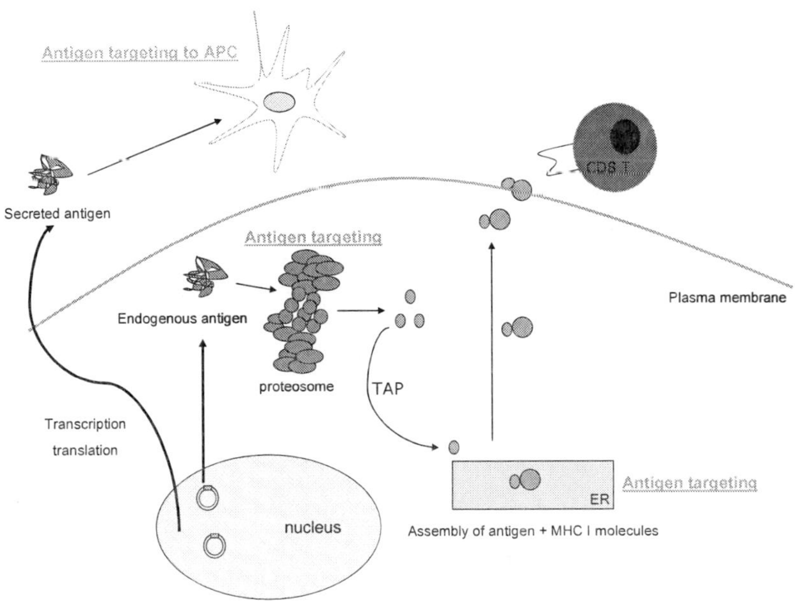

Figure 5. MHC I antigen processing and presentation pathway. Endogenous proteins are proteolytically degraded in proteosomes and transported to the endoplasmic reticulum to form a complex with MHC I molecules. Once they are transported to the cell surface the antigen is presented to $CD8^+$T cells.

Promoting protein transport to the proteosomes and to the endoplasmic reticulum results in a more effective presentation to the immune system. (TAP – Transporter Associated Protein).

Promoting antigen secretion had also proved to be a successful way to improve the DNA vaccine immunogenicity. Following intramuscular injection most of the transfected cells are non-professional presenting cells that cannot trigger the immune response alone. If these cells were to act as factories, producing a secreted protein, the antigen would be available in the extracellular space, increasing the probabilities of an encounter with an APC (Figure 5). Therefore, this strategy also allows the antigen to be processed through the MHC II pathway, with improved results regarding the potency of the immune response detected after immunization [61-63].

Maximizing antigen loading to the MHC I pathway has also been used in DNA vaccine development to increase the $CD8^+$ T-cell response, including the use of ubiquitin-fused proteins to promote its transport to the proteasomes where degradation occurs [64]. The resulting epitopes are then transported to the ER where they are assembled with the MHC I molecules and transported to the cell surface to promote $CD8^+$ T cell activation. Directing the antigen to the ER also seems to result in the increase of the cytotoxic response and induction of protective immunity [65, 66].

A final approach to improve the immunological potency of DNA vaccines is to promote antigen targeting to the antigen-presenting cells [67-69], i.e., the cells responsible for the generation of the immune response following DNA vaccination.

Adjuvants

A vaccine adjuvant is used to enhance, accelerate or prolong the immune response. The expected outcome of their use includes the increase of the immunogenicity of the antigen or the change of the nature of the immune response [70].

Plasmid DNA is rapidly degraded by nucleases present in the plasma or trapped in lysosomes and endosomes during trafficking to the nucleus. Therefore, adjuvants such as liposomes, polymers and nano and micro particles were designed to promote the increase of transfection efficiencies. DNA encapsulation using liposomes is an effective way of protecting DNA from nuclease degradation. Liposomes form an aqueous compartment encapsulated by a layer of organic molecules with a charged DNA-binding head group and an aliphatic tail. By varying the electric charge of the lipids used to construct the liposomes both cationic and anionic liposomes can be prepared [71]. Examples of liposome-based vaccines have proved to elicit stronger immune responses [72] and protective immunity [73].

Additionally, several cationic polymers have been developed for gene delivery, such as polyethyleneimine (PEI) [74]. Cationic polymers, once associated with DNA, make its surface charge less negative, which facilitates plasmid DNA transport across membranes. While promoting plasmid transfection, the complexes formed also protect DNA from degradation and facilitate its release from endosomes [75].

A third approach for gene delivery is the use of nano or micro particles that encapsulate or allow DNA to adsorb onto their surfaces. These particles are effective at attracting APC to the site of injection and facilitate plasmid DNA uptake, while protecting DNA from degradation [76]. Enhancement of DNA vaccine immunogenicity has been observed using cationic particles made of polymers such as poly(lactide-co-glycolide) (PLG) [77], poly-L-lysine (PLL) or stabilized plasmid lipid particles [78].

Other chemical adjuvants, such as aluminum salts, traditionally used in conventional vaccines, have also been applied to DNA vaccination [79].

Immunologic adjuvants, such as cytokines, when co-administrated with DNA result in a more potent immune response [80, 81] and allows to manipulate of the immune system towards the desired response [82]. Recombinant cytokines can be administered alongside with the DNA vaccine, or alternatively, their coding genes can be inserted into the DNA vaccine plasmid and co-expressed with the therapeutic gene.

The European Agency for the Evaluation of Medical Products (EMEA) has issued a guideline for the quality, non-clinical and clinical topics to be considered when using a new or established adjuvant for vaccine development. This guideline also includes a list of some of the approved adjuvants, namely mineral salts, oil emulsions and surfactant based formulations, particulate adjuvants, microbial derivatives (natural or synthetics), endogenous human immunomodulators and inert vehicles [83].

Prime Boost Strategy

As mentioned before, immunization with DNA alone might not result in a strong immune response, especially in the case of larger animals or humans. Special immunization protocols have been designed in order to overcome this limitation. For example, prime-boost strategies use in a first instance DNA to prime the immune system for the encoded antigen and a then a second vector (e.g. a recombinant viral vector) expressing the same antigen to boost the immunological reaction. Although the immunogenicity of antigens delivered by viral vectors is much higher when compared with antigens delivered by plasmid DNA, when used alone much of the immunogenicity of the recombinant virus is targeted at the vector components. In the prime boost strategy, the DNA vaccine induces the production of T-cells specific to the antigen of interest. Following the boost, a second contact with the antigen promotes the proliferation of these cells [84]. The most widely used virus in this strategy is the modified vaccinia Ankara [85], but others such as recombinant adenovirus [86] or recombinant poxvirus [87] have been tested. Generically speaking, prime-boost strategy elicits the production of high antibodies titters and an improved cellular response, associated with protective immunity [88].

Boosts of recombinant antigens can also be used following DNA priming. With this approach, stronger cellular and humoral immune responses are obtained when compared with the ones induced by vaccination with DNA or with the antigenic protein alone [89, 90].

Improved Delivery Systems

Improved delivery devices have been licensed for human use, such as the Biojector 2000 needle free system for intramuscular injection [25] or the PowerJet XR1 [23] to propel DNA into epithelial cells. Administration of DNA by gene gun is a very efficient delivery method that uses a regulated burst of helium gas to propel pDNA-coated gold beads into the epithelial cells of the skin. Plasmid-coated gold beads penetrate cells of the epidermis and consequently deposit plasmid DNA intracellularly. Cells are directly transfected and therefore a lower amount of DNA is required to induce a strong immune response compared with other delivery methods. Although 0.1-1.0µg of plasmid DNA is usually necessary to induce a strong immune response [91], doses as little as 0.4ng have been shown to generate a measurable immune response using the gene gun delivery [92].

Potent immune responses were also observed following electroporation [81]. This method uses an electric charge, applied after injection of naked pDNA, to facilitate the entrance of macromolecules such as DNA. The technique has shown to substantially increase the efficiency of transfection of cells, leading to increase of vaccine potency in several animal models, including non-human primates [93]. A phase I trial using interleukin 12 plasmid delivered by *in vivo* electroporation for the treatment of metastic melanoma was recently published, reporting this method to be safe, effective and reproducible [81]. Furthermore, one of the licensed DNA products uses this immunization method [32].

Multivalent DNA Vaccines by Bicistronic Plasmid Design

For DNA vaccine formulation, pDNA expression vectors are usually monocistronic (i.e., the transcribed mRNA encodes a single ORF). However, in recent times various regulatory elements that allow the coexpression of antigens and/or antigen-adjuvant became instrumental in the generation of multivalent vaccines. These multivalent vaccines are based on bicistronic vectors encoding two antigens, or an immunomodulator adjuvant (for example cytokines) encoded together with an antigen, on a single pDNA formulation [94].

The term bicistronic pDNA is used to name a eukaryotic expression vector that expresses a single transcriptional unit, in which the first gene is translated in a cap-dependent manner and the second one in an internal ribosome entry site (IRES)-dependent manner, thereby permitting simultaneous expression of two proteins. Faithful evidences of the usefulness of bicistronic pDNA to improve the immunogenicity of DNA vaccines came from two line of experimental approaches. In the first one, the pDNA is designed to generate a multivalent vaccine in which two viral or parasite antigens are encoded in a single transcriptional unit [95, 96]. The second strategy is aimed to enhance the effectiveness and antigen-specific immune response of vaccination, through the use of various cytokines and co-stimulatory genes as molecular adjuvants [97-99]. The basis of these approaches has in common the use of bicistronic pDNA and co-expression of two antigenic genes is required in the multivalent vaccine approach, whereas simultaneous expression of an antigen with an adjuvant gene is needed in the immunoenhancer approach.

Limitations and Safety Concerns of DNA Vaccines

Although DNA vaccination can elicit strong immune responses, the results obtained in non-human primates and humans revealed weak immune responses when compared to the ones observed in other animal models. Vaccination with DNA seems less immunogenic when compared with vaccination with recombinant viral vectors, whereas the use of recombinant protein-based vaccines appears to be more effective at the induction of antibodies responses [100].

Additionally, four main concerns were pointed out when DNA vaccines appeared as an alternative method for immunization: 1) integration of genes into the host's genome; 2) induction of tolerance 3) autoimmunity and 4) development of antibiotic resistance. Pre-clinical studies and clinical trials have dismissed these safety risks and presently DNA vaccines are considered to be a safe and reliable platform. These potential concerns regarding the use of DNA vaccines are listed and debated in table 2.

Table 2. Potential concerns associated with DNA vaccination

Low immunogenicity	Optimization strategies (vector and gene optimization, use of adjuvants and improved immunization protocols).
Integration	Probability of integration into host genome is lower than the risk of spontaneous mutation.
Tolerance	Immunization protocols using amounts as low as 10μg of plasmid DNA trigger the immune system and do not result in the induction of tolerance. It is possible to induce the immune system of neonates, even though tolerance following DNA vaccination has been observed.
Autoimmunity	Phenomenon is unlikely to occur. Anti-DNA antibodies are not induced by plasmid DNA.
Antibiotic resistance	Use of antibiotics as selection markers that are not used to treat human infections. Methods of plasmid selection antibiotic-free have been developed.

Integration may occur randomly or as a result of homologous recombination between sequences present in both plasmid and host. This might result in the occurrence of insertional mutagenesis that, in turn, could be responsible for the activation of oncogenes or the inactivation of tumour suppressor genes [101]. Studies using PCR methodologies revealed that no integration could be observed for the detection limit of 1-7.5 plasmid copies per 150,000 nuclei, below the probability of occurrence of a spontaneous mutation [101]. Presently, FDA recommends that integration studies should be performed only if the plasmid persists above 30,000 copies per μg host DNA [102]. Nevertheless, plasmids are constructed with only a very limited number of homologous sequences with the human genome.

Also plasmids do not have an operational origin of replication in eukaryotes which ensures that plasmid replication does not occur. Furthermore, most of the transfected cells such as myocytes do not undergo division, which prevents plasmid propagation.

A second issue to be addressed is the possibility of induction of tolerance following immunization with plasmid DNA. Protein expression is detected for several weeks or even months after DNA immunization, which could hypothetically lead to immunologic un-responsiveness due to constant exposure of the immune system to the antigen, especially if present in small amounts. Pre-clinical studies listed above clearly demonstrate that DNA vaccination is capable of eliciting the immune system, while protective immunity is also reported. Immunizations with doses as low as 10μg are immunogenic [103] and elicit protective immunity [103].

Newborns present an immature immune system and are considered to have a higher susceptibility to tolerance and potential negative interference of maternal antibodies [104].

Even though tolerance has been reported in neonates [105], other studies demonstrated that DNA immunization lead to the induction of cytotoxic T cell response [106]. The presence of CpG motifs and their ability of activating immature antigen-presenting cells appear to be critical for the neonatal immunogenicity of DNA vaccines. Further, the endogenous production of the antigen might explain the lack of inhibition by maternal antibodies of cellular responses. These properties of DNA vaccination make this immunization methodology appropriate for neonates (reviewed in [104]).

The question of whether auto-immunity could result from exposure to plasmid DNA has also been raised. In DNA vaccines, a nonself molecule of DNA is introduced in the cell, where it will be expressed in a nonself protein. This can trigger an autoimmune response against transfected cells [36] or can lead to the development of antibodies anti-DNA [107]. Moreover, DNA can convert cells in antigen presenting cells (APCs), which become subject of destruction. However, the expression of nonself antigens and hence the destruction of cells of the host occur also in viral and bacterial infections, and so the risk of autoimmunity induced by DNA vaccines is not significant when compared with that induced by natural infections [108]. Indeed, following DNA vaccination the production of anti-DNA antibodies is not detected [103, 109]. Furthermore, no reports exist describing the development of an immune response against cells expressing the plasmid [110]. Since the risk of auto-immunity is an unlikely event, FDA guidelines no longer suggest that pre-clinical studies should be performed specifically to address this safety issue [102]

Finally, there were concerns about the use of antibiotics for plasmid selection. Antibiotics such as penicillin and ampicillin are not recommended due to the risk of causing hypersensitivity reactions in patients. The gene that confers resistance to kanamycin is one of the most used resistance markers in vectors for DNA vaccination, since it does not elicit allergic reactions in humans. Further, this antibiotic is not used to treat human infections. In fact, FDA encourages the choice of selection markers that are not extensively used in therapeutic regimens, in order to avoid unnecessary risk of spread of antibiotic resistance traits to environmental microbes [111]. New alternatives for plasmid selection have been explored such as the use of strains of bacteria that can be selected by repressor titration [112].

DNA Purification Methods and Quality Control

A production and purification method capable of delivering high yields of pharmaceutical-grade plasmid DNA is necessary to respond to the increasing demand of DNA for ongoing clinical trials, where milligrams of DNA might be necessary to elicit a protective immune response. A process for plasmid production and purification based on hydrophobic chromatography was developed, capable of delivering high yields of plasmid with elevated degree of purity [113, 114]. The purification process relies on the low hydrophobic properties of DNA compared to those of characterizing RNA, genomic DNA and other contaminants. Plasmid DNA can be eluted using a phenyl-sepharose matrix buffered with a high salt concentration solution, while the removal of contaminants is achieved by decreasing the ionic strength of the buffer. For *in vivo* applications, DNA must be dissolved in a compatible physiological buffer, so an additional step in the purification process is necessary. The buffer can be changed performing a gel filtration chromatography, again using a phenyl-sepharose matrix. Finally, plasmid DNA should be eluted in a suitable buffer for the desired downstream applications.

Production methods have also to meet the quality standard specifications required by the regulatory agencies [115] and be able to consistently deliver a vaccine free of any contamination and impurities that could potentially be harmful to the target organism. In Europe each country has its own regulatory agencies, whereas in the United States the Food and Drug Administration (FDA) establishes the standards procedures for production and licensing of new vaccines [102, 116, 117]. The World Health Organisation (WHO) [118] and the European Community [119] also publishes guidelines to support the application of new products for license. FDA stipulations for plasmid production necessary for the licensing of a new vaccine are shown on table 3 [102].

Table 3. Quality specifications and required information for the license of a new vaccine [102]

Quality controls	Specification and data required
Product manufacture	• Description of gene (sequence and origin) and cloning protocol • Annotated plasmid map • Genotype of the bacterial cells used for production
Bulk Plasmid Product release testing	
Quality	• >80% supercoiled plasmid content
	• < 1% macromolecules (DNA, RNA and protein)
	• Test for pyrogenic substances
	• < 40U/mg plasmid of bacterial endotoxins and endotoxin contamination
Product identity	• Agarose gel electrophoreses following restriction analysis
Potency	• *In vitro* measurements of transfection efficiencies • Verify transcription and translation of the encoded gene • *In vivo* assays of DNA vaccine immunogenicity • Potency assays that evaluate biological activity • If available data of correlation between the potency assays and the immunogenicity and protection results obtained in clinical trials should be provided
Final Product release	• Test the final product for potency, general safety, sterility, purity, quantity and identity (same methods and specifications for the intermediate products)

Detailed information about the immunogenicity and safety data of the new product is also required before submitting a request to proceed with a clinical trial [102] (Table 4). The product license is dependent of the successful completion of the clinical trials, where the new vaccine has to prove to be efficient, safe, while presenting quality and consistent results [117].

Table 4. Preclinical data requirements to request the permission to conduct a clinical trial [102]

Preclinical immunogenicity and safety	Information required
Immunogenicity	• Assessed in a relevant animal model • Antibodies titers • Cell mediated immune responses
Cytokines	• Studies should be performed in animal models responsive to human cytokines or using the homologous animal gene • Assess the occurrence of immunosuppression, chronic inflammation, autoimmunity and other immunopathologies
Prime/Boost strategies	• Safety and tolerability of the dose, schedule and route of administration of each component
Autoimmunity	• This phenomena is unlikely following DNA vaccination, therefore only the general welfare of the animals should be monitored
Local reactogenicity and systemic toxicity studies	• Studies should include at least one more immunization than planned for clinical trials, using the higher dose • Toxicity studies to potential target organs • Detailed clinical pathology and histopathology assessments
Biodistribution, Persistence and integration analysis	• <100 copies of plasmid per microgram of host DNA detect by Q-PCR using a panel of tissues at several time points post administration • Biodistribution studies are only required for non documented vectors, formulations, methods of delivery and routes of administration • Integration studies are only necessary if the plasmid persists in any tissue > 30,000 copies per μg of host DNA

Advantages of DNA Vaccines

DNA vaccines are capable of eliciting protective immunity in animals against several diseases without some of the disadvantages associated with conventional vaccines. The ideal vaccine candidate should be efficient at eliciting a protective immune response, it has to be safe and preferably associated with a low production cost. Furthermore, it would be advantageous if the vaccine could provide simultaneous protection against different pathogens. A comparison with conventional vaccines is outlined on table 5.

The purification and production methods of DNA vaccines are generic and independent of the encoded antigen, which configures an important advantage compared with conventional vaccines. Attenuated and protein based vaccines are not associated with a single purification method and each case presents itself specific challenges that need to be overcome. Usually this is a time and cost consuming process. On the contrary, the design of a DNA vaccine is straightforward. It starts by the choice of the appropriate antigen(s), whose gene is then amplified by PCR and inserted into a plasmid using its multiple cloning site.

Furthermore, the methods for plasmid DNA purification are reproducible and allow large scale production with high yields, while maintaining the required quality standards.

Plasmids are easy to store and more stable at room temperature than conventional vaccines, therefore there is no need for the use of a cold chain. The commercial benefits of this flexibility are huge, dramatically reducing the costs associated with vaccine distribution. This is especially relevant for developing countries.

DNA vaccines design is simple and associated with a standard production protocol. However, this is not sufficient to establish DNA vaccines as an attractive vaccination protocol since it has also to present therapeutic benefits compared to conventional approaches.

One of the most important therapeutic advantages of DNA vaccination is the fact that both arms of the immune response are activated. Following DNA vaccination, the antigen is synthesized and processed within the cell and then can be presented to the immune system not only by MHC class II pathway, as occurs with the majority of conventional vaccines, but also by MHC class I molecules to the development of cellular immunity, as explained in detailed previously. Attenuated viruses based vaccines have also this ability, but in this case there is the risk of reversion to the virulent form.

On the other hand, protein vaccines are only able to present the antigen to the immune system in the context of the MHC II pathway, eliciting the production of antibodies. However, recombinant proteins may present a different conformation compared to the wild-type, if they are produced in prokaryotes. Also the folded protein structure may be disrupted during the protein purification process. Therefore, the administration of this misshapen protein leads to the production of antibodies that fail to recognise the native form of the protein, and so the vaccine is not effective in inducing a correct immune response and the vaccinated organism is not protected against the pathogen. In the case of DNA vaccines, the encoded protein is produced by the organism and it is usually processed with the correct folded structure, leading to the production of antibodies that specifically recognise the wild-type protein.

This advantage is especially relevant for the development of a vaccine against a viral disease, in which viral proteins are expressed by eukaryotic cells of the host. DNA vaccines mimics a viral infection, since the protein is also expressed by the host, leading to proper folding, posttranslational modifications and intracellular transport, as in a natural infection. On the other hand, DNA vaccines developed for diseases caused by prokaryotes will encode a protein with a different conformation compared to the native form and might not be as be effective [110].

So far, clinical trials have demonstrated that DNA vaccines are safe for human use, with no indications of adverse effects. Plasmid vaccines encode an antigen or, in some cases, the epitope of the protein. Therefore, unlike live or attenuated vaccines, there is no risk of infection or reversion to the virulent form.

The design of plasmid vectors is made according to regulations established for gene therapy, in order to minimize safety risks. For example, the origin of replication should not be compatible with the mammalian host in order to preclude plasmid replication in the vaccinated individual. Further, homologous regions with the human genome are deleted to limit the probability of integration into the host genome. Additionally, the choice of kanamycin as the preferential antibiotic for plasmid selection is made to prevent allergy complications and antibiotic resistance that could happen with other antibiotics such as ampicillin.

Table 5. Comparison between DNA immunization and conventional vaccines (adapted from [120])

	Vaccine type			
	Live attenuated	**Killed microorganism**	**Protein subunit**	**DNA vaccine**
Humoral immunity	+++	+++	+++	+++
Cellular immunity	+++	-	-	++
Antigen presentation	MHC I & II	MHC II	MHC II	MHC I & II
Ease of development and production	+	+	+	+++
Ease of storage and transport	+	+	+	+++
Cost	+	+	+	+++
Safety	++	+++	+++	+++
Antibodies against the natural protein	+++	++	+	+++
Several proteins included	+++	+++	+	+++
Risk of reversion of pathogenicity	+++	+	-	-

DNA vaccination is a more versatile approach compared to conventional vaccines. They can be used either as prophylactic or therapeutic means, and allow the combination of several encoding genes in one plasmid. Simultaneous immunization and protection against several pathogens can also be achieved using more than one plasmid administrated in a single formulation. Furthermore, immunomodulation can be accomplished simply by changing the localization of antigen expression (secreted protein *vs* cytoplasmic) or by using an adequate delivery method to the desired immune response (intramuscular *vs* gene gun). Finally, the vector has immunogenic and adjuvant properties. The plasmid includes sequences of unmethylated CpG motifs from bacterial origin. Several reports revealed that these sequences contribute to increase the immune response due to their ability to activate the innate immune system [121] and B lymphocytes [122], contributing for the increase of DNA vaccine potency [14].

The above discussed advantages of DNA vaccination are summarized on table 6.

Table 6. Advantages of DNA vaccination technology

Design	• Vaccine design includes a simple step of cloning where the therapeutic gene is inserted into an appropriate vector.
Production	• Purification protocol independent of gene of interest. • Low cost production compared to conventional vaccines. • No need for cold chain.
Mechanism of action	• Induction of both arms of the immune response. • Humoral response raised against the natural form of the protein.
Safety	• No risk of reversion to the infectious phenotype. • Potential safety concerns not observed in preclinical studies and clinical trials. • Well tolerated by patients, no adverse reactions following immunization.
Versatility	• One plasmid can include several antigens from different pathogens, so a broad protection scope can be achieved in one immunization. • Is easy to modulate the immune response simply by changing the administration route or by using cytokines in conjunction with the DNA. • Vector has adjuvant properties

Conclusions

The proof of concept that DNA vaccination works was achieved by the licensing of four products for veterinary medicine. In 2008, plasmid DNA vectors accounted for 18% of the total of clinical trials using gene-carrier vectors. However, there is still a long way to go before the first DNA vaccine can be licensed for human use. Optimized approaches have been pursued, including the development of improved vectors, use of adjuvants, efficient delivery systems and other immunization protocols such as the prime-boost strategy. Hopefully, this second generation of vaccines will prove to be more immunogenic. The outcome of the ongoing clinical trials will be decisive for the establishment of DNA vaccination as an attractive and reliable immunization strategy.

Acknowledgments

The authors thank the Portuguese Fundação para a Ciência e a Tecnologia for fund support (POCTI/CVT/47260/2002; POCTI/CVT/61090/2004; POCI/BIO/55799/2004; PTDC/CVT/72624/2006), and doctoral grant to J.A. Carvalho (SFRH/BD/21423/2005) and post-doctoral grants to M.S. Silva (SFRH/BPD/26491/2006) and A.M. Henriques (SFRH/BD/2904/2000).

References

[1] Wolff, JA; Malone, RW; Williams, P; et al. Direct gene transfer into mouse muscle in vivo. *Science.*, 1990, 247(4949 Pt 1), 1465-8.

[2] Tang, Dc; DeVit, M; Johnston, SA. Genetic immunization is a simple method for eliciting an immune response. *Nature.*, 1992, 356(6365), 152-4.

[3] Robinson, HL; Hunt, LA; Webster, RG. Protection against a lethal influenza virus challenge by immunization with a haemagglutinin-expressing plasmid DNA. *Vaccine.*, 1993, 11(9), 957-60.

[4] Okura, Y; Miyakoshi, A; Kohyama, K; Park, IK; Staufenbiel, M; Matsumoto, Y. Nonviral Abeta DNA vaccine therapy against Alzheimer's disease: long-term effects and safety. *Proc Natl Acad Sci.*, U.S.A., 2006, 103(25), 9619-24.

[5] Li, G; Liu, Z; Zhong, N; Liao, B; Xiong, Y. Therapeutic effects of DNA vaccine on allergen-induced allergic airway inflammation in mouse model. *Cell Mol Immunol.*, 2006, 3(5), 379-84.

[6] Andersson, A; Isaksson, M; Wefer, J; et al. Impaired autoimmune T helper 17 cell responses following DNA vaccination against rat experimental autoimmune *encephalomyelitis.*, *PLoS ONE*, 2008, 3(11), e3682.

[7] Yan, J; Harris, K; Khan, AS; Draghia-Akli, R; Sewell, D; Weiner, DB. Cellular immunity induced by a novel HPV18 DNA vaccine encoding an E6/E7 fusion consensus protein in mice and rhesus macaques. *Vaccine.*, 2008, 26(40), 5210-5.

[8] Lu, B; Tao, L; Wang, T; et al. Humoral and Cellular Immune Responses Induced by 3a DNA Vaccines against Severe Acute Respiratory Syndrome (SARS) or SARS-Like Coronavirus in Mice. *Clin Vaccine Immunol.*, 2009, 16(1), 73-7.
[9] Santra, S; Korber, BT; Muldoon, M; et al. A centralized gene-based HIV-1 vaccine elicits broad cross-clade cellular immune responses in rhesus monkeys. *Proc Natl Acad Sci.*, U.S.A., 2008, 105(30), 10489-94.
[10] Huber, VC; Thomas, PG; McCullers, JA. A multi-valent vaccine approach that elicits broad immunity within an influenza subtype. *Vaccine* In Press, Corrected Proof.
[11] Ahmed, SBH; Touihri, L; Chtourou, Y; Dellagi, K; Bahloul, C. DNA based vaccination with a cocktail of plasmids encoding immunodominant Leishmania (Leishmania) major antigens confers full protection in BALB/c mice. *Vaccine.*, 2009, 27(1), 99-106.
[12] Daubersies, P; Ollomo, B; Sauzet, JP; et al. Genetic immunisation by liver stage antigen 3 protects chimpanzees against malaria despite low immune responses., *PLoS ONE*, 2008, 3(7), e2659
[13] Liang, Y; Wu, X; Zhang, J; et al. The treatment of mice infected with multi-drug-resistant Mycobacterium tuberculosis using DNA vaccines or in combination with rifampin. *Vaccine.*, 2008, 26(35), 4536-40.
[14] Hoft, DF; Eickhoff, CS; Giddings, OK; Vasconcelos, JRC; Rodrigues, MM. Trans-Sialidase Recombinant Protein Mixed with CpG Motif-Containing Oligodeoxynucleotide Induces Protective Mucosal and Systemic Trypanosoma cruzi Immunity Involving CD8+ CTL and B Cell-Mediated Cross-Priming. *J Immunol.*, 2007, 179(10), 6889-900.
[15] Henriques, AM; Fevereiro, M; Prazeres, DMF; Monteiro, GA. Development of a candidate DNA vaccine against Maedi-Visna virus. *Veterinary Immunology and Immunopathology.*, 2007, 119(3-4), 222-32.
[16] Tagawa, ST; Lee, P; Snively, J; et al. Phase I study of intranodal delivery of a plasmid DNA vaccine for patients with Stage IV melanoma. *Cancer.*, 2003, 98(1), 144-54.
[17] Trimble, CL; Peng, S; Kos, F; et al. A Phase I Trial of a Human Papillomavirus DNA Vaccine for HPV16+Cervical Intraepithelial Neoplasia 2/3. *Clin Cancer Res.*, 2009, 15(1), 361-7.
[18] Mincheff, M; Tchakarov, S; Zoubak, S; et al. Naked DNA and adenoviral immunizations for immunotherapy of prostate cancer: a phase I/II clinical trial. *Eur Urol.*, 2000, 38(2), 208-17.
[19] Garren, H; Robinson, WH; Krasulova, E; et al. Phase 2 trial of a DNA vaccine encoding myelin basic protein for multiple sclerosis. *Ann Neurol.*, 2008, 63(5), 611-20.
[20] Le, TP; Coonan, KM; Hedstrom, RC; et al. Safety, tolerability and humoral immune responses after intramuscular administration of a malaria DNA vaccine to healthy adult volunteers. *Vaccine.*, 2000, 18(18), 1893-901.
[21] Catanzaro, AT; Roederer, M; Koup, RA; et al. Phase I clinical evaluation of a six-plasmid multiclade HIV-1 DNA candidate vaccine. *Vaccine.*, 2007, 25(20), 4085-92.
[22] Gilbert, PB; Chiu, YL; Allen, M; et al. Long-term safety analysis of preventive HIV-1 vaccines evaluated in AIDS vaccine evaluation group NIAID-sponsored Phase I and II clinical trials. *Vaccine.*, 2003, 21(21-22), 2933-47.
[23] Tacket, CO; Roy, MJ; Widera, G; Swain, WF; Broome, S; Edelman, R. Phase 1 safety and immune response studies of a DNA vaccine encoding hepatitis B surface antigen delivered by a gene delivery device. *Vaccine.*, 1999, 17(22), 2826-9.

[24] Alvarez-Lajonchere, L; Shoukry, NH; Gra, B; et al. Immunogenicity of CIGB-230, a therapeutic DNA vaccine preparation, in HCV-chronically infected individuals in a Phase I clinical trial. *J Viral Hepat.*, 2009, 16(3), 156-67.

[25] Martin, J; Pierson, T; Hubka, S; et al. A West Nile Virus DNA Vaccine Induces Neutralizing Antibody in Healthy Adults during a Phase 1 Clinical Trial. *The Journal of Infectious Diseases.*, 2007, 196(12), 1732-40.

[26] Martin, JE; Louder, MK; Holman, LA; et al. A SARS DNA vaccine induces neutralizing antibody and cellular immune responses in healthy adults in a Phase I clinical trial. *Vaccine.*, 2008, 26(50), 6338-43.

[27] Martin, JE; Sullivan, NJ; Enama, ME; et al. A DNA Vaccine for Ebola Virus Is Safe and Immunogenic in a Phase I Clinical Trial. *ClinVaccine Immunol.*, 2006, 13(11), 1267-77.

[28] Cattamanchi, A; Posavad, CM; Wald, A; et al. Phase I Study of a Herpes Simplex Virus Type 2 (HSV-2) DNA Vaccine Administered to Healthy, HSV-2-Seronegative Adults by a Needle-Free Injection System. *Clin Vaccine Immunol.*, 2008, 15(11), 1638-43.

[29] Davidson, AH; Traub-Dargatz, JL; Rodeheaver, RM; et al. Immunologic responses to West Nile virus in vaccinated and clinically affected horses. *J Am Vet Med Assoc.*, 2005, 226(2), 240-5.

[30] Bergman, PJ; Camps-Palau, MA; McKnight, JA; et al. Development of a xenogeneic DNA vaccine program for canine malignant melanoma at the Animal Medical Center. *Vaccine.*, 2006, 24(21), 4582-5.

[31] Garver, KA; LaPatra, SE; Kurath, G. Efficacy of an infectious hematopoietic necrosis (IHN) virus DNA vaccine in Chinook Oncorhynchus tshawytscha and sockeye O. nerka salmon. *Dis Aquat Organ.*, 2005, 64(1), 13-22.

[32] Thacker, EL; Holtkamp, DJ; Khan, AS; Brown, PA; Draghia-Akli, R. Plasmid-mediated growth hormone-releasing hormone efficacy in reducing disease associated with Mycoplasma hyopneumoniae and porcine reproductive and respiratory syndrome virus infection. *J Anim Sci.*, 2006, 84(3), 733-42.

[33] Davis, HL; Michel, ML; Whalen, RG. DNA-based immunization induces continuous secretion of hepatitis B surface antigen and high levels of circulating antibody. *Hum Mol Genet.*, 1993, 2(11), 1847-51.

[34] Levy, MY; Barron, LG; Meyer, KB; Szoka, FC; Jr. Characterization of plasmid DNA transfer into mouse skeletal muscle: evaluation of uptake mechanism, expression and secretion of gene products into blood. *Gene Ther.*, 1996, 3(3), 201-11.

[35] Tripathy, SK; Svensson, EC; Black, HB; et al. Long-term expression of erythropoietin in the systemic circulation of mice after intramuscular injection of a plasmid DNA vector. *Proc Natl Acad Sci.*, U.S.A., 1996, 93(20), 10876-80.

[36] Ulmer, JB; Deck, RR; DeWitt, CM; Donnhly, JI; Liu, MA. Generation of MHC class I-restricted cytotoxic T lymphocytes by expression of a viral protein in muscle cells: antigen presentation by non-muscle cells. *Immunology.*, 1996, 89(1), 59-67.

[37] Gurunathan, S; Klinman, DM; Seder, RA. DNA Vaccines: Immunology, Application, and Optimization*. *Annual Review of Immunology.*, 2000, 18(1), 927-74.

[38] Corr, M; Lee, DJ; Carson, DA; Tighe, H. Gene vaccination with naked plasmid DNA: mechanism of CTL priming. *J Exp Med.*, 1996, 184(4), 1555-60.

[39] Doe, B; Selby, M; Barnett, S; Baenziger, J; Walker, CM. Induction of cytotoxic T lymphocytes by intramuscular immunization with plasmid DNA is facilitated by bone marrow-derived cells. *Proc Natl Acad Sci.*, U.S.A., 1996, 93(16), 8578-83.

[40] Iwasaki, A; Torres, CA; Ohashi, PS; Robinson, HL; Barber, BH. The dominant role of bone marrow-derived cells in CTL induction following plasmid DNA immunization at different sites. *J Immunol.*, 1997, 159(1), 11-4.

[41] Albert, ML; Sauter, B; Bhardwaj, N. Dendritic cells acquire antigen from apoptotic cells and induce class I-restricted CTLs. *Nature.*, 1998, 392(6671), 86-9.

[42] Harding, CV; Song, R. Phagocytic processing of exogenous particulate antigens by macrophages for presentation by class I MHC molecules. *J Immunol.*, 1994, 153(11), 4925-33.

[43] Rocha, N; Neefjes, J. MHC class II molecules on the move for successful antigen presentation. *EMBO J.*, 2008, 27(1), 1-5.

[44] Feltquate, DM; Heaney, S; Webster, RG; Robinson, HL. Different T helper cell types and antibody isotypes generated by saline and gene gun DNA immunization. *J Immunol.*, 1997, 158(5), 2278-84.

[45] Watts, AM; Kennedy, RC. DNA vaccination strategies against infectious diseases. International *Journal for Parasitology.*, 1999, 29(8), 1149-63.

[46] Koide, Y; Nagata, T; Yoshida, A; Uchijima, M. DNA vaccines. *Jpn J Pharmacol.*, 2000, 83(3), 167-74.

[47] Moreau, P; Hen, R; Wasylyk, B; Everett, R; Gaub, MP; Chambon, P. The SV40 72 base repair repeat has a striking effect on gene expression both in SV40 and other chimeric recombinants. *Nucl Acids Res.*, 1981, 9(22), 6047-68.

[48] Boshart, M; Weber, F; Jahn, G; Dorsch, H; Fleckenstein, B; Schaffner, W. A very strong enhancer is located upstream of an immediate early gene of human cytomegalovirus. *Cell.*, 1985, 41(2), 521-30.

[49] Cazeaux, N; Bennasser, Y; Vidal, PL; Li, Z; Paulin, D; Bahraoui, E. Comparative study of immune responses induced after immunization with plasmids encoding the HIV-1 Nef protein under the control of the CMV-IE or the muscle-specific desmin promoter. *Vaccine.*, 2002, 20(27-28), 3322-31.

[50] Montgomery, DL; Shiver, JW; Leander, KR; et al. Heterologous and homologous protection against influenza A by DNA vaccination: optimization of DNA vectors. *DNA Cell Biol.*, 1993, 12(9), 777-83.

[51] Azzoni, AR; Ribeiro, SC; Monteiro, GA; Prazeres, DM. The impact of polyadenylation signals on plasmid nuclease-resistance and transgene expression. *J Gene Med.*, 2007, 9(5), 392-402.

[52] Barouch, DH; Yang, Zy; Kong, Wp; et al. A Human T-Cell Leukemia Virus Type 1 Regulatory Element Enhances the Immunogenicity of Human Immunodeficiency Virus Type 1 DNA Vaccines in Mice and Nonhuman Primates. *J Virol.*, 2005, 79(14), 8828-34.

[53] Garg, S; Oran, AE; Hon, H; Jacob, J. The Hybrid Cytomegalovirus Enhancer/Chicken {beta}-Actin Promoter along with Woodchuck Hepatitis Virus Posttranscriptional Regulatory Element Enhances the Protective Efficacy of DNA Vaccines. *J Immunol.*, 2004, 173(1), 550-8.

[54] Ramakrishna, L; Anand, KK; Mohankumar, KM; Ranga, U. Codon Optimization of the Tat Antigen of Human Immunodeficiency Virus Type 1 Generates Strong Immune Responses in Mice following Genetic Immunization. *J Virol.*, 2004, 78(17), 9174-89.
[55] Cheung, YK; Cheng, SCS; Sin, FWY; Xie, Y. Plasmid encoding papillomavirus Type 16 (HPV16) DNA constructed with codon optimization improved the immunogenicity against HPV infection. *Vaccine.*, 2004, 23(5), 629-38.
[56] Wu, X; J÷rnvall, H; Berndt, KD; Oppermann, U. Codon optimization reveals critical factors for high level expression of two rare codon genes in Escherichia coli: RNA stability and secondary structure but not tRNA abundance. *Biochemical and Biophysical Research Communications.*, 2004, 313(1), 89-96.
[57] Kozak, M. Recognition of AUG and alternative initiator codons is augmented by G in position +4 but is not generally affected by the nucleotides in positions +5 and +6. *EMBO J.*, 1997, 16(9), 2482-92.
[58] Ruff, AL; Guarnieri, FG; Staveley-O'Carroll, K; Siliciano, RF; August, JT. The Enhanced Immune Response to the HIV gp160/LAMP Chimeric Gene Product Targeted to the Lysosome Membrane Protein Trafficking Pathway. *J Biol Chem.*, 1997, 272(13), 8671-8.
[59] Chikhlikar, P; de Arruda, LB; Maciel, M; et al. DNA encoding an HIV-1 Gag/human lysosome-associated membrane protein-1 chimera elicits a broad cellular and humoral immune response in Rhesus macaques. *PLoS ONE.*, 2006, 1:e135. e135
[60] Niazi, KR; Ochoa, MT; Sieling, PA; et al. Activation of human CD4+ T cells by targeting MHC class II epitopes to endosomal compartments using human CD1 tail sequences. *Immunology.*, 2007, 122(4), 522-31.
[61] Kim, SJ; Suh, D; Park, SE; et al. Enhanced immunogenicity of DNA fusion vaccine encoding secreted hepatitis B surface antigen and chemokine RANTES. *Virology.*, 2003, 314(1), 84-91.
[62] Qiu, JT; Liu, B; Tian, C; Pavlakis, GN; Yu, XF. Enhancement of Primary and Secondary Cellular Immune Responses against Human Immunodeficiency Virus Type 1 Gag by Using DNA Expression Vectors That Target Gag Antigen to the Secretory Pathway. *J Virol.*, 2000, 74(13), 5997-6005.
[63] Fan, H; Xiao, S; Tong, T; et al. Immunogenicity of porcine circovirus type 2 capsid protein targeting to different subcellular compartments. *Molecular Immunology.*, 2008, 45(3), 653-60.
[64] Brandsma, JL; Shlyankevich, M; Zelterman, D; Su, Y. Therapeutic vaccination of rabbits with a ubiquitin-fused papillomavirus E1, E2, E6 and E7 DNA vaccine. *Vaccine.*, 2007, 25(33), 6158-63.
[65] Xu, W; Chu, Y; Zhang, R; Xu, H; Wang, Y; Xiong, S. Endoplasmic reticulum targeting sequence enhances HBV-specific cytotoxic T lymphocytes induced by a CTL epitope-based DNA vaccine. *Virology.*, 2005, 334(2), 255-63.
[66] Sherritt, M; Cooper, L; Moss, DJ; Kienzle, N; Altman, J; Khanna, R. Immunization with tumor-associated epitopes fused to an endoplasmic reticulum translocation signal sequence affords protection against tumors with down-regulated expression of MHC and peptide transporters. *Int Immunol.*, 2001, 13(3), 265-71.
[67] Deliyannis, G; Boyle, JS; Brady, JL; Brown, LE; Lew, AM. A fusion DNA vaccine that targets antigen-presenting cells increases protection from viral challenge. *Proc Natl Acad Sci.*, U.S.A., 2000, 97(12), 6676-80.

[68] Nchinda, G; Kuroiwa, J; Oks, M; et al. The efficacy of DNA vaccination is enhanced in mice by targeting the encoded protein to dendritic cells. *J Clin Invest.*, 2008, 118(4), 1427-36.

[69] Fredriksen, AB; Sandlie, I; Bogen, B. DNA Vaccines Increase Immunogenicity of Idiotypic Tumor Antigen by Targeting Novel Fusion Proteins to Antigen-Presenting *Cells. Mol Ther.*, 2006, 13(4), 776-85.

[70] Vogel, FR. Improving vaccine performance with adjuvants. *Clin Infect Dis.*, 2000, 30 Suppl 3, S266-70.:S266-S270.

[71] Greenland, JR; Letvin, NL. Chemical adjuvants for plasmid DNA vaccines. *Vaccine.*, 2007, 25(19), 3731-41.

[72] Wang, D; Christopher, ME; Nagata, LP; et al. Intranasal immunization with liposome-encapsulated plasmid DNA encoding influenza virus hemagglutinin elicits mucosal, cellular and humoral immune responses. *Journal of Clinical Virology.*, 2004, 31(Supplement 1), 99-106.

[73] Zhou, WZ; Kaneda, Y; Huang, S; Morishita, R; Hoon, D. Protective immunization against melanoma by gp100 DNA-HVJ-liposome vaccine. *Gene Ther.*, 1999, 6(10), 1768-73.

[74] Reina, R; Barbezange, C; Niesalla, H; et al. Mucosal immunization against ovine lentivirus using PEI-DNA complexes and modified vaccinia Ankara encoding the gag and/or env genes. *Vaccine.*, 2008, 26(35), 4494-505.

[75] Bieber, T; Meissner, W; Kostin, S; Niemann, A; Elsasser, HP. Intracellular route and transcriptional competence of polyethylenimine-DNA complexes. *Journal of Controlled Release.*, 2002, 82(2-3), 441-54.

[76] Denis-Mize, KS; Dupuis, M; MacKichan, ML; et al. Plasmid DNA adsorbed onto cationic microparticles mediates target gene expression and antigen presentation by dendritic cells. *Gene Ther.*, 2000, 7(24), 2105-12.

[77] Otten, GR; Schaefer, M; Doe, B; et al. Enhanced Potency of Plasmid DNA Microparticle Human Immunodeficiency Virus Vaccines in Rhesus Macaques by Using a Priming-Boosting Regimen with Recombinant Proteins. *J Virol.*, 2005, 79(13), 8189-200.

[78] Wilson, KD; de Jong, SD; Kazem, M; et al. The combination of stabilized plasmid lipid particles and lipid nanoparticle encapsulated CpG containing oligodeoxynucleotides as a systemic genetic vaccine. *J Gene Med.*, 2009, 11(1), 14-25.

[79] Kwissa, M; Lindblad, EB; Schirmbeck, R; Reimann, J. Codelivery of a DNA vaccine and a protein vaccine with aluminum phosphate stimulates a potent and multivalent immune response. *J Mol Med.*, 2003, 81(8), 502-10.

[80] Xu, R; Megati, S; Roopchand, V; et al. Comparative ability of various plasmid-based cytokines and chemokines to adjuvant the activity of HIV plasmid DNA vaccines. *Vaccine.*, 2008, 26(37), 4819-29.

[81] Daud, AI; DeConti, RC; Andrews, S; et al. Phase I Trial of Interleukin-12 Plasmid Electroporation in Patients With Metastatic Melanoma. *J Clin Oncol.*, 2008, JCO.

[82] Kim, JJ; Simbiri, KA; Sin, JI; et al. Cytokine molecular adjuvants modulate immune responses induced by DNA vaccine constructs for HIV-1 and SIV. *J Interferon Cytokine Res.*, 1999, 19(1), 77-84.

[83] European Agency for the Evaluation of Medical Products. 2005. Guideline on adjuvants in vaccines. Available from URL: http://www.emea.europa.eu/pdfs/human/vwp/13471604en.pdf

[84] Dunachie, SJ; Hill, AVS. Prime-boost strategies for malaria vaccine development. *J Exp Biol.*, 2003, 206(21), 3771-9.

[85] Carson, C; Antoniou, M; Ruiz-Arg³ello, MBa; et al. A prime/boost DNA/Modified vaccinia virus Ankara vaccine expressing recombinant Leishmania DNA encoding TRYP is safe and immunogenic in outbred dogs, the reservoir of zoonotic visceral leishmaniasis. *Vaccine.*, 2009, 27(7), 1080-6.

[86] Yu, S; Feng, X; Shu, T; et al. Potent specific immune responses induced by prime-boost-boost strategies based on DNA, adenovirus, and Sendai virus vectors expressing gag gene of Chinese HIV-1 subtype B. *Vaccine.*, 2008, 26(48), 6124-31.

[87] Weiss, WR; Kumar, A; Jiang, G; et al. Protection of rhesus monkeys by a DNA prime/poxvirus boost malaria vaccine depends on optimal DNA priming and inclusion of blood stage antigens. *PLoS ONE.*, 2007, 2(10), e1063

[88] Ramsay, AJ; Kent, SJ; Strugnell, RA; Suhrbier, A; Thomson, SA; Ramshaw, IA. Genetic vaccination strategies for enhanced cellular, humoral and mucosal immunity. *Immunol Rev.*, 1999, 171:27-44.:27-44.

[89] Abel, K; Strelow L; Yue, Y; Eberhardt, MK; Schmidt, KA; Barry, PA. A heterologous DNA prime/protein boost immunization strategy for rhesus cytomegalovirus. *Vaccine.*, 2008, 26(47), 6013-25.

[90] Arankalle, VA; Lole, KS; Deshmukh, TM; Srivastava, S; Shaligram, US. Challenge studies in Rhesus monkeys immunized with candidate hepatitis E vaccines: DNA, DNA-prime-protein-boost and DNA-protein encapsulated in liposomes. *Vaccine.*, 2009, 27(7), 1032-9.

[91] Dunham, SP. The application of nucleic acid vaccines in veterinary medicine. *Research in Veterinary Science.*, 2002, 73(1), 9-16.

[92] Feltquate, DM. DNA vaccines: vector design, delivery, and antigen presentation. *J Cell Biochem Suppl.*, 1998, 30-31, 304-11.:304-11.

[93] Otten, G; Schaefer, M; Doe, B; et al. Enhancement of DNA vaccine potency in rhesus macaques by electroporation. *Vaccine.*, 2004, 22(19), 2489-93.

[94] Tang, M; Wang, H; Zhou, S; Tian, G. Enhancement of the immunogenicity of an infectious bronchitis virus DNA vaccine by a bicistronic plasmid encoding nucleocapsid protein and interleukin-2. *Journal of Virological Methods.*, 2008, 149(1), 42-8.

[95] Manoj, S; Babiuk, LA; van Drunen Littel-van den Hurk S. Immunization with a dicistronic plasmid expressing a truncated form of bovine herpesvirus-1 glycoprotein D and the amino-terminal subunit of glycoprotein B results in reduced gB-specific immune responses. *Virology.*, 2003, 313(1), 296-307.

[96] Rainczuk, A; Scorza, T; Spithill, TW; Smooker, PM. A Bicistronic DNA Vaccine Containing Apical Membrane Antigen 1 and Merozoite Surface Protein 4/5 Can Prime Humoral and Cellular Immune Responses and Partially Protect Mice against Virulent Plasmodium chabaudi adami DS Malaria. *Infect Immun.*, 2004, 72(10), 5565-73.

[97] Zhang, HY; Sun, SH; Guo, YJ; et al. Optimization strategy for plasmid DNAs containing multiple-epitopes of foot-and-mouth disease virus by cis-expression with IL-2. *Vaccine.*, 2008, 26(6), 769-77.

[98] Kim, JJ; Simbiri, KA; Sin, JI; et al. Cytokine molecular adjuvants modulate immune responses induced by DNA vaccine constructs for HIV-1 and SIV. *J Interferon Cytokine Res.*, 1999, 19(1), 77-84.

[99] Iwasaki, A; Stiernholm, BJ; Chan, AK; Berinstein, NL; Barber BH. Enhanced CTL responses mediated by plasmid DNA immunogens encoding costimulatory molecules and cytokines. *J Immunol.*, 1997, 158(10), 4591-601.

[100] Kutzler, MA; Weiner, DB. DNA vaccines: ready for prime time? *Nat Rev Genet.*, 2008, 9(10), 776-88.

[101] Nichols, WW; Ledwith, BJ; Manam, SV; Troilo, PJ. Potential DNA vaccine integration into host cell genome. *Ann N Y Acad Sci.*, 1995, 772, 30-9. 30-9.

[102] FDA. Guidance for Industry - Considerations for plasmid DNA vaccines for infectious disease Indications. 2007.

[103] Available from URL: http://www.fda.gov/CbER/gdlns/plasdnavac.pdf

[104] Liu, MA; McClements, W; Ulmer, JB; Shiver, J; Donnelly, J. Immunization of non-human primates with DNA vaccines. *Vaccine.*, 1997, 15(8), 909-12.

[105] Bot, A; Bona, C. Genetic immunization of neonates. *Microbes and Infection.*, 2002, 4(4), 511-20.

[106] Mor, G; Yamshchikov, G; Sedegah, M; et al. Induction of neonatal tolerance by plasmid DNA vaccination of mice. *J Clin Invest.*, 1996, 98(12), 2700-5.

[107] Sarzotti, M; Dean, TA; Remington, MP; Ly, CD; Furth, PA; Robbins, DS. Induction of cytotoxic T cell responses in newborn mice by DNA immunization. *Vaccine.*, 1997, 15(8), 795-7.

[108] Davis, HL; McCluskie, MJ. DNA vaccines for viral diseases. *Microbes and Infection.*, 1999, 1(1), 7-21.

[109] Harriet, L. Robinson PhDHSGMDHLDPhDSAJPhDaMAL. *The Scientific Future of DNA for Immunization*. American Academy of Microbiology, 1997.

[110] MacGregor, RR; Boyer, JD; Ugen, KE; et al. First human trial of a DNA-based vaccine for treatment of human immunodeficiency virus type 1 infection: safety and host response. *J Infect Dis.*, 1998, 178(1), 92-100.

[111] Donnelly, JJ; Ulmer, JB; Shiver, JW; Liu, MA. DNA VACCINES. *Annual Review of Immunology.*, 1997, 15(1), 617-48.

[112] FDA. Guidance for Industry - Guidance for human *Somatic Cell Therapy and Gene Therapy.*, 1998.

[113] Available from URL: http://www.fda.gov/CBER/gdlns/somgene.pdf

[114] Cranenburgh, RM; Hanak, JAJ; Williams, SG; Sherratt, DJ. Escherichia coli strains that allow antibiotic-free plasmid selection and maintenance by repressor titration. *Nucl Acids Res.*, 2001, 29(5), e26

[115] Diogo, MM; Queiroz, JA; Monteiro, GA; Martins, SA; Ferreira, GN; Prazeres, DM. Purification of a cystic fibrosis plasmid vector for gene therapy using hydrophobic interaction chromatography. *Biotechnol Bioeng.*, 2000, 68(5), 576-83.

[116] Diogo, MM; Ribeiro, SC; Queiroz, JA; et al. Production, purification and analysis of an experimental DNA vaccine against rabies. *J Gene Med.*, 2001, 3(6), 577-84.

[117] Prazeres, DMF; Ferreira, GNM; Monteiro, GA; Cooney, CL; Cabral, JMS. Large-scale production of pharmaceutical-grade plasmid DNA for gene therapy: problems and bottlenecks. *Trends in Biotechnology.*, 1999, 17(4), 169-74.

[118] Medjitna, TD; Stadler, C; Bruckner, L; Griot, C; Ottiger, HP. DNA vaccines: safety aspect assessment and regulation. *Dev Biol (Basel)*., 2006, 126:261-70; discussion, 327, 261-70.

[119] Robertson, JS; Griffiths, E. Assuring the quality, safety, and efficacy of DNA vaccines. *Mol Biotechnol*., 2001, 17(2), 143-9.

[120] WHO. *Guidelines for assuring the quality and nonclinical safety evaluation of DNA vaccines*. 2007.

[121] Available from URL: http://www.who.int/entity/biologicals/publications/trs/areas/vaccines/dna/Annex%201_DNA%20vaccines.pdf

[122] European Agency for the Evaluation of Medical Products. *Note for Guidance on clinical evaluation of vaccines*., 2005.

[123] Available from URL: http://www.emea.europa.eu/pdfs/human/vwp/16465305en.pdf

[124] Hassett, DE; Lindsay Whitton, J. DNA immunization. *Trends in Microbiology*., 1996, 4(8), 307-12.

[125] Klinman, DM; Yamshchikov, G; Ishigatsubo, Y. Contribution of CpG motifs to the immunogenicity of DNA vaccines. *J Immunol*, 1997, 158(8), 3635-9.

[126] Krieg, AM; Yi, AK; Matson, S; et al. CpG motifs in bacterial DNA trigger direct B-cell activation. *Nature*., 1995, 374(6522), 546-9.

In: Veterinary Immunology and Immunopathology
Editors: L. Nermann, S. Meier
ISBN: 978-1-60876-342-9

Chapter 5

NATURAL RESISTANCE AND INNATE IMMUNITY IN BRUCELLOSIS

Tatiane A. Paixão, Erica A. Costa, Mariana N. Xavier, Teane M. A. Silva and Renato L. Santos
Escola de Veterinária, Universidade Federal de Minas Gerais. Av. Antônio Carlos, 6627; 30123-970 Belo Horizonte, MG, Brazil

ABSTRACT

Brucellosis is one of the most important zoonotic diseases worldwide. Genetic resistance against brucellosis has been characterized in several host species. Recent studies have also shown that *Brucella* has evolved mechanisms for evading the host immune system. This chapter is focused on host natural resistance and innate immunity against *Brucella* infection.

INTRODUCTION

Brucellosis is a chronic disease that affects a variety of mammalian hosts including man. The disease is caused by the intracellular gram-negative bacteria *Brucella* spp.. Each *Brucella* species has a preferential host, including sheep and goats (*Brucella melitensis*), pigs (*B. suis*), cattle (*B. abortus),* dogs (*B. canis*), sheep (*B. ovis*), and desert rat (*B. neotomae*). In addition to these six classical species, two new *Brucella* species were isolated from marine mammals, *B. ceti* and *B. pinnipedialis*, with predilection to cetaceans and pinipeds, respectively [51], and novel species named *B. microti* has been recently isolated from rodent *Microtus arvalis* [113].

Brucellosis in domestic animals is a chronic infection that affects lymphoid tissues and the reproductive tract causing abortion and placentitis in females [32, 99] or orchitis and epididymitis in males [32]. *B. melitensis*, *B suis*, and *B. abortus* are considered the most pathogenic species for humans [32]. *B. ceti*, *B. pinnipedialis*, and *B. canis* can also infect humans [51, 56], whereas *B. ovis* and *B. neotomae* do not cause human brucellosis. Human brucellosis is a debilitating disease characterized by fever, weakness, malaise, arthritis, osteomyelitis, abscesses in the liver and spleen, endocarditis, and meningoencephalitis [56].

Interaction between host immunity and pathogen may or may not result in disease manifestation. Both innate and adaptive immune systems are important for controlling *Brucella* infection. It is well established that cell-mediated adaptive immunity is essential for controlling brucellosis [90]. However, the innate immune response also plays a critical role. As the first line of defense, innate immunity is the major component of natural resistance against brucellosis, which has been recognized in different host species. In addition, innate immunity is triggered by recognition of *Brucella* by the host during the early stages of infection, and it is essential to drive the adaptive immune response. In this review we discuss the innate immunity in brucellosis emphasizing aspects related to natural resistance, effector mechanisms and pathogen recognition.

I. Natural Resistance to Brucellosis

By definition, natural resistance to infectious disease depends upon the host capacity to resist to disease when exposed to the infectious agent without a previous contact or immunization [119]. Natural resistance against *B. abortus* infection in cattle was initially observed in a selection and breeding study including phenotypically resistant or susceptible cattle. Abortion and re-isolation of *B. abortus* did not occur in resistant cows after experimental infection. The frequency of resistant phenotype increased from 20 to 58.6% through selection and controlled breeding. Genetic analysis suggested the involvement of more than two genes controlling these phenotypes [2]. Additional studies demonstrated that this resistance *in vivo* correlates with macrophage bactericidal activity *in vitro*. Mammary gland or peripheral blood monocyte-derived macrophages from phenotypically resistant cattle have the ability to control intracellular replication of *B. abortus in vitro* and to generate higher oxidative burst activity after phagocytosis of opsonized bacteria when compared to those from phenotypically susceptible cows [17, 66, 102, 103].

Genes involved in natural resistance against different intracellular pathogens, including *Mycobacterium bovis*, *Salmonella enterica* serotype Typhimurium, and *Leishmania donovani* have been identified in mice based on antimicrobial activity of macrophages [15, 60, 80]. Later on these genes were recognized as one single locus identified as *Nramp1* (natural resistance associated macrophage protein 1), which has been more recently renamed as *Slc11a1* (solute carrier family 11 member 1) [122]. *Nramp1* is a gene located in chromosome 1 of the mouse that encodes a 90-100 kD protein with several transmembrane domains and phosphorylation sites. Nramp1 locates in the membrane of late endosomos and lysosomos of macrophages, and it is recruited to the membrane of phagolysosomos after phagocytosis [64]. *Nramp1* expression is induced by lipopolysaccharide (LPS) or cytokines such as interferon gamma (IFNγ), tumor necrosis factor alpha (TNFα), and interleukin 1 (IL-1). Although resistant mice have higher levels of *Nramp1* expression than susceptible mouse strains [114, 127], resistance or susceptibility of mice against *Brucella* infection is not linked to *Nramp1* since macrophages from mice resistant and susceptible to other intracellular pathogens have similar intracellular growth rates of *B. abortus* [112]. Additionally, *Nramp1* is not associated with the control *B. melitensis* infection in mice [65].

Nramp1 is a pH-dependent divalent cation transporter that transports cations such as iron ($Fe2^{+}$) and manganese ($Mn2^{+}$) from the lumen of the phagosome to the cytosol, preventing iron acquisition by intracellular pathogens [5, 50, 58, 70]. However, there is another hypothesis suggesting that the movement of cations transported by Nramp1 occurs from cytosol to phagosome, resulting in increased levels of iron inside the phagosomo thus favoring pathogen killing by production of ROI through the Fenton reaction [59, 77, 132]. Furthermore, there are evidences that Nramp1 can modulate cytokine and chemokine production, respiratory burst, nitric oxide (NO) production, and neutrophil recruitment to the infection site [12, 127]. A human ortholog of the murine *Nramp1* gene has been identified [26], and several polymorphisms have been located in the human *Nramp1* [115]. However, these polymorphisms in the human *Nramp1* apparently are not associated with resistance to brucellosis [16]. Furthermore, an experimental infection of susceptible and resistant cows from controlled breeding demonstrated a significant association between a polymorphism at the 3'UTR (3´untranslated region) of bovine *Nramp1* with natural resistance to brucellosis [119]. The bovine *Nramp1* gene encodes a 60kD protein that is expressed mostly in macrophages and cells of reticuloendothelial system. These 3'UTR polymorphisms are due to a variation in the number of (GT) repeats in a microsatellite [45, 67]. The resistant allele (GT13) controls the replication of *B. abortus* in murine macrophages line cells transfected with bovine *Nramp1* [9]. A recent study detected novel GTn microsatellite polymorphisms (GT10 and GT12), which appears to be associated with resistance or susceptibility based on intracellular growth of *Brucella* in cultured bovine macrophages [84]. Conversely, we have recently performed a systematic study aiming to asses the influence of *Nramp1* 3'UTR polymorphisms on natural resistance to bovine brucellosis, assessing the link of particular *Nramp1* 3'UTR genotypes with resistance in naturally and experimentally infected cows as well as macrophages *in vitro*. We found no correlation between the putative resistant allele GT13 and natural resistance against *B. abortus* infection under any of the experimental conditions [94]. Interestingly, we demonstrated a significant difference in allelic frequencies of *Nramp1* 3'UTR polymorphisms between *Bos taurus taurus* (Hostein cattle) and *Bos taurus indicus* (Zebu cattle). *Bos taurus taurus* has extremely homogeneous genotype, with 100% of the individuals with a GT13 genotype [93]. Similar results were found comparing the frequency of polymorphism among different bovine breeds [78, 83]. Several other polymorphisms of the bovine *Nramp1* gene have been identified, which are located in introns, exons or promoter region [1, 33, 83]. However, the importance of these polymorphisms on natural resistance to bovine brucellosis has not yet been established.

Natural resistance against brucellosis has also been characterized in water buffalo (*Bubalus bubalis*). Apparently, approximately 20% of animals remain negative by serological tests in water buffalo herds heavily infected with *B. abortus* [14]. The polymorphisms in 3'UTR of the bubaline *Nramp1* has also been associated to natural resistance to brucellosis under natural and *in vitro* conditions [14, 19, 20, 53]. However, this polymorphism is characterized by a different number of GT repeats in a microsatellite, a single insertion, or a mutation in one allele [14]. Monocyte-derived macrophage from animals with a resistant genotype have higher levels of *Nramp1* expression and limit more efficiently intracellular replication of *B. abortus* and *B. melitensis* during early stages of infection [14, 19, 20]. In addition a second gene, coding for a Mannose Binding Lectin (MBL), has been associated to natural resistance against *B. abortus* infection in water buffalo [21]. MBL is an acute phase protein that binds to mannose and other sugars expressed by a variety of pathogenic microorganisms. This binding can activates complement, favor opsonization and regulate pro-inflammatory cytokine production by phagocytes, favoring natural resistance to pathogens [120].

II. Macrophage Activity during *Brucella* Infection

Phagocytes like macrophages and dendritic cells (DCs) and non-phagocytic cells such as trophoblasts are target cells for *Brucella* replication (Figure 1) [24, 25, 109]. This predilection reflects the disease manifestation, which is characterized by persistent infection in lymphoid tissues and placentitis [123]. The macrophage is an important component of the innate immunity that can directly control or eliminate pathogens working as the first line of defense. In addition, macrophages can also act as antigen-presenting cells along with DCs. Their cytokines production after contact with pathogens directs the type of a subsequent adaptive immune response.

After phagocytosis by macrophages, the majority of intracellular *Brucella* is killed, and less than 10% of the original bacterial load remains viable intracellularly in macrophages. However, these viable organisms survive bactericidal mechanisms and are capable of re-direct intracellular trafficking of the *Brucella*-containing vacuole towards the rough endoplasmic reticulum, which is the intracellular replication niche of *Brucella* [4, 25, 98]. Conversely, attenuated *Brucella* mutant strains are completely inactivated within phagolysosomes because they are not capable of direct intracellular trafficking towards the replication niche [25, 101].Various mechanisms are employed by macrophages to control or eliminate pathogens after phagocytosis [87]. *Brucella* has evolved several mechanisms to evade macrophage killing and to establish infection as summarized Table 1 [3, 25, 81, 101].

Reduction of intraphagosomal pH is an important killing mechanism of macrophages, but *Brucella* can survive in acidified intracellular compartments under pH as low as 4. Indeed, neutralization of phagosomal pH impairs intracellular replication of *B. suis.* Therefore, *Brucella* not just survives acidification, but early acidification within the phagosome is essential for intracellular bacterial replication [75, 100].

Two other important bactericidal mechanisms of macrophages are nitric oxide (NO) production as a result of induction of iNOS (inducible Nitric Oxide Synthase) and production of reactive oxygen intermediate (ROI) species generated by the oxidative burst. NO contributes to kill intracellular *Brucella* just during the first 24 hours of infection, affecting particularly opsonized bacteria [61, 62, 124]. Apparently *Brucella* may limit NO production by macrophage to ensure its intracellular survival since deletion of the *Brucella* gene that encodes NO reductase renders an attenuated *B. suis* strain in mice [81]. Importantly, *Brucella* infection weakly stimulates iNOS production in comparison to *Escherichia coli* LPS or IFNγ [62, 124]. Oxidative burst activated by NADPH oxidase, is important for *B. abortus* killing by IFNγ-activated macrophages [73, 116]. IFNγ-activated macrophage supplemented with iron efficiently kills intracellular *Brucella* due to the ability of iron to catalyze oxygen radicals, thus confirming the importance of ROI [6]. However, ROI are not critical for limiting bacterial growth in mice [116]. Interestingly, under *in vitro* conditions *B. suis* does not induce an oxidative burst when internalized by phagocytes [75].

Table 1. Summary of mechanisms employed by *Brucella* to evade these host defenses mediated by efector cells

Efectors cells	*Brucella* resistance pathways	Virulence factors involved
Macrophages	Survive intracellular acidification	-
	Limit NO production	NO reductase
	Induce weakly iNOS production	-
	Inhibit oxidative burst	-
	Inhibit endososomo maturation	cyclic β-1,2-glucan and LPS
	Inhibit phagosome-lysosome fusion	T4SS
	Direct bacteria to replication niche	T4SS
	Inhibit apoptosis	LPS
Neutrophils	Decrease PMN recruitment	-
	Inhibit and resist to PMN degranulation	LPS
	Induce weakly oxidative burst	-
NK cell	Depletion of IFNγ production (mouse)	-
	Impair NK activity (human)	-

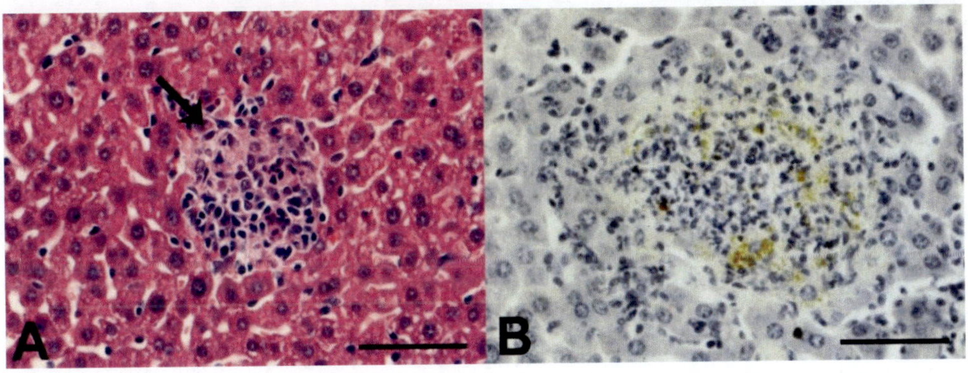

Figure 1. Mouse. Liver. (A) Microgranuloma (arrow) due to *B. abortus* 2308 infection in C57BL/6 mouse, 21 days after inoculation. HE. (B) The bacteria are localized inside of macrophages in microgranulomas. Immunohistochemistry. Bar = 50 μm.

Macrophages can induce its own death by apoptosis, signaling to other cells the presence of pathogens when its antimicrobial mechanisms fail to control bacterial growth. It is noteworthy that virulent smooth *Brucella* is able to inhibit apoptosis of infected macrophages [44, 49, 63, 95], whereas mutant rough *Brucella* induces macrophage cell death [95, 97].

Some virulence factors of *Brucella* have been identified as essential for surviving and establishing the intracellular replication niche within macrophages. The cyclic β-1,2-glucan and LPS are required for early intracellular survival of smooth *Brucella* [3, 101]. *Brucella* cyclic β-1, 2-glucan may interfere with maturation of early endosome by disrupting the lipid rafts and consequently prevent phagosome-lysosome fusion [3]. Entry of *B. suis* occurs via lipid raft and it is dependent on the LPS O-chain. Consequently, lipid raft-mediated entry prevents phagosome-lysosome fusion [101]. The type IV secretion system (T4SS) is required at later stages of intracellular infection but is not for early survival stages [25]. *Brucella* T4SS is encoded by the *virB* operon, and *virB* mutants do not survive inside macrophage [35, 36, 125]. The T4SS is required for inhibiting phagosome-lysosome fusion and directing the

Brucella-containing vacuole towards the acquisition of endoplasmic reticulum membranes, which results in the establishment of the *Brucella* replication niche [25, 29].

III. Other Innate Immunity Effector Cells

In addition to macrophages, others cells associated with innate immunity such as polymorphonuclear leukocytes (PMN), natural killer cells (NK) and γδ T cells play a role during *Brucella* infection. PMN recruitment at early stages of infection represents the first-line of cellular defense particularly against gram-negative bacterial infections [111]. The lipid A and LPS from gram-negative bacteria bind to neutrophils, inducing oxygen radical and lysozyme production that are responsible for bactericidal activity of these cells [105]. However, *Brucella* has peculiar mechanisms to prevent a significant recruitment of PMN, and to resist to their bactericidal extracts [7, 105]. It is known that smooth strains of *Brucella* induce lower PMN degranulation and are more resistant to intracellular killing than rough strains [7, 76,106, 107]. The presence of LPS O-chain in *Brucella* smooth strains is responsible to impair the induction of PMN respiratory burst and lysozyme secretion [79, 105]. Furthermore, the virulence factors located on the *B. abortus* membrane inhibit PMN degranulation even by non viable organisms [105].

Oxygen-independent reactions of PMN do not play a significant role in killing *B. abortus* [106]. However, oxidative-dependent mechanisms mediated by myeloperoxidase may have effective bactericidal activity when associated with hydrogen peroxide and halides, such as chloride and iodine [76, 106,107]. Thus, *Brucella*-infected neutrophils do trigger significant oxygen metabolites production by respiratory burst, which limits PMN killing capacity [106]. Moreover, recent studies emphasize that *B. abortus* infection induces mild respiratory burst and significantly lower PMN degranulation compared to other pathogenic intracellular bacteria, such as *Salmonella* [7]. These results suggest that PMN do not have an important role in controlling *B. abortus* infection possibly serving as a pathogen carrier within the host [7].

NK cells are one of the most important resources of the immune system in the first line of defense against pathogens because they are able to directly kill infected target cells and to produce IFNγ. The production of IFNγ is important when a cellular immune response (Th1) is needed, especially against intracellular pathogens such as *Brucella* [57]. In addition to its capacity of killing microbe-infected cells, NK cells can directly cause lysis of gram-negative bacteria [54]. These antimicrobial properties could potentially contribute for controlling *Brucella* infection as demonstrated in the case of other intracellular bacteria such as *Listeria* and *Salmonella* [39, 47]. Conversely, NK cells do not play an important role in controlling of *B. abortus* during the early stages of infection in mice [47]. Although infected mice are able to properly stimulate NK cells, it is not sufficient for clearing bacteria in the infected animals. NK cells apparently produce small amounts of IFNγ during the early stages of infection when compared to CD4+ T lymphocytes [47]. Therefore, other effectors of the innate immune system seem to play a more significant role in eliminating *Brucella* and activating a Th1 immune response in the mouse. In contrast, an *in vitro* study using *B. suis* and a co-culture of human macrophages and NK cells demonstrated that NK cells do have an important role in controlling the *Brucella* infection [39]. Differently from mice, human NK cells are able to kill infected macrophages and, therefore, to prevent proliferation of *Brucella*. NK cells cytotoxic effect is mostly mediated by cell-to-cell contact by releasing perforin and granzymes into infected-macrophages rather than through cytokine production such as INFγ [39]. However, although these results support the notion that NK cells may be important in controlling human brucellosis by direct cytotoxic effects, NK activity is impaired in human patients with acute brucellosis [110].

A putative role of γδ T cells during *Brucella* infection has been proposed in human brucellosis since the number of γδ T cells in peripheral blood greatly increase during the acute phase of infection with *B. melitensis* [11]. γδ T cells are effector cells associated primarily with adaptive immunity, but they also act in innate immune responses with significant contribution in IFNγ production [13]. Importantly, γδ T cell activation and proliferation are stimulated by *B. suis* infected-macrophages [38, 91, 92], and supposedly γδ T cells may contribute for controlling *Brucella* infection in humans due to a combination of mechanisms, including lyses of infected macrophages via cytotoxic effects, and reduction in bacterial loads by secreting antimicrobial peptides and cytokines [40, 88, 92].

IV. Other Components of the Innate Immune Response

Complement and cytokines are effector components of innate immunity that participate during acute *Brucella* infection. Importantly, *Brucella* is able to escape or resist the action of these effector mechanisms. Outer membrane proteins (OMPs) and LPS of gram-negative bacteria are involved in serum sensitivity, and induction of both classical and alternative pathways of complement activation [79]. Complement activation is a mechanism for killing extracellular pathogens, and therefore could be important during the early stages of infection with *Brucella* or in cases of bacteremia. However, apparently *Brucella* has the ability to impair serum bactericidal activity by preventing activation of the alternative complement pathway [7, 22, 79]. In addition, *Brucella* prevents formation of stable membrane attack complexes because C1 does not bind to the bacterial surface [43]. Apparently, serum sensitive strains of *B. abortus* induce the classical pathway of complement cascade in cattle [31, 42], leading to C1q binding, then C3b binding and activation, followed by C5b-9 [42, 43]. However, serum resistant strains of *B. abortus* do not activate the complement cascade, and the organisms that survive direct killing are opsonized by C3b and phagocytosed [43]. Smooth strains of *Brucella* seem to be more resistant to complement killing than rough strains [31, 42, 79]. The presence of LPS O-chain is one of the major factors that determine smooth strains resistance [79, 105]. It is known that *B. abortus* LPS O-chain blocks C1q access to the outer membrane proteins and that long O-chain molecules prevent C3b attachment [43, 105]. Additionally, serum resistance of smooth strains may be due to lack of C9 deposition on bacterial surface when compared to rough strains [42].

Microarray analyses demonstrated that *Brucella* infection *in vitro* up-regulates expression of pro-inflammatory cytokines as TNFα, IL-1β, and modulator molecules associated with Th1 responses, such as IL-12 and MHCII in macrophages during early stages of infection [44]. In addition, *Brucella* infection *in vivo* induces an increase of gene expression induced by type I and II interferon in splenocytes from infected mice [108]. Indeed, it has been demonstrated that *Brucella* infection can also induce IFNα and IFNβ expression *in vitro* [69, 109].

Although type I IFN is recognized to be essential against viruses, it is also induced in response to bacterial infections [34]. However, the protective role of type I IFN signaling during infection with *Brucella* virulent strains is still not clear [108]. Conversely, resistance to intracellular bacterial pathogens such as *Brucella* is strongly dependent on the ability of the organism to develop a cellular type of immune response (Th1), which is mediated by cytokines such as INFγ (a type II IFN), IL-12, TNFα, among others. These cytokines activate macrophages, lymphocytes, and other effectors of the innate or adaptive immune system to optimize their capacity of pathogen clearance [71].

In innate immunity, IFNγ enhances the NK cells activity and activates macrophages inducing expression of iNOS. IFNγ is produced by Th1 T cells, CD8+ T cells, and NK cells, acting on B cells, T cells, NK cells, and macrophages [85]. The critical role of IFNγ in controlling *Brucella* infection in mice has been well established through *in vitro* and *in vivo* studies [38, 72, 74, 129]. *In vitro* studies have demonstrated that IFNγ-treated macrophages are more capable of inhibiting growth of *B. abortus* when compared to untreated macrophages [72, 74]. Antibody depletion of endogenous IFNγ in mice results in increased numbers of *B. abortus* in the spleen and liver, which is associated with a reduction in splenomegaly and lower numbers of macrophages in the spleen of infected mice. Furthermore, when the ability to produce endogenous IFNγ is restored, clearance of *Brucella* is promptly recommenced [129]. In humans, one study demonstrated that levels of IFNγ in serum from patients with brucellosis decreases throughout the course of infection [104]. Therefore, the implications of this pattern of IFNγ production in human brucellosis have yet to be clarified.

In the murine model of *Brucella* infection, endogenous IL-12 stimulates IFNγ production and clearance of the bacteria *in vivo* [130,131]. Antibody depletion of IL-12 results in reduced IFNγ production by splenocytes, decrease number of macrophages in spleen, and lower levels of NO, indicating an impairment of macrophage activation [130]. Recent *in vitro* and *in vivo* studies demonstrated that induction of both TNFα and IL-12 p40 requires an intact MyD88 pathway [69, 82]. IL-12 is a Th1 cytokine produced by activated macrophages, dendritic cells, and B cells. Thus, IL-12 induces secretion of IL-2 and IFNγ, enhancing T and NK cells cytotoxicity [71].

Endogenous TNFα may play an important role in controlling *Brucella* infection in mice, mainly during the early stages of the infection [131]. TNFα is a cytokine produced by various cell types, including macrophages and B cells. Among other functions, TNFα activates macrophages, stimulates production of IL-1, IL-6, M-CSF (macrophage colony stimulating-factor) and GM-CSF (granulocyte macrophage colony-stimulating factor) as well as further induction of TNFα secretion [71]. Antibody depletion of TNFα prior to *Brucella* infection does not abrogate IFNγ production by murine T cells in response to *Brucella* antigens. Therefore, TNFα in mice is involved in resistance to brucellosis via a direct action on effector cells, but it is not essential for induction of acquired cellular resistance [131]. Conversely, human macrophages were not able to produce TNFα when inoculated with live *B. suis*, although heat killed *B. suis* triggered TNFα production [38]. Failure of human *Brucella*-infected cells to produce TNFα is due to a direct active bacterial mechanism which impairs cell signaling pathways that leads to TNFα synthesis [23]. Therefore, during the phagocytic process of *Brucella* by human macrophages, the absence of TNFα may result in an impairment of early macrophage killing mechanisms as well as failure in NK cell activation.

Disruption of these mechanisms may favor the initial establishment of the bacteria in human host cells [38].

Production and action of other pro-inflammatory cytokines during *Brucella* infection has been investigated. IL-10 is known to be produced in mice in response to immunization with heat-killed *B. abortus* strain 19 [117]. IL-10 is produced by CD4+ Th2 cells, activated macrophages, and B cells [86], and it down-regulates Th1 responses and/or increases susceptibility to a number of bacterial and parasitic infections [46, 86]. Endogenous IL-10 has a detrimental effect on the control of *B. abortus* infections in BALB/c mice even in the presence of IFNγ [46]. In addition, *in vitro* data demonstrated that IL-10 can decrease the production by spleen cells of IFNγ in response to stimulation by *Brucella* antigens and inhibit the anti-*Brucella* effector functions of macrophages [46].

Mice experimentally infected with *B. abortus* have a significant decrease in IL-18 synthesis [48]. IL-18 has a direct pro-inflammatory effect on T and NK cells, enhancing proliferation and cytotoxicity and stimulating the production of cytokines such as TNFα [37, 48]. IL-18 alone does not induce significant IFN production, but it can synergistically enhance IL-12-stimulated IFNγ production [128]. A study demonstrated that *B. abortus* and L-Omp19 stimulate macrophages via TLR-2 to produce IL-6 [55]. Secretion of IL-6 inhibits IFNγ induced MHC-II expression and antigen presentation by the macrophage infected with *Brucella* [8]. These new evidence suggests that IL-6 has inhibitory effects on macrophage functions during *Brucella* infection *in vitro*. The role of IL-6 *in vivo* studies has yet to be determined.

Recent studies from our lab and others have demonstrated that *B. abortus* suppresses pro-inflammatory cytokine and chemokine expression during early stages of infection, which is followed by a delayed and mild expression of pro-inflammatory chemokine *in vitro* and *in vivo* [7, 24, 109]. Apparently, this ability to inhibit expression of pro-inflammatory mediators may be a strategy of *B. abortus* to favor persistent infection.

V. Pathogen Recognition by the Host

The first line of host defense mechanisms against infection includes the pattern recognition receptors (PRRs) that recognize pathogen-associated molecular patterns (PAMPs) from a variety of pathogens. This recognition activates intracellular signaling pathways that culminate in a rapid induction of cytokines, chemokines, and co-stimulatory molecules [89]. There are several functionally distinct classes of PRRs and the best characterized are the Toll-like receptors (TLRs). These receptors recognize a broad range of microbial products [121]. The TLR is a single-pass type I transmembrane receptor with an extra cellular domain, and an intracellular Toll/interleukin-1 receptor homology (TIR) domain, and it is expressed in macrophages, dendritic cells, leucocytes, and epithelial cells [118]. The others important PRRs are proteins of the nucleotide-binding oligomerization domain (NOD) subfamily, NOD1 and NOD2 as well as other similar intracellular receptors generically named NOD-like receptors (NLRs). These innate immune receptors are involved in bacterial intracellular detection through the recognition of distinct motifs of peptidoglycan fragments within the host cell cytosol [27, 52]. Although there are various studies investigated the role of TLRs during *Brucella* infection, so far there is no evidence of the participation of NLRs.

Brucella is recognized by TLR-2, TLR-4, and TLR-9 [7, 89, 109]. LPS and lipoproteins are detected by TLR-4 and TLR-2, respectively. TLR-9 has restricted expression in endosomal compartments and recognizes intracellular bacterial DNA [121]. The first evidence that *Brucella* PAMPs can interact with TLRs was demonstrated using heat-killed *B. abortus* (HKBA) [68], which stimulates cells via MyD88 (Myeloid differentiation primary response gene 88), resulting in NF-κB (transcription factor-kappaB) translocation and secretion of cytokines such as TNFα, IL-6, and IL-12. *Brucella* outer membrane lipoproteins are active components in HKBA triggering a TLR-2 response [55, 68]. In macrophages, TNFα induction is TLR-4 and TLR-2 dependent, and it is triggered by live but not killed *B. abortus* [7]. TNFα induction is TLR-2 dependent and TLR-4 independent in DCs [82]. IL-12 secretion is neither TLR-2 nor TLR-4 dependent in macrophages, whereas it is TLR-2 dependent in DCs [82]. Conversely, TLR-9 that recognizes *Brucella* DNA in endosomes is not required for TNFα induction in macrophages and DCs, but it mediates IL-12 secretion [69, 82]. Importantly, other Th1 cytokines such as IFNγ are also induced by HKBA via TLR-9 [69].

Brucella LPS has a non-classical structure, which is not a strong agonist of TLR-4, and consequently has low immunostimulatory activity and low endotoxicity [79]. Accordingly, *Brucella* LPS is a hundred-times less active and toxic than *Escherichia coli* LPS and it has reduced and late ability to activate pro-inflammatory cytokines compared to *Salmonella* LPS [7, 41,126]. When different *Brucella* LPS phenotypes are compared, a rough LPS *B. abortus* mutant can stimulate TLR-4 and consequently macrophage activation, whereas the smooth LPS virulent strain cannot. In addition, it has been demonstrated that interaction of *B. abortus* LPS and TLR-4 plays a role in bacterial uptake by macrophages [96]. Interestingly, there are evidences that *Brucella* molecules other than LPS can stimulate DCs via TLR-4. The enzyme lumazine synthase (BLS) interacts with TLR-4 and stimulates expression of co-stimulatory molecules and production of IL-6, TNFα, and IL-12 [10].

The role of TLR-4 during *Brucella* clearance *in vivo* is controversial. Some studies indicate a significant role of TLR-4 in controlling *Brucella* infection in mice [18, 30], whereas other studies did not demonstrate any involvement of TLR-4 in resistance [7, 126]. Although *Brucella* lipoproteins are responsible for a pro-inflammatory response induced via TLR-2, this receptor is not important for controlling *Brucella* infection in mice [7, 126]. Conversely, it has been demonstrated that the TLR-9 is required for controlling *Brucella* infection *in vivo* [30, 82]. Therefore, it is still not clear how exactly TLRs confer resistance to *Brucella*. However, it is well established that the adaptor molecule MyD88 is required for controlling *Brucella* infection *in vivo* independently of which specific TLR is involved. MyD88 is critical for TLR-mediated activation of the NF-κB, and consequently induction of pro-inflammatory cytokines such as TNFα during *Brucella* infection [68, 69, 82]. MyD88 also affects both DC maturation and macrophage function [30, 82, 126]. Additionally, DC maturation and production of IL-12 is dependent on the adaptor molecule MyD88, regulating activation of *Brucella*-specific IFNγ producing T cells [82].

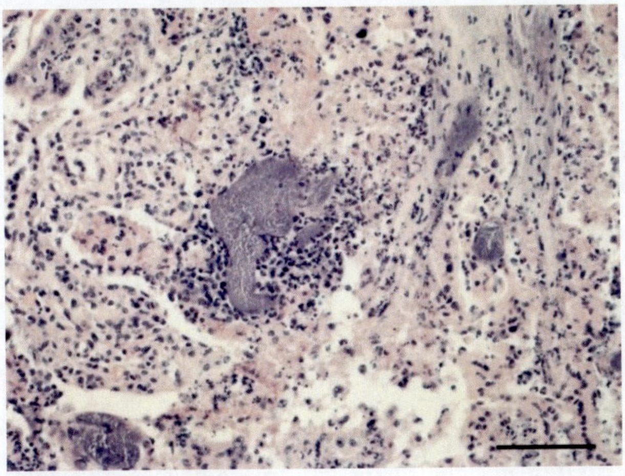

Figure 2. Cow. Caruncula. Necrotic placentitis associated with neutrophilic infiltrate and myriad of *Brucella abortus*. HE. Bar = 100 μm.

Various studies have shown that live and killed *Brucella* induces pro-inflammatory cytokines [18, 126]. However, *in vivo* levels of IL-10, IL-1β, IL-6 and TNFα were comparatively insignificant after infection with *B. abortus* in mice when compared with *Salmonella* [7]. It is worth emphasizing that we demonstrated suppression of pro-inflammatory cytokines and mediators during the early stages of infection in bovine trophoblastic cells during *in vitro* and *in vivo* infection with *B. abortus* [24]. This suppression mechanism may be essential to *B. abortus* colonize the pregnant uterus and cause necrotic placentitis (Figure 2) and late gestation abortion observed in infected cattle [123].

Interestingly, recent studies demonstrated that *Brucella* interferes in TLR signaling by producing a protein that contains a Toll/interleukin-1 receptor (TIR) domain, which interferes with TLR/MyD88 signaling [28, 109]. Two independent groups have recently identified *Brucella* proteins, named Btp1 in *B. abortus* [109] and TcpB in *B. melitensis* [28]. These are hortholog genes whose products interfere with TLR signaling. *B. abortus* Btp1 affects TLR-2 signaling, down-modulating DCs maturation, affecting DC function and inhibiting secretion of pro-inflammatory cytokines [109]. Similarly, *B. melitensis* TcpB was shown to interact with MyD88 impairing TLR-2 and TLR-4 signaling, preventing secretion of pro-inflammatory cytokines [28]. Therefore, these proteins appear to constitute a new class of virulence factors that has ability to inhibit TLRs and MyD88-specific signaling, thus suppressing innate immunity.

In summary, *Brucella* can interact with TLR-2, TLR-4, and TLR-9, which signals via MyD88. However, the pathogen is capable of hampering or inhibiting this mechanism, resulting in low levels of cytokines. Therefore, activation of host inflammatory mechanisms is inhibited, favoring long-term intracellular survival of *Brucella*. Apparently survival and replication within phagocytic cells with minimal signaling of host defenses is a critical step in pathogenesis of *Brucella*.

REFERENCES

[1] Ables, G. P., Nishibori, M., Kanemaki, M. & Watanabe, T. (2002). Sequence analysis of the NRAMP1 genes from different bovine and buffalo breeds. *Journal of Veterinary Medical Science.*, *64*, 1081-1083.

[2] Adams, L. G. & Templeton, J. W. (1998). Genetic resistance to bacterial diseases of animals. *Revue Scientifique et Technique.*, *17*, 200-219.

[3] Arellano-Reynoso, B., Lapaque, N., Salcedo, S., Briones, G., Ciocchini, A. E., Ugalde, R., Moreno, E., Moriyon, I. & Gorvel, J. P. (2005). Cyclic beta-1,2-glucan is a *Brucella* virulence factor required for intracellular survival. *Nature Immunology.*, *6*, 618-625.

[4] Arenas, G. N., Staskevich, A. S., Aballay, A. & Mayorga, L. S. (2000). Intracellular trafficking of *Brucella abortus* in J774 macrophages. *Infection and Immunity.*, *68*, 4255-4263.

[5] Atkinson, P. G. & Barton, C. H. (1998). Ectopic expression of NRAMP1 in COS-1 cells modulates iron accumulation. *Federation of European Biochemical Societies Letters.*, *425*, 239-242.

[6] Baldwin, C. L., Jiang, X. & Fernandes, D. M. (1993). Macrophage control of *Brucella abortus*: influence of cytokines and iron. *Trends in Microbiology.*, *1*, 99-104.

[7] Barquero-Calvo, E., Chaves-Olarte, E., Weiss, D. S., Guzmán-Verri, C., Chacón-Díaz, C., Rucavado, A., Moriyón, I. & Moreno, E. (2007). *Brucella abortus* uses a stealthy strategy to avoid activation of the innate immune system during the onset of infection. *PLoS ONE.*, *2*, 1-14.

[8] Barrionuevo, P., Cassataro, J., Delpino, M. V., Zwerdling, A., Pasquevich, K. A., García Samartino, C., Wallach, J. C., Fossati, C. A. & Giambartolomei, G. H. (2008). *Brucella abortus* inhibits major histocompatibility complex class II expression and antigen processing through interleukin-6 secretion via Toll-like receptor 2. *Infection and Immunity.*, *76*, 250-262.

[9] Barthel, R., Feng, J., Piedrahita, J. A., McMurray, D. N., Templeton, J. W. & Adams, L. G. (2001). Stable transfection of the bovine NRAMP1 gene into murine RAW264.7 cells: effect on *Brucella abortus* survival. *Infection and Immunity.*, *69*, 3110-3119.

[10] Berguer, P. M., Mundiñano, J., Piazzon, I., Goldbaum, F. A. (2006). A polymeric bacterial protein activates dendritic cells via TLR4. *Journal of Immunology.*, *176*, 2366-2372.

[11] Bertotto, A., Gerli, R., Spinozzi, F., Muscat, C., Scalise, F., Castellucci, G., Sposito, M., Candio, F. & Vaccaro, R. (1993). Lymphocytes bearing the γδ T cell receptor in acute *Brucella melitensis* infection. *European Journal of Immunology.*, *23*, 1177-1180.

[12] Blackwell, J. M., Searle, S., Goswami, T. & Miller, E. N. (2000). Understanding the multiple functions of NRAMP1. *Microbes and Infection.*, *2*, 317-321.

[13] Boismenu, R. & Havran, W. L. (1997). An innate view of γδ T cells. *Current Opinion in Immunology.*, *9*, 57-63.

[14] Borriello, G., Capparelli, R., Bianco, M., Fenizia, D., Alfano, F., Capuano, F., Ercolini, D., Parisi, A., Roperto, S. & Iannelli, D. (2006). Genetic resistance to *Brucella abortus* in the water buffalo (*Bubalus bubalis*). *Infection and Immunity.*, *74*, 2115-2120.

[15] Bradley, D. J., Taylor, B. A., Blackwell, J., Evans, E. P. & Freeman, J. (1979). Regulation of *Leishmania* populations within the host. III. Mapping of the locus controlling susceptibility to visceral leishmaniasis in the mouse. *Clinical and Experimental Immunology.*, *37*, 7-14.

[16] Bravo, M. J., Colmenero, J. de, D., Martín, J., Alonso, A. & Caballero González, A. (2006). Variation in the NRAMP1 gene does not affect susceptibility or protection in human brucellosis. *Microbes and Infection.*, *8*, 154-156.

[17] Campbell, G. A. & Adams, L. G. (1992) The long-term culture of bovine monocyte-derived macrophages and their use in the study of intracellular proliferation of *Brucella abortus*. *Veterinary Immunology and Immunopathology.*, *34*, 291-305.

[18] Campos, M. A., Rosinha, G. M., Almeida, I. C., Salgueiro, X. S., Jarvis, B. W., Splitter, G. A., Qureshi, N., Bruna-Romero, O., Gazzinelli, R. T. & Oliveira, S. C. (2004). Role of Toll-like receptor 4 in induction of cell-mediated immunity and resistance to *Brucella abortus* infection in mice. *Infection and Immunity.*, *72*, 176-186.

[19] Capparelli, R., Alfano, F., Amoroso, M. G., Borriello, G., Fenizia, D., Bianco, A., Roperto, S., Roperto, F. & Iannelli, D. (2007). Protective effect of the NRAMP1 BB genotype against *Brucella abortus* in the water buffalo (*Bubalus bubalis*). *Infection and Immunity.*, *75*, 988-996.

[20] Capparelli, R., Borriello, G., Marabelli, R., Roperto, S., Roperto, F. & Iannelli, D. (2007). The NRAMP1 AA genotype confers susceptibility to *Brucella abortus* in water buffalo. *Mammalian Genome.*, *18*, 137-143.

[21] Capparelli, R., Parlato, M., Amoroso, M. G., Roperto, S., Marabelli, R., Roperto, F. & Iannelli, D. (2008). Mannose-binding lectin haplotypes influence *Brucella abortus* infection in the water buffalo (*Bubalus bubalis*). *Immunogenetics.*, *60*, 157-165.

[22] Cardoso, P. G., Macedo, G. C., Azevedo, V. & Oliveira, S. C. (2006). *Brucella* spp noncanonical LPS: structure, biosynthesis, and interaction with host immune system. *Microbial Cell Factories.*, *5*, 13.

[23] Caron, E., Peyrard, T., Kohler, S., Cabane, S., Liautard, J. P. & Dornand, J. (1994). Live *Brucella* spp. fail to induce TNF-α excretion upon infection of U-937-derived phagocytes. *Infection and Immunity.*, *62*, 5267-5274.

[24] Carvalho Neta, A. V., Stynen, A. P., Paixão, T. A., Miranda, K. L., Silva, F. L., Roux, C. M., Tsolis, R. M., Everts, R. E., Lewin, H. A., Adams, L. G., Carvalho, A. F., Lage, A. P. & Santos, R. L. (2008). Modulation of the bovine trophoblastic innate immune response by *Brucella abortus*. *Infection and Immunity.*, *76*, 1897-1907.

[25] Celli, J., de Chastellier, C., Franchini, D. M., Pizarro-Cerda, J., Moreno, E. & Gorvel, J. P. (2003). *Brucella* evades macrophage killing via VirB-dependent sustained interactions with the endoplasmic reticulum. *The Journal of Experimental Medicine.*, *198*, 545-556.

[26] Cellier, M., Govoni, G., Vidal, S., Kwan, T., Groulx, N., Liu, J., Sanchez, F., Skamene, E., Schurr, E. & Gros, P. (1994). Human natural resistance-associated macrophage protein: cDNA cloning, chromosomal mapping, genomic organization,

and tissue-specific expression. *The Journal of Experimental Medicine.*, *180*, 1741-1752.

[27] Chen, G., Shaw, M. H., Kim, Y. G. & Nuñez, G. (2009). Nod-like Receptors: Role in Innate Immunity and Inflammatory Disease. *Annual Review of Pathology.*, *4*, 365-398.

[28] Cirl, C., Wieser, A., Yadav, M., Duerr, S., Schubert, S., Fischer, H., Stappert, D., Wantia, N., Rodriguez, N., Wagner, H., Svanborg, C. & Miethke, T. (2008). Subversion of Toll-like receptor signaling by a unique family of bacterial Toll/interleukin-1 receptor domain-containing proteins. *Nature Medicine.*, *14*, 399-406.

[29] Comerci, D. J., Martinez-Lorenzo, M. J., Sieira, R., Gorvel, J. P. & Ugalde, R. A. (2001). Essential role of the VirB machinery in the maturation of the *Brucella abortus*-containing vacuole, *Cellular Microbiology.*, *3*, 159-168.

[30] Copin, R., De Baetselier, P., Carlier, Y., Letesson, J. J., Muraille, E. (2007). MyD88-dependent activation of B220-CD11b+LY-6C+ dendritic cells during *Brucella melitensis* infection. *Journal of Immunology.*, *178*, 5182-5191.

[31] Corbeil, L. B., Blau, K., Inzana, T. J., Nielsen, K. H., Jacobson, R. H., Corbeil, R. R. & Winter, A. J. (1988). Killing of *Brucella abortus* by bovine serum. *Infection and Immunity.*, *56*, 3251-3261.

[32] Corbel, M. J. *Brucellosis human and animals*. Geneva: World health organization Press; 2006.

[33] Coussens, P. M., Coussens, M. J., Tooker, B. C. & Nobis, W. (2004). Structure of the bovine natural resistance associated macrophage protein (NRAMP1) gene and identification of a novel polymorphism. *DNA Sequence.*, *15*, 15-25.

[34] Decker, T., Müller, M. & Stockinger, S. (2005). The yin and yang of type I interferon activity in bacterial infection. *Nature Reviews Immunology.*, *5*, 675-687.

[35] Den Hartigh, A. B., Rolán, H. G., De Jong, M. F. & Tsolis, R. M. (2008). VirB3 to VirB6 and VirB8 to VirB11, but not VirB7, are essential for mediating persistence of *Brucella* in the reticuloendothelial system. *Journal of Bacteriology.*, *190*, 4427-4436.

[36] Den Hartigh, A. B., Sun, Y. H., Sondervan, D., Heuvelmans, N., Reinders, M. O., Ficht, T. A. & Tsolis, R. M. (2004). Differential requirements for VirB1 and VirB2 during *Brucella abortus* infection. *Infection and Immunity.*, *72*, 5143-5149.

[37] Dinarello, C. A. & Fantuzzi, G. (2003). Interleukin-18 and host defense against infection. *Journal of Infectious Diseases.*, *187*, Suppl 2, S370-384.

[38] Dornand, J., Gross, A., Lafont, V., Liautard, J., Oliaro, J. & Liautard, J. P. (2002). The innate immune response against *Brucella* in humans. *Veterinary Microbiology.*, *90*, 383-394.

[39] Dornand, J., Lafont, V., Oliaro, J., Terraza, A., Castaneda-Roldan, E. & Liautard, J. P. (2004). Impairment of intramacrophagic *Brucella suis* multiplication by human natural killer cells through a contact-dependent mechanism. *Infection and Immunity.*, *72*, 2303-2311.

[40] Dudal, S., Turriere, C., Bessoles, S., Fontes, P., Sanchez, F., Liautard, J., Liautard, J. P. & Lafont, V. (2006). Release of LL-37 by activated human Vgamma9Vdelta2 T cells: a microbicidal weapon against *Brucella suis*. *Journal of Immunology.*, *177*, 5533-5539.

[41] Dueñas, A. I., Orduña, A., Crespo, M. S. & García-Rodríguez, C. (2004). Interaction of endotoxins with Toll-like receptor 4 correlates with their endotoxic potential and may explain the proinflammatory effect of *Brucella* spp. LPS. *International Immunology*., *16*, 1467-1475.

[42] Eisenschenk, F. C., Houle, J. J. & Hoffmann, E. M. (1995). Serum sensitivity of field isolates and laboratory strains of *Brucella abortus*. *American Journal of Veterinary Research*., *56*, 1592-1598.

[43] Eisenschenk, F. C., Houle, J. J. & Hoffmann, E. M. (1999). Mechanism of serum resistance among *Brucella abortus* isolates. *Veterinary Microbiology*., *68*, 235-244.

[44] Eskra, L., Mathison, A. & Splitter, G. (2003). Microarray analysis of mRNA levels from RAW264.7 macrophages infected with *Brucella abortus*. *Infection and Immunity*., *71*, 1125-1133.

[45] Feng, J., Li, Y., Hashad, M., Schurr, E., Gros, P., Adams, L. G. & Templeton, J. W. (1996). Bovine natural resistance associated macrophage protein 1 (NRAMP1) gene. *Genome Research*., *6*, 956-964.

[46] Fernandes, D. M. & Baldwin, C. L. (1995). Interleukin-10 downregulates protective immunity to *Brucella abortus*. *Infection and Immunity*., *63*, 1130-1133.

[47] Fernandes, D. M., Benson, R. & Baldwin, C. L. (1995). Lack of a role for natural killer cells in early control of *Brucella abortus* 2308 infections in mice. *Infection and Immunity*., *63*, 4029-4033.

[48] Fernández-Lago, L., Orduña, A. & Vizcaíno, N. (2005). Reduced interleukin-18 secretion in *Brucella abortus* 2308-infected murine peritoneal macrophages and in spleen cells obtained from *B. abortus* 2308-infected mice. *Journal of Medical Microbiology*., *54*, 527-531.

[49] Fernandez-Prada, C. M., Zelazowska, E. B., Nikolich, M., Hadfield, T. L., Roop, R. M., Robertson, G. L. & Hoover, D. L. (2003). Interactions between *Brucella melitensis* and human phagocytes: bacterial surface O-Polysaccharide inhibits phagocytosis, bacterial killing, and subsequent host cell apoptosis. *Infection and Immunity*., *71*, 2110-2119.

[50] Forbes, J. R. & Gros, P. (2003). Iron, manganese, and cobalt transport by NRAMP1 (Slc11a1) and NRAMP2 (Slc11a2) expressed at the plasma membrane. *Blood*., *102*, 1884-1892.

[51] Foster, G., Osterman, B. S., Godfroid, J., Jacques, I. & Cloeckaert, A. (2007). *Brucella ceti* sp. nov. and *Brucella pinnipedialis* sp. nov. for *Brucella* strains with cetaceans and seals as their preferred hosts. *International Journal of Systematic and Evolutionary Microbiology*., *11*, 2688-2693.

[52] Franchi, L., Park, J. H., Shaw, M. H., Marina-Garcia, N., Chen, G., Kim, Y. G. & Núñez, G. (2008). Intracellular NOD-like receptors in innate immunity, infection and disease. *Cellular Microbiology*., *10*, 1-8.

[53] Ganguly, I., Sharma, A., Singh, R., Deb, S. M., Singh, D. K. & Mitra, A. (2008). Association of microsatellite (GT)n polymorphism at 3'UTR of NRAMP1 with the macrophage function following challenge with *Brucella* LPS in buffalo (*Bubalus bubalis*). *Veterinary Microbiology*., *129*, 188-196.

[54] Garcia-Penarrubia, P., Foster, F. T., Kelley, R. O., McDowell, T. D., Bankhurst, A. D. (1989). Antibacterial activity of natural killer cells. *The Journal of Experimental Medicine*., *169*, 99-113.

[55] Giambartolomei, G. H., Zwerdling, A., Cassataro, J., Bruno, L., Fossati, C. A. & Philipp, M. T. (2004). Lipoproteins, not lipopolysaccharide, are the key mediators of the proinflammatory response elicited by heat-killed *Brucella abortus*. *Journal of Immunology.*, *173*, 4635-4642.

[56] Godfroid, J., Cloeckaert, A., Liautard, J. P., Kohler, S., Fretin, D., Walravens, K., Garin-Bastuji, B. & Letesson, J. J. (2005). From the discovery of the Malta fever's agent to the discovery of a marine mammal reservoir, brucellosis has continuously been a re-emerging zoonosis. *Veterinary Research.*, *36*, 313-326. Review.

[57] Golding, B., Scott, D. E., Scharf, O., Huang, L. Y., Zaitseva, M., Lapham, C., Eller, N. & Golding, H. (2001). Immunity and protection against *Brucella abortus*. *Microbes and Infection.*, *3*, 43-48.

[58] Gomes, M. S. & Appelberg, R. (1998). Evidence for a link between iron metabolism and NRAMP1 gene function in innate resistance against *Mycobacterium avium*. *Immunology.*, *95*, 165-168.

[59] Goswami, T., Bhattacharjee, A., Babal, P., Searle, S., Moore, E., Li, M., Blackwell, J. M. (2001). Natural-resistance-associated macrophage protein 1 is an H+/bivalent cation antiporter. *The Biochemical Journal.*, *354*, 511-519.

[60] Gros, P., Skamene, E. & Forget, A. (1981). Genetic control of natural resistance to *Mycobacterium bovis* (BCG) in mice. *Journal of Immunology.*, *127*, 2417-2421.

[61] Gross, A., Bertholet, S., Mauel, J. & Dornand, J. (2004). Impairment of *Brucella* growth in human macrophagic cells that produce nitric oxide. *Microbial Pathogenesis.*, *36*, 75-82.

[62] Gross, A., Spiesser, S., Terraza, A., Rouot, B., Caron, E. & Dornand, J. (1998). Expression and bactericidal activity of nitric oxide synthase in *Brucella suis*-infected murine macrophages. *Infection and Immunity.*, *66*, 1309-1316.

[63] Gross, A., Terraza, A., Ouahrani-Bettache, S., Liautard, J. P. & Dornand, J. (2000). In vitro *Brucella suis* infection prevents the programmed cell death of human monocytic cells. *Infection and Immunity.*, *68*, 342-351.

[64] Gruenheid, S., Pinner, E., Desjardins, M. & Gros, P. (1997). Natural resistance to infection with intracellular pathogens: the NRAMP1 protein is recruited to the membrane of the phagosome. *The Journal of Experimental Medicine.*, *185*, 717-730.

[65] Guilloteau, L. A., Dornand, J., Gross, A., Olivier, M., Cortade, F., Vern, Y. L. & Kerboeuf, D. (2003). NRAMP1 is not a major determinant in the control of *Brucella melitensis* infection in mice. *Infection and Immunity.*, *71*, 621-628.

[66] Harmon, B. G., Adams, L. G., Templeton, J. W. & Smith, R. (1989). Macrophage function in mammary glands of *Brucella abortus*-infected cows and cows that resisted infection after inoculation of *Brucella abortus*. *American Journal of Veterinary Research.*, *50*, 459-465.

[67] Horín, P., Rychlík, I., Templeton, J. W. & Adams, L. G. (1999). A complex pattern of microsatellite polymorphism within the bovine NRAMP1 gene. *European Journal of Immunogenetics.*, *26*, 311-313.

[68] Huang, L. Y., Aliberti, J., Leifer, C. A., Segal, D. M., Sher, A., Golenbock, D. T. & Golding, B. (2003). Heat-killed *Brucella abortus* induces TNF and IL-12p40 by distinct MyD88-dependent pathways: TNF, unlike IL-12p40 secretion, is Toll-like receptor 2 dependent. *Journal of Immunology.*, *171*, 1441-1446.

[69] Huang, L. Y., Ishii, K. J., Akira, S., Aliberti, J. & Golding, B. (2005). Th1-like cytokine induction by heat-killed Brucella abortus is dependent on triggering of TLR9. *Journal of Immunology.*, *175*, 3964-3970.

[70] Jabado, N., Jankowski, A., Dougaparsad, S., Picard, V., Grinstein, S. & Gros, P. (2000). Natural resistance to intracellular infections: natural resistance-associated macrophage protein 1 (NRAMP1) functions as a pH-dependent manganese transporter at the phagosomal membrane. *The Journal of Experimental Medicine.*, *192*, 1237-1248.

[71] Janeway, C. A., Travers, P., Walport, M. & Shlomchik, M. (2001). *Immunobiology: The immune system in health and disease*. 5th edition. New York: Garland publishing;.

[72] Jiang, X. & Baldwin, C. L. (1993). Effects of cytokines on intracellular growth of *Brucella abortus*. *Infection and Immunity.*, *61*, 124-134.

[73] Jiang, X., Leonard, B., Benson, R. & Baldwin, C. L. (1993). Macrophage control of *Brucella abortus*: role of reactive oxygen intermediates and nitric oxide. *Cellular Immunology..*, *151*, 309-319.

[74] Jones, S. M. & Winter, A. J. (1992). Survival of virulent and attenuated strains of *Brucella abortus* in normal and gamma interferon-activated murine peritoneal macrophages. *Infection and Immunity.*, *60*, 3011-3014.

[75] Köhler, S., Porte, F., Jubier-Maurin, V., Ouahrani-Bettache, S., Teyssier, J. & Liautard, J.P. (2002). The intramacrophagic environment of Brucella suis and bacterial response. *Veterinary Microbiology.*, *90*, 299-309.

[76] Kreutzer, D. L., Dreyfus, L. A. & Robertson, D. C. (1979). Interaction of polymorphonuclear leukocytes with smooth and rough strains of *Brucella abortus*, *Infection and Immunity.*, *23*, 737-742.

[77] Kuhn, D. E., Baker, B. D., Lafuse, W. P. & Zwilling, B. S. (1999). Differential iron transport into phagosomes isolated from the RAW264.7 macrophage cell lines transfected with NRAMP1 Gly169 or NRAMP1 Asp169. *Journal of Leukocyte Biology.*, *66*, 113-119.

[78] Kumar, N., Mitra, A., Ganguly, I., Singh, R., Deb, S. M., Srivastava, S. K. & Sharma, A. (2005). Lack of association of brucellosis resistance with (GT)(13) microsatellite allele at 3'UTR of NRAMP1 gene in Indian zebu (Bos indicus) and crossbred (Bos indicus xBos taurus) cattle. *Veterinary Microbiology.*, *111*, 139-143.

[79] Lapaque, N., Moriyon, I., Moreno, E. & Gorvel, J. P. (2005). *Brucella* lipolysaccharide acts as a virulence factor. *Current Opinion in Microbiology.*, *8*, 60-66.

[80] Lissner, C. R., Swanson, R. N. & O'Brien, A. D. (1983). Genetic control of the innate resistance of mice to *Salmonella typhimurium*: expression of the Ity gene in peritoneal and splenic macrophages isolated in vitro. *Journal of Immunology.*, *131*, 3006-3013.

[81] Loisel-Meyer, S., Jimenez de Bagues, M. P., Basseres, E., Dornand, J., Kohler, S., Liautard, J. P. & Jubier-Maurin, V. (2006). Requirement of norD for *Brucella suis* virulence in a murine model of in vitro and in vivo infection. *Infection and Immunity.*, *74*, 1973-1976.

[82] Macedo, G. C., Magnani, D. M., Carvalho, N. B., Bruna-Romero, O., Gazzinelli, R. T. & Oliveira, S. C. (2008). Central role of MyD88-dependent dendritic cell

maturation and proinflammatory cytokine production to control *Brucella abortus* infection. *Journal of Immunology.*, *180*, 1080-1087.

[83] Martínez, R., Dunner, S., Barrera, G. & Cañon, J. (2008a). Novel variants within the coding regions of the Slc11A1 gene identified in *Bos taurus* and *Bos indicus* breeds. *Journal of Animal Breeding and Genetics.*, *125*, 57-62.

[84] Martínez, R., Toro, R., Montoya, F., Burbano, M., Tobón, J., Gallego, J., Dunner, S. & Cañón, J. (2008b). Bovine SLC11A1 3' UTR SSCP genotype evaluated by a macrophage in vitro killing assay employing a *Brucella abortus* strain. *Journal of Animal Breeding and Genetics.*, *125*, 271-279.

[85] Medzhitov, R. & Janeway, C. A. Jr. (1997). Innate immunity: impact on the adaptive immune response. *Current Opinion of Immunology.*, *9*, 4-9.

[86] Moore, K. W., O'Garra, A., Malefyt, R. D. W., Vieira, P. & Mosmann, T. R. (1993). Interleukin-10. *Annual Review of Immunology.*, *11*, 165-190.

[87] Mosser, D. M. (2003). The many faces of macrophage activation. *Journal of Leukocyte Biology.*, *73*, 209-212.

[88] Oliaro, J., Dudal, S., Liautard, J., Andrault, J. B., Liautard, J. P. & Lafont, V. (2005). Vgamma9Vdelta2 T cells use a combination of mechanisms to limit the spread of the pathogenic bacteria *Brucella*. *Journal of Leukocyte Biology.*, *77*, 652-660.

[89] Oliveira, S. C., De Oliveira, F. S., Macedo, G. C., De Almeida, L. A. & Carvalho, N. B. (2008). The role of innate immune receptors in the control of *Brucella abortus* infection: Toll-like receptors and beyond. *Microbes and Infection.*, *10*, 1005-1009.

[90] Oliveira, S. C., Soeurt, N. & Splitter, G. (2002). Molecular and cellular interactions between *Brucella abortus* antigens and host immune responses. *Veterinary Microbiology.*, *90*, 417-424.

[91] Ottones, F., Liautard, J., Gross, A., Rabenoelina, F., Liautard, J. P. & Favero, J. (2000a). Activation of human Vgamma9Vdelta2 T cells by a *Brucella suis* non-peptidic fraction impairs bacterial intracellular multiplication in monocytic infected cells. *Immunology.*, *100*, 252-258.

[92] Ottones, F., Dornand, J., Naroeni, A., Liautard, J. P. & Favero, J. (2000b). Vgamma9Vdelta2 T cells impair intracellular multiplication of *Brucella suis* in autologous monocytes through soluble factor release and contact-dependent cytotoxic effect. *Journal of Immunology.*, *165*, 7133-7139.

[93] Paixão, T. A., Ferreira, C., Borges, A. M., Oliveira, D. A., Lage, A. P. & Santos, R. L. (2006). Frequency of bovine NRAMP1 (Slc11a1) alleles in Holstein and Zebu breeds. *Veterinary Immunology and Immunopathology.*, *109*, 37-42.

[94] Paixão, T. A., Poester, F. P., Carvalho Neta, A. V., Borges, A. M., Lage, A. P. & Santos, R. L. (2007). NRAMP1 3' untranslated region polymorphisms are not associated with natural resistance to *Brucella abortus* in cattle. *Infection and Immunity.*, *75*, 2493-2499.

[95] Pei, J. & Ficht, T. A. (2004). *Brucella abortus* rough mutants are cytopathic for macrophages in culture. *Infection and Immunity.*, *72*, 440-450.

[96] Pei, J., Turse, J. E. & Ficht, T. A. (2008). Evidence of *Brucella abortus* OPS dictating uptake and restricting NF-kappaB activation in murine macrophages. *Microbes and Infection.*, *10*, 582-590.

[97] Pei, J., Turse, J. E., Wu, Q. & Ficht, T. A. (2006). *Brucella abortus* rough mutants induce macrophage oncosis that requires bacterial protein synthesis and direct interaction with the macrophage. *Infection and Immunity.*, *74*, 2667-2675.

[98] Pizarro-Cerda, J., Moreno, E., Sanguedolce, V., Mege, J.L. & Gorvel, J. P. (1998). Virulent *Brucella abortus* prevents lysosome fusion and is distributed within autophagosome-like compartments. *Infection and Immunity.*, *66*, 2387-2392.

[99] Poester, F. P., Gonçalves, V. S., Paixão, T. A., Santos, R. L., Olsen, S. C., Schurig, G. G. & Lage, A. P. (2006). Efficacy of strain RB51 vaccine in heifers against experimental brucellosis. *Vaccine.*, *24*, 5327-5334.

[100] Porte, F., Liautard, J. P. & Köhler, S. (1999). Early acidification of phagosomes containing *Brucella suis* is essential for intracellular survival in murine macrophages. *Infection and Immunity.*, *67*, 4041-4047.

[101] Porte, F., Naroeni, A., Ouahrani-Bettache, S. & Liautard, J. P. (2003). Role of the *Brucella suis* lipopolysaccharide O antigen in phagosomal genesis and in inhibition of phagosome-lysosome fusion in murine macrophages. *Infection and Immunity.*, *71*, 1481-1490.

[102] Price, R. E., Templeton, J. W., Smith III, R. & Adams, L. G. (1990). Ability of mononuclear phagocytes from cattle naturally resistant or susceptible to brucellosis to control in vitro intracellular survival of *Brucella abortus*. *Infection and Immunity.*, *58*, 879-886.

[103] Qureshi, T., Templeton, J. W. & Adams, L. G. (1996). Intracellular survival of *Brucella abortus*, *Mycobacterium bovis* BCG, S*almonella dublin*, and *Salmonella typhimurium* in macrophages from cattle genetically resistant to *Brucella abortus*. *Veterinary Immunology and Immunopathology.*, *50*, 55-65.

[104] Rafiei, A., Ardestani, S. K., Kariminia, A., Keyhani, A., Mohraz, M. & Amirkhani, A. (2006). Dominant Th1 cytokine production in early onset of human brucellosis followed by switching towards Th2 along prolongation of disease. *The Journal of Infection.*, *53*, 315-324.

[105] Rasool, O., Freer, E., Moreno, E. & Jarstrand, C. (1992). Effect of *Brucella abortus* lipopolysaccharide on oxidative metabolism and lysozyme release by human neutrophils. *Infection and Immunity.*, *60*, 1699-1702.

[106] Riley, L. K. & Robertson, D. C. (1984a). Brucellacidal activity of human and bovine polymorphonuclear leukocyte granule extracts against smooth and rough strains of Brucella abortus. *Infection and Immunity.*, *46*, 231-236.

[107] Riley, L. K. & Robertson, D. C. (1984). Ingestion and intracellular survival of *Brucella abortus* in Human. *Infection and Immunity.*, *46*, 224-230.

[108] Roux, C. M., Rolán, H. G., Santos, R. L., Beremand, P. D., Thomas, T. L., Adams, L. G. & Tsolis, R. M. (2007). *Brucella* requires a functional Type IV secretion system to elicit innate immune responses in mice. *Cellular Microbiology.*, *9*, 1851-1869.

[109] Salcedo, S. P., Marchesini, M. I., Lelouard, H., Fugier, E., Jolly, G., Balor, S., Muller, A., Lapaque, N., Demaria, O., Alexopoulou, L., Comerci, D. J., Ugalde, R. A., Pierre, P. & Gorvel, J. P. (2008). *Brucella* control of dendritic cell maturation is dependent on the TIR-containing protein BTP1. *PLoS Pathogens.*, *4*, 1-16.

[110] Salmeron, I., Rodriguez-Zapata, M., Salmeron, O., Manzano, L., Vaquer, S. & Alvarez-Mon, M. (1992). Impaired activity of natural killer cells in patients with acute brucellosis. *Clinical Infectious Diseases.*, *15*, 764-770.

[111] Sansonetti, P. & Di Santo, J. (2007). Debugging how bacteria manipulate the immune response . *Immunity.*, *26*, 149-161.

[112] Sathiyaseelan, J., Jiang, X. & Baldwin, C. L. (2000). Growth of Brucella abortus in macrophages from resistant and susceptible mouse strains. *Clinical and Experimental Immunology.*, *121*, 289-294.

[113] Scholz, H. C., Hubalek, Z., Sedlácek, I., Vergnaud, G., Tomaso, H., Al Dahouk, S., Melzer, F., Kämpfer, P., Neubauer, H., Cloeckaert, A., Maquart, M., Zygmunt, M.S., Whatmore, A. M., Falsen, E., Bahn, P., Göllner, C., Pfeffer, M., Huber, B., Busse, H. J. & Nöckler, K. (2008). *Brucella microti* sp. nov., isolated from the common vole *Microtus arvalis. International Journal of Systematic and Evolutionary Microbiology.*, *58*, 75-82.

[114] Searle, S., Bright, N. A., Roach, T. I., Atkinson, P. G., Barton, C. H., Meloen, R. H. & Blackwell, J. M. (1998). Localization of NRAMP1 in macrophages: modulation with activation and infection. *Journal of Cell Science.*, *111*, 2855-2866.

[115] Skamene, E., Schurr, E. & Gros, P. (1998). Infection genomics: NRAMP1 as a major determinant of natural resistance to intracellular infections. *Annual Review of Medicine.*, *49*, 275-287.

[116] Sun, Y. H., Den Hartigh, A. B., Santos, R. L., Adams, L. G. & Tsolis, R. M. (2002). virB-Mediated survival of *Brucella abortus* in mice and macrophages is independent of a functional inducible nitric oxide synthase or NADPH oxidase in macrophages. *Infection and Immunity.*, *70*, 4826-4832.

[117] Svetic, A., Jian, Y. C., Finkleman, F. D. & Gause, W. C. (1993). *Brucella abortus* induces a novel cytokine gene expression pattern characterized by elevated IL-10 and IFN-g in CD41 T cells. *International Immunology.*, *5*, 877-883.

[118] Takeda, K., Kaisho, T. & Akira, S. (2003). Toll-like receptors. Annual Review of Immunology, *21*, 335-376.

[119] Templeton, J. W., Smith III, R. & Adams, L. (1988). Natural disease resistance in domestic animals. *Journal of the American Veterinary Medical Association.*, *192*, 1306-1315.

[120] Turner, M. W. (2003). The role of Mannose-binding lectin in health and disease. *Molecular Immunology.*, *40*, 423-429.

[121] Uematsu, S. & Akira, S. (2006). Toll-like receptors and innate immunity. *Journal of Molecular Medicine.*, *84*, 712-725.

[122] Vidal, S., Gros, P. & Skamene, E. (1995). Natural resistance to infection with intracellular parasites: molecular genetics identifies NRAMP1 as the Bcg/Ity/Lsh locus. *Journal of Leukocyte Biology.*, *58*, 382-390. Review.

[123] Xavier, M. N., Paixão, T. A., Poester, F. P., Lage, A. P. & Santos, R. L. (2009). Pathological, immunohistochemical, and bacteriological study of tissues and milk of cows and fetuses experimentally infected with *Brucella abortus. Journal of Comparative Pathology.*, *140*, 149-157.

[124] Wang, M., Qureshi, N., Soeurt, N. & Splitter, G. (2001). High levels of nitric oxide production decrease early but increase late survival of *Brucella abortus* in macrophages. *Microbial Pathogenesis.*, *31*, 221-230.

[125] Watarai, M., Makino, S. & Shirahata, T. (2002). An essential virulence protein of *Brucella abortus*, VirB4, requires an intact nucleoside-triphosphate-binding domain. *Microbiology.*, *148*, 1439-1446.

[126] Weiss, D. S., Takeda, K., Akira, S., Zychlinsky, A. & Moreno, E. (2005). MyD88, but not toll-like receptors 4 and 2, is required for efficient clearance of *Brucella abortus*. *Infection and Immunity.*, *73*, 5137-5143.

[127] Wyllie, S., Seu, P. & Goss, J. A. (2002). The natural resistance-associated macrophage protein 1 Slc11a1 (formerly NRAMP1) and iron metabolism in macrophages. *Microbes and Infection.*, *4*, 351-359.

[128] Yoshimoto, T., Nagai, N., Ohkusu, K., Ueda, H., Okamura, H. & Nakanishi, K. (1998). IL-12 up-regulates IL-18 receptor expression on T cells, Th1 cells, and B cells: synergism with IL-18 for IFN-α production. *Journal of Immunology.*, *161*, 3400-3407.

[129] Zhan, Y. & Cheers, C. (1993). Endogenous gamma interferon mediates resistance to *Brucella abortus* infection. *Infection and Immunity.*, *61*, 4899-4901.

[130] Zhan, Y. & Cheers, C. (1995). Endogenous interleukin-12 is involved in resistance to *Brucella abortus* infection. *Infection and Immunity.*, *63*, 1387-1390.

[131] Zhan, Y., Liu, Z. & Cheers, C. (1996). Tumor necrosis factor alpha and interleukin-12 contribute to resistance to the intracellular bacterium *Brucella abortus* by different mechanisms. *Infection and Immunity.*, *64*, 2782-2786.

[132] Zwilling, B. S., Kuhn, D. E., Wikoff, L., Brown, D. & Lafuse, W. (1999). Role of iron in NRAMP1-mediated inhibition of mycobacterial growth. *Infection and Immunity.*, *67*, 1386-1392.

In: Veterinary Immunology and Immunopathology
Editors: L. Nermann, S. Meier

ISBN: 978-1-60876-342-9

Chapter 6

EXPERIMENTAL PULMONARY TUBERCULOSIS AND IMMUNE INHIBITORY SIGNALS: A POTENTIAL ROLE FOR PD1-PD1-LIGAND INTERPLAY

T. Santos-Mendoza,[1,3] A. Silva-Sánchez,[1,3] S. Meza-Pérez,[1,3] D. Aguilar-León,[2] I. Estrada-García,[1] S. Estrada-Parra,[1] R. Hernández-Pando,[2] and L. Flores-Romo.[3] [1]

[1]Department of Immunology, National School of Biological Sciences ENCB-IPN, México, DF;

[2] Department of Experimental Pathology National Institute of Medical Science and Nutrition "Salvador Zubiran", México, DF;

[3] Department of Cell Biology, Center for Advanced Research CINVESTAV-IPN, México, D.F.

ABSTRACT

Immune responses may substantially differ between conventional laboratory antigens and microbial ones. Pathogens like Mycobacterium tuberculosis (Mtb) have evolved complex mechanisms to evade, divert or subvert immune responses. Tuberculosis (TB) is a major cause of death worldwide whose etiology, *M. tuberculosis*, produces latent asymptomatic infections and chronic progressive diseases, being pulmonary TB its most important manifestation. TB immunity is mediated by Th1-type responses, nevertheless this response seems not enough to definitely eradicate the chronic infection, nor the latent bacilli.

The experimental pulmonary tuberculosis is a convenient *in vivo* model to study the disease because it provides important advantages. Mice infection is established through the airways and murine immune responses are well known. Different aspects of the disease can be explored using different mice strains with different susceptibilities. Another crucial advantage is the suitability to infect with different mycobacterial strains, including clinical isolates and mutant bacteria.

Mycobacteria induce strong Th1-type responses, however, infected immunocompetent hosts are somehow rendered unable to generate effective sterilizing immunity and bacilli can persist long periods in an ill-understood dormant state.

Although the molecule Programmed Death-1 (PD1) belongs to the CD28 co-stimulatory family, unlike CD28, PD1 is an inhibitory receptor inducible upon activation of T, B and NKT cells, with two ligands described (PDL-1/PDL-2). The relevance of the PD1/PDL-1 pathway to induce and maintain immunologic tolerance is well documented, and dysfunction in this pathway is associated with autoimmunity. Recently, the PD1/PDL-1 pathway was shown as a suitable target for exploitation by chronic viral diseases provoking T cell exhaustion (a phenotype that can be reverted *in vivo* by blocking the PD1/PDL1 interaction). It is thus possible that other chronic infectious agents may use this strategy too.

We therefore evaluated the *in vivo* PD1 expression on CD4+ and CD8+ T cells, both in the TB target organ, the lung, and the mediastinal lymph nodes, along the infection (days 1-60) in Balb/c mice. Results revealed that compared to the less virulent *M. canetti*, airborne infection with the virulent strain *M. tuberculosis* H37Rv, triggered an increase in PD1+ T cells, especially in the lungs, although with different kinetics for CD4+ and CD8+ lymphocytes. Conceivably, virulent Mtb might elicit this inhibitory molecule in effector T cells to facilitate both the establishing and the persistence of the infection.

INTRODUCTION

Tuberculosis (TB) is a major cause of death worldwide estimated to kill over two million people annually. TB etiology *Mycobacterium tuberculosis* (Mtb) has evolved complex mechanisms to evade, divert or subvert immune responses avoiding the destruction by innate and adaptive mechanisms of immunity (1). Mtb is capable to induce latent asymptomatic infections and chronic progressive diseases, being pulmonary TB its most important manifestation (2, 3). Although one third of human population is infected with Mtb, less than 10% of immunocompetent individuals develop the clinical disease and 85% out of these develop exclusively pulmonary TB (4).

TB immunity is mediated by Th1-type responses, nevertheless this response seems not enough to definitely eradicate the chronic infection, nor the latent bacilli. Moreover, it would seem that the elicited Th1-type response is able to control the mycobacteria proliferation in all organs except the lung, resulting in the disease transmission by infective aerosols (4-6).

Much of what is known about the immunology of TB has come from studies in mice. Animal models are essential for investigating the immune response *in vivo* where the ability to control infection can be assessed in a more physiological setting. Animal models offer several advantages for studying TB immunity. Of capital interest is the capacity to monitor *in vivo* the pathogenicity and immunogenicity of different mycobacteria strains such as clinical isolates, mutant bacteria generated in the laboratory, or even isolated mycobacterial components, like for instance certain lipids. It is well recognized that the immunology of mouse tuberculosis is comparable to that in humans and it has been proposed that mouse is an appropriate experimental model to study pulmonary TB. Like in humans, mouse TB is restricted to lungs and also develops in immunocompetent intact hosts, not only in immunodefficient animals. There is a variety of mice strains with different susceptibilities for the same pathogen, and this generates the possibilities for using defined transgenic (TG) or knock out (KO) animals to study the contribution of specific pathways for the establishing and development of the disease. Additional advantages of the mouse model includes the vast arsenal of reagents to assess murine immune responses and also the possibilities to explore *in vivo* different therapeutic strategies (1, 3, 4).

Once a pathogen enters through the airways of an organism, concomitantly to other innate immune responses elicited, local antigen presenting cells (APC) can uptake the pathogen, migrate to regional lymph nodes (in this case the mediastinal ones, MLN), and can present the processed antigens (Ags) to T cells. Activation and expansion of Ag-specific T lymphocytes ensues, leading to the onset of the adaptive cellular immune response. Once this starts in regional LN, Ag-specific T lymphocytes traffic to the site of infection and can eliminate the intruding Ags. This process takes a few days for most of the pathogens (7, 8), nevertheless there are compelling new data describing an important delay in this process for pulmonary Mtb infection (9). Indeed, different groups using different Mtb inoculation amounts and routes, and different mice strains, coincide in that mycobacteria progressively grows within the lung apparently unimpeded during the first 3 weeks of infection. This is precisely the time that T cell adaptive response takes to appear in the lungs to control mycobacterial growth, although without ensuring its elimination. The reason for this exceptionally long delay for a T cell response to appear (in the lungs) remains unknown. In line with this, our group has reported a rather prolonged delay of about three weeks for DC to increase and get activated during airborne murine TB infection (9-12).

In mice and humans, Mtb infection is controlled at the beginning mainly by macrophage activation induced through Th-1 type cytokines. In the model of progressive pulmonary TB in Balb/c mice, airways Mtb inoculation induces a disease with an initial phase of approximately three weeks of partial apparent resistance, dominated by a Th1 cytokine profile. Interferon-gamma (IFN-γ) induces macrophage activation that in turn produces tumor necrosis factor-α (TNF-α) and expresses inducible nitric oxide synthase (iNOS). Granulomas concomitantly develop at this stage. Then, the phase of chronic progressive disease starts after one month of infection. This late phase is characterized both by a drop in the number of cells expressing INF-γ, IL-2, TNF-α, and iNOS, and by a progressive pneumonia, extensive tissue damage, high bacillary counts and an increase in Th-2 type cytokines such as IL-4. Pneumonia and a high bacillary burden, causes the animal's death (13, 14).

The model described above, where different phases of the disease are well defined for the virulent strain H37Rv is suitable to compare the pathogenicity and immunogenicity exerted by different mycobacterial strains. For example, mice infected with *M. canettii* strains (belonging to the MTb complex but much less virulent than H37Rv) exhibit longer survival compared to animals infected with H37Rv, lower bacillary loads and significantly fewer areas of pneumonia, together with constant and stable expression of IFN-γ, TNF-α and iNOS (15).

Although it is well recognized that immunity to TB is primarily dependent on a conventional Th1-type response, debate on its effectiveness is currently emerging. Altered genetic pathways for both the IFN-γ and IL-12 axis have clearly unraveled the importance of Th1 response for TB control (16), nevertheless the presence of high numbers of specific, IFN-γ-secreting T cells rarely if ever results in the complete mycobacteria eradication and thus in sterilizing immunity (9, 10). In this scenario, much investigation is needed to determine additional factors that contribute to the establishing and the permanence of the pathogen in the host.

Appropriate T cell responses are affected by a complex balance between activating and inhibitory molecules. Programmed death 1 (PD1) belongs to the co-stimulatory CD28 family, but in contrast to CD28, PD1 is a co-inhibitory receptor (17). PD1 is inducibly expressed in T and B lymphocytes as well as NKT cells upon activation. Two ligands for PD1 have been described to date, PDL1 and PDL2 (18) . The relevance of PD1/PDL1 pathway for the induction and maintenance of immunologic tolerance is well documented, and dysfunction in this pathway is associated with autoimmune manifestations (19-21).

Recently, during chronic murine LCMV infection, an important increase of PD1 on T cells has been described, resulting in the so called "exhausted" T cell phenotype, where specific T cells cannot put to work their effector functions, a situation that can be reverted by blocking PD1-PDL1 interactions using anti-PD1 or anti-PDL1 antibodies (22). In this context it has been suggested that the PD1/PDL1 pathway might be a potential target exploited by a variety of viruses during chronic infections. Indeed, humans infected with HIV show increased PD1 expression on specific T cells, which coincides with an increase in viral titers and disease progression, the same is observed during Hepatitis B and HTLV viral infections in humans (23-25).

Persistent infections with pathogens other than viruses may also be associated with suboptimal T cell responsiveness due to PD1/PDL interaction. Conceivably, this could be the case of TB where the sole presence of activated, IFN-γ producing T cells seems not enough to control and eradicate the infection.

In this chapter we explore the *in vivo* modulation of PD1 on the two major T cell subsets (CD4+ and CD8+) during TB. Both, the pulmonary draining lymph nodes (MLN) and the target organ, the lung, were analyzed in the experimental murine model of pulmonary TB. The virulent strain H37Rv *vs* the less virulent *M. canetti* are compared.

Materials and Methods

Experimental Model of Progressive Pulmonary Tuberculosis

The experimental model of progressive pulmonary TB has been already described (13). Briefly, male Balb/c mice aged 6-8 weeks were infected with 2.5 x 10^5 viable bacteria suspended in 100 μl of endotoxin-free saline solution, at least five mice per group and time point were used. Mice were anesthetized with sevofluorane and infected by inoculating the bacilli in the trachea. Infected mice were maintained in groups of five in cages fitted with microisolators connected to negative pressure. All procedures were performed in a biological security cabinet, P3. The protocol was approved by the Ethics Committee for Experimentation in Animals of the National Institute of Medical Sciences and Nutrition.

Lung and Regional Lymph Node Cell Suspensions

Animals were sacrificed at 3, 14, 21, 28 and 60 days postinfection. Lung and MLN from at least five mice per experimental group and time point were pooled, lung was cut into small pieces and both organs were digested for 60 min in complete medium (RPMI-1640 plus 10% fetal calf serum, 2 mM glutamine, 100 U/ml of penicillin G, and 100 mg/ml of streptomycin) containing 370UI/ml of collagenase type II (Gibco, BRL, Carlsbad, CA). The small tissue pieces were passed through a 100-μm cell strainer (Becton Dickinson, San Jose, CA) to obtain cell suspensions. Red blood cells were lysed with NH_4Cl lysis buffer. Pooled cell suspensions were thoroughly washed with PBS, and finally resuspended in FACS buffer (PBS containing 1% bovine serum albumin and 0.01% sodium azide).

Flow Cytometry Analysis

Cells were blocked with Universal Blocking Reagent (BioGenex, San Ramon, CA) for 15 min to reduce nonspecific binding. 1 x 10^6 cells were incubated with specific antibodies for 15 min on ice. Monoclonal antibodies used were the following: Anti-CD3-fluoresceinisothiocyanate (CD3-FITC), anti-CD4-allophycocyanin (CD4-APC), anti-PD1-phycoerythrin (PD1-PE; all from Pharmingen, BD, San Jose, CA). The acquisition was performed in a FACScalibur flow cytometer using the Cell-Quest program (Becton Dickinson Immunocytometry Systems) recording at least 50 000 events per assessment. For data analysis flow jo program was used. Lymphocyte region was determined on the forward/scatter plot where CD3+ cells were gated. FL4+ corresponds to CD4+ lymphocytes whereas FL4- (but CD3+) was considered the CD8+ lymphocyte population. FL2+ represents the PD1+ cells for each population.

RESULTS

Differential T Cell Expression of PD1 Inhibitory Receptor in Lungs of Mice Infected with Virulent *vs* Non-Virulent Mycobacteria

In vivo Expression of PD1 on CD4+ T Cells during Airways Mtb Infection

We infected mice via the airways with the virulent strain MtbH37Rv and the non virulent strain *M. canetti* to compare de expression of PD1 in T cells obtained from the regional (mediastinal) lymph nodes and lungs. Control mice were inoculated the same way with sterile, endotoxin-free saline solution (SS). The kinetics of PD1 expression on CD3+CD4+ T lymphocytes is shown in Figure 1. The percentage of double positive CD4+PD1+ T cells from MLN shows similar behavior in both, MtbH37Rv and *M. canetti* infected mice. There is a clear increase at three days postinfection that in the case of H37Rv returns to the control values of SS mice (dotted line) on day 14, and decreases afterwards until day 28 maintained steadily until 60 days. *M. canetti* shows a more pronounced decrement of PD1 on day 14 with a little recovery on day 21 (that does not reach the control SS values) and diminishes again on day 28 to keep steadily until day 60. These results indicate that except for a little fluctuation seen with *M. canetti,* both mycobacteria strains behave rather similar regarding the induction of PD1 in T cells in the regional lymph nodes.

In contrast, the percentage of double positive CD4+PD1+ T cells is quite different in the lung compared with MLN and within the two mycobacteria strains. Both mycobacteria display a sort of biphasic behavior which coincides with the well defined first (early) and second (chronic) phases of the TB infection according to the pattern of cytokine microenvironment and lung pathology. Since basal levels of PD1 expression on pulmonary T cells are virtually negative, H37Rv infection shows a considerable increment of CD4+PD1+ T cells starting on day 3 postinfection, reaching a peak on day 14 to decline on day 21 (although not to basal levels), increasing again from day 28 to day 60. The kinetics of PD1 expression with *M. canetti* infection shows an increment on day 3 similar to that of H37Rv. In contrast, under *M. canetti* there is a decline by day 14 reaching basal levels on day 21, increasing again and reaching its maximum (although more modest compared with H37Rv) on day 28, declining then almost to the basal levels (dotted line) by day 60. In this regard, it is interesting to note that while the virulent H37Rv triggers a more pronounced increment of CD4+PD1+ T cells on lung during the first phase (day 14) of the infection, the non-virulent *M. canetti* increment is induced earlier and lower (day 3). Of note, under both mycobacteria the minimum PD1 expression is reached at day 21, which coincides with the point of Th-1 type cytokines decrement and the concomitant increase in Th-2 type environment for H37Rv. In the second, chronic phase of the disease, pulmonary PD1 expression was again different: both mycobacteria augmented PD1, however, while virulent mycobacteria triggered a progressive increase till day 60, *M. canetti* declined almost to the basal level by day 60.

In vivo expression of PD1 on CD8+ T cells during airways Mtb infection

The kinetics of PD1 expression on CD8+ lymphocytes did not differ much from that of CD4+, but there were peculiarities. First, percentage of CD8+PD1+ T cells (Figure 2) is lower compared with CD4+ in MLN whereas the contrary happens in lung where percentage of CD8+PD1+ lymphocytes is higher compared with CD4+ cells. Both mycobacteria behave very much alike regarding PD1 expression in CD8+ lymphocytes from MLN. There is an increment of CD8+PD1+ T cells since day 3 with a peak on day 14, declining even below basal levels from day 21 onwards. Interestingly, although the precise role of CD8+ T cells in TB is still unclear, this population is deemed important regarding intracellular pathogens.

In contrast with MLN, *M. canetti* triggers in the lungs an increment of CD8+PD1+ T cells on day 3 declining on days 14 and 21, increasing again reaching its maximum (even higher than with H37Rv) by day 28, whereas on day 60 diminishes almost to basal levels. CD8+PD1+ T cells under H37Rv infection show a different pattern; there is a considerable increase since day 3 reaching a peak on day 14 to decrease abruptly on day 21. On day 28 there is again an increment although not as high as for *M. canetti*, which is maintained steadily until day 60.

It is interesting to note the different effects that the two mycobacteria had upon the PD1 expressing T cells. Especially between the MLN where the Ag-specific T cells are activated and expanded, and the lungs where these activated T cells arrive to exert their functions against the infecting pathogen. In MLN both mycobacteria increased PD1 expression in the first phase of infection, though earlier for CD4+ T cells (day 3) than for CD8+ cells (day 14). It has been reported that after Mtb inoculation, T cell proliferation in MLN is delayed until day 12 (11). At first glance, this early increment in the inhibitory molecule PD1 might be interpreted as precedent for the delayed response in the regional LN, nevertheless the same pattern is displayed by both, virulent and non-virulent mycobacteria.

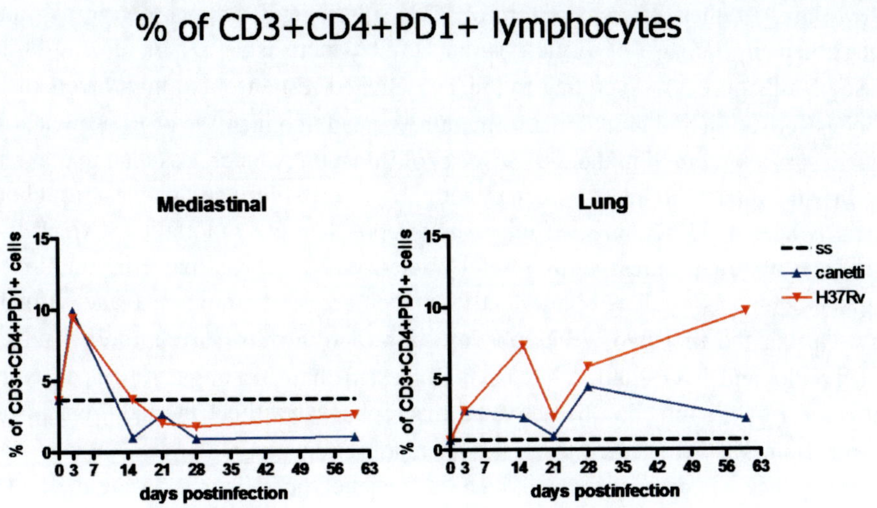

Figure 1. Mice were intratracheally inoculated with virulent MtbH37Rv or non virulent *M.canetti*, while mice inoculated with endotoxin-free sterile saline solution (SS) were used as controls (dotted line). At least five mice per group and time point were sacrificed on days 3, 14, 21, 28 and 60 post-infection and MLN and lung cell suspensions were prepared for flow cytometry. Pooled cell suspensions from these organs were stained with anti-CD3-FITC, anti-CD4-APC and anti-PD1-PE. The percentage of CD4+PD1+ lymphocytes previously gated on CD3+, is represented for each mycobateria strain and time point indicated.

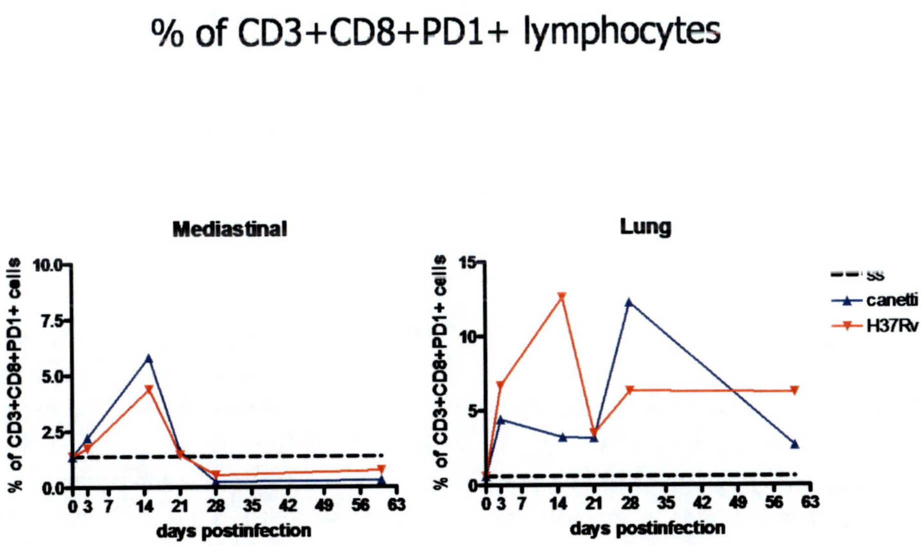

Figure 2. Mice were intratracheally inoculated with virulent MtbH37Rv or non virulent *M.canetti*, while mice inoculated with endotoxin-free sterile saline solution (SS) were used as controls (dotted line). At least five mice per group and time point were sacrificed on days 3, 14, 21, 28 and 60 post-infection. Lungs and MLN were obtained and respectively pooled to prepare cell suspensions. These were then stained with anti-CD3-FITC, anti-CD4-APC and anti-PD1-PE for flow cytometry analysis. Lymphocytes gated first on CD3+ which were CD4- are deemed the CD8+ and were co-labelled for PD1. Percentages of this CD8+PD1+ population are represented for each mycobateria strain and time point.

Noteworthy, a rather different pattern of PD1-expressing T cells is seen in the lung, the TB target organ. In the case of virulent strain H37Rv, there is a PD1 increment during the first phase of infection corresponding to the Th1 elicited response and high levels of IFN-γ, TNFα, iNOs activation and granuloma formation. Instead, the increase of PD1 elicited by the non virulent strain *M canetti* in the first 3 weeks of infection is much less than that exerted by H37Rv. In this model, animals infected with *M. canetti* survive longer throughout the experiment, whereas H37Rv infected mice show more than 50% mortality at week five. *M. canetti* induces a slow progressive disease characterized by delayed bacterial multiplication, limited pneumonia, as well as significantly smaller areas of granuloma and virtually no mortality. During the first two weeks postinfection there are low levels of IFN-γ but high levels of TNF-α and iNOS and it has been suggested that this might reflect an early and efficient control of the infection by activated macrophages, without the requirement of high IFN-γ expression (15). Altogether, these results might reflect one of the virulent mycobacteria strategies to induce an unresponsive ("exhausted") phenotype in T cells during the first phase of infection, which later results in the subsequent host incapacity to eliminate the mycobacteria.

On the second phase (after weeks 2-3) of infection, *M. canetti* importantly increases PD1 expression, more pronounced for CD8+ than for CD4+ T cells. The increment of CD8+PD1+ T cells by day 28 is higher than that under H37Rv. One likely explanation is to prevent the extensive tissue damage in the advanced disease. The interaction of PD1 with its PDL1-ligand has been implicated in the development and maintenance of immune tolerance, concomitantly, this interaction plays a crucial role controlling autoimmunity and tissue damage (21). In this regard, it is plausible that PD1 in the lung microenvironment during the chronic phase of infection with *M. canetti*, might partially explain for counteracting tissue damage by excessive T cell reactivity. Indeed, *M. canetti* infection displays much less lung pathology than H37Rv (15). In the second, late phase of H37Rv infection, tissue damage is quite advanced with lot of granulomas within great lung areas and progressive pneumonia, this is the disease´s chronic phase and PD1 increment is maintained. It could be that in this case the increment of PD1+ ("exhausted") T cells might participate in the evasion of the immune response by mycobacteria, favoring its persistence within these tissues.

In keeping with our work, it was recently reported that in the pleural fluid of patients with active TB, more than 90% of IFN-γ secreting cells after Mtb stimulation *in vitro*, co-express PD1, and it has been suggested that PD1 might exert an important role *in vivo* as inhibitor of T cell responses against mycobacteria (26). In line with this, we found higher PD1 expression but in the lung itself as the TB target organ of H37Rv infected mice, compared with *M. canetti* infection. To the best of our knowledge, these are previously unrecognized *in vivo* effects of mycobacterial infection upon PD1 expression on both T cell populations within the TB target tissue itself, the lung, and also within the nearest lymphoid organ, the mediastinal nodes.

Conclusion

Many pathogens have evolved complex mechanisms to evade or alter the efficacy of immune responses. *M. tuberculosis* seems one of the best examples, infecting one third of human population, persisting latent in most of them, but inducing clinical manifestations in less than 10% of immunocompetent individuals. The mouse model of experimental pulmonary TB is suitable for studying TB pathogenesis, immunity or therapeutics. In this chapter we analyzed the *in vivo* expression of the inhibitory molecule PD1 (responsible for the "exhausted" T cell phenotype) on the two main T cell populations: CD4+ and CD8+. We assessed both, the MLN (nearest lymphoid tissue) and the lungs (the TB target organ) in mice infected with virulent (H37Rv) or non-virulent (*M. canetti*) mycobacteria strains. Depending on the mycobacterial strain, PD1+CD4+ and PD1+CD8+ T cells increased differently in the TB target tissue the lungs; the former did it rather progressively along the infection while the latter mostly in the early phase, when the increase was more pronounced with the virulent than with the non-virulent strain. The PD1+CD8+ T cells increment in the second phase under Canetti might well be related to the attenuated tissue pathology observed.

Conceivably, virulent Mtb might elicit these responses to ensure, first the infection and then the persistence, especially in lungs. The model of experimental pulmonary TB is a suitable tool to examine *in vivo* the detailed contribution of inhibitory pathways for T cells during immune evasion and disease development by mycobacteria.

ACKNOWLEDGMENTS

This work was supported by National Council for Science and Technology of Mexico (CONACYT). T Santos-Mendoza, A Silva-Sanchez and S Meza-Perez are CONACYT fellows. T Santos-Mendoza, I Estrada-Garcia, D Aguilar, R Hernandez-Pando, S Estrada-Parra and L Flores-Romo are members of the National System of Researchers of Mexico, SNI.

REFERENCES

[1] Rook, G. A. & R. Hernandez Pando. (1996). The pathogenesis of tuberculosis. *Annu Rev Microbiol, 50*, 259-284.

[2] Rook, G. A., Dheda, K. & Zumla, A. (2005). Immune responses to tuberculosis in developing countries: implications for new vaccines. *Nat Rev Immunol, 5*, 661-667.

[3] Casanova, J. L. & Abel, L. (2002). Genetic dissection of immunity to mycobacteria: the human model. *Annu Rev Immunol* 20:581-620.

[4] North, R. J. & Y. J. Jung. 2004. Immunity to tuberculosis. *Annu Rev Immunol 22*,599-623.

[5] Mogues, T., Goodrich, M. E., Ryan, L., LaCourse, R. & North, R. J. (2001). The relative importance of T cell subsets in immunity and immunopathology of airborne Mycobacterium tuberculosis infection in mice. *J Exp Med., 193*, 271-280.

[6] Jung, Y. J., LaCourse, R., Ryan, L. & North, R. J. (2002). Virulent but not avirulent Mycobacterium tuberculosis can evade the growth inhibitory action of a T helper 1-dependent, nitric oxide Synthase 2-independent defense in mice. *J Exp Med., 196*, 991-998.

[7] Xia, W., Pinto, C. E. & Kradin. R. L. (1995). The antigen-presenting activities of Ia+ dendritic cells shift dynamically from lung to lymph node after an airway challenge with soluble antigen. *J Exp Med., 181*, 1275-1283.

[8] Bozza, S., Gaziano, R., Spreca, A., Bacci, A., Montagnoli, C., di Francesco, P. & Romani, L. (2002). Dendritic cells transport conidia and hyphae of Aspergillus fumigatus from the airways to the draining lymph nodes and initiate disparate Th responses to the fungus. *J Immunol., 168*, 1362-1371.

[9] Jung, Y. J., Ryan, L., LaCourse, R. & North, R. J. (2005). Properties and protective value of the secondary versus primary T helper type 1 response to airborne Mycobacterium tuberculosis infection in mice. *J Exp Med., 201*, 1915-1924.

[10] Jung, Y. J., LaCourse, R., Ryan, L. & North. R. J. (2008). 'Immunization' against airborne tuberculosis by an earlier primary response to a concurrent intravenous infection. *Immunology., 124*, 514-521.

[11] Wolf, A. J., Desvignes, L., Linas, B., Banaiee, N., Tamura, T., Takatsu, K. & Ernst, J. D. (2008). Initiation of the adaptive immune response to Mycobacterium tuberculosis depends on antigen production in the local lymph node, not the lungs. *J Exp Med., 205*, 105-115.

[12] Garcia-Romo, G. S., Pedroza-Gonzalez, A., Aguilar-Leon, D., Orozco-Estevez, H., Lambrecht, B. N., Estrada-Garcia, I., Flores-Romo, L. & Hernandez-Pando, R. (2004). Airways infection with virulent Mycobacterium tuberculosis delays the influx of dendritic cells and the expression of costimulatory molecules in mediastinal lymph nodes. *Immunology,, 112*, 661-668.

[13] Hernandez-Pando, R., Orozcoe, H., Sampieri, A., Pavon, L., Velasquillo, C., Larriva-Sahd, J., Alcocer, J. M. & Madrid, M. V. (1996). Correlation between the kinetics of Th1, Th2 cells and pathology in a murine model of experimental pulmonary tuberculosis. *Immunology., 89*, 26-33.

[14] Hernandez-Pando, R., Orozco, H., Arriaga, K., Sampieri, A., Larriva-Sahd, J. & Madrid-Marina, V. (1997). Analysis of the local kinetics and localization of interleukin-1 alpha, tumour necrosis factor-alpha and transforming growth factor-beta, during the course of experimental pulmonary tuberculosis. *Immunology., 90*, 607-617.

[15] Lopez, B., Aguilar, D., Orozco, H., Burger, M., Espitia, C., Ritacco, V., Barrera, L., Kremer, K., Hernandez-Pando, R., Huygen, K. & van Soolingen, D. (2003). A marked difference in pathogenesis and immune response induced by different Mycobacterium tuberculosis genotypes. *Clin Exp Immunol., 133*, 30-37.

[16] Jouanguy, E., Doffinger, R., Dupuis, S., Pallier, A., Altare, F. & Casanova, J. L. (1999). IL-12 and IFN-gamma in host defense against mycobacteria and salmonella in mice and men. *Curr Opin Immunol., 11*, 346-351.

[17] Chen, L. (2004). Co-inhibitory molecules of the B7-CD28 family in the control of T-cell immunity. *Nat Rev Immunol., 4*, 336-347.

[18] Keir, M. E., Butte, M. J., Freeman, G. J. & Sharpe, A. H. (2008). PD-1 and its ligands in tolerance and immunity. *Annu Rev Immunol., 26*, 677-704.

[19] Nishimura, H., Nose, M., Hiai, H., Minato, N. & Honjo, T. (1999). Development of lupus-like autoimmune diseases by disruption of the PD-1 gene encoding an ITIM motif-carrying immunoreceptor. *Immunity., 11*, 141-151.

[20] Nishimura, H., Okazaki, T., Tanaka, Y., Nakatani, K., Hara, M., Matsumori, A., Sasayama, S., Mizoguchi, A., Hiai, H., Minato, N. & Honjo, T. (2001). Autoimmune dilated cardiomyopathy in PD-1 receptor-deficient mice. *Science., 291*, 319-322.

[21] Keir, M. E., Liang, S. C., Guleria, I., Latchman, Y. E., Qipo, A., Albacker, L. A., Koulmanda, M., Freeman, G. J., Sayegh, M. H. & Sharpe, A. H. (2006). Tissue expression of PD-L1 mediates peripheral T cell tolerance. *J Exp Med., 203*, 883-895.

[22] Barber, D. L., Wherry, E. J., Masopust, D., Zhu, B., Allison, J. P., Sharpe, A. H., Freeman, G. J. & Ahmed, R. (2006). Restoring function in exhausted CD8 T cells during chronic viral infection. *Nature., 439*, 682-687.

[23] Day, C. L., Kaufmann, D. E., Kiepiela, P., Brown, J. A., Moodley, E. S., Reddy, S., Mackey, E. W., Miller, J. D., Leslie, A. J., DePierres, C., Mncube, Z., Duraiswamy, J., Zhu, B., Eichbaum, Q., Altfeld, M., Wherry, E. J., Coovadia, H. M., Goulder, P. J., Klenerman, P., Ahmed, R., Freeman, G. J. & Walker, B. D. (2006). PD-1 expression on HIV-specific T cells is associated with T-cell exhaustion and disease progression. *Nature., 443*, 350-354.

[24] Kozako, T., Yoshimitsu, M., Fujiwara, H., Masamoto, I., Horai, S., White, Y., Akimoto, M., Suzuki, S., Matsushita, K., Uozumi, K., Tei, C. & Arima, N. (2009). PD-1/PD-L1 expression in human T-cell leukemia virus type 1 carriers and adult T-cell leukemia/lymphoma patients. *Leukemia., 23*, 375-382.

[25] Boni, C., Fisicaro, P., Valdatta, C., Amadei, B., Di Vincenzo, P., Giuberti, T., Laccabue, D., Zerbini, A., Cavalli, A., Missale, G., Bertoletti, A. & Ferrari, C. (2007). Characterization of hepatitis B virus (HBV)-specific T-cell dysfunction in chronic HBV infection. *J Virol., 81*, 4215-4225.

[26] Jurado, J. O., Alvarez, I. B., Pasquinelli, V., Martinez, G. J., Quiroga, M. F., Abbate, E., Musella, R. M., Chuluyan, H. E. & Garcia, V. E. (2008). Programmed death (PD)-1:PD-ligand 1/PD-ligand 2 pathway inhibits T cell effector functions during human tuberculosis. *J Immunol., 181*, 116-125.

In: Veterinary Immunology and Immunopathology
Editors: L. Nermann, S. Meier
ISBN: 978-1-60876-342-9

Chapter 7

SOCIAL STRESS, IMMUNE FUNCTIONS AND IMMUNO-ENDOCRINE DISORDERS IN RODENTS AND OTHER ANIMALS

Paola Palanza, Stefano Parmigiani, and Alessandro Bartolomucci
Department of Evolutionary and Functional Biology, University of Parma, V.le G.P. Usberti 11A, 43100 Parma, Italy.

ABSTRACT

The link between stress, health and social factors has been the focus of many studies and the preferred subject of investigation of researchers from different disciplines. In particular rodent-based models modulating the social environment are becoming increasingly popular in the scientific literature as realistic models of human disease. A growing number of studies now prove that social factors in both early phases and adult life do have a widespread effect on neuroimmune functions. Accordingly, here we will provide an updated review of studies conducted prominently on laboratory rodents. However a number of studies have also been conducted in non-human primates, farm animals and birds. Results from different disciplines will be integrated to reach a clear conclusion: an allostatic overload, including neuroimmune abnormalities, is more likely to develop when unpredictable stressors of social nature, chronically induce physiological and behavioral adjustments that may 'wear and tear' the underlying physiological functions.

INTRODUCTION

The link between stress, health and social factors has been the focus of many studies and the preferred subject of investigation of researchers from different disciplines. In particular rodent-based models modulating the social environment are becoming increasingly popular in the scientific literature as realistic models of human disease. This review will focus on recent advancements at the intersection between immunology, ethology, psychoneuroimmunolgy, neuroendocrinology and neuroscience proving that social factors in rodents and other animals are indeed causally linked with immune disorders/disease susceptibility. First, I will briefly describe sociality and its meaning as well as the modern view of stress and psychoneuroimmunology, then I will discuss available evidences proving that social factors are among the most potent activator of the stress response and among the major causes of stress-related disorders. Because of the great amount of literature, most of the analysis will be devoted to rodents. In reviewing the rodent literature I will distinguish between papers dealing with dyadic social interaction, individual housing and differential group housing. In addition even if much less information are available I will discuss a few examples of studies on social factors and neuroimmune functions in non-human primates, farm animals and birds. I will integrate recent advancements within the framework devised by Bohus and Koolhaas in their landmark paper on social factors and immune functions (1991).

SOCIALITY, STRESS AND PSYCHONEUROIMMUNOLOGY

Sociality is one of the more widespread phenomenons in animal kingdom (Alexander, 1974; Lindberg 2001). Living in a group, as any other behavioral trait, has costs and benefits. Within a social group individuals are generally not all equal and genetic-environmental-experiential differences will interact to determine the position of an individual inside a dominance hierarchy. The evolutionary meaning of the emergence of hierarchies is generally considered to be the benefit of priority or exclusivity of access to resources for high rank individuals and theoretical models predict that a given social setting evolved because of the net advantage of the components to live in that condition (Hemelrijk and Wantia, 2005; Conradt & Roper, 2003; Drummond 2006). Different conditions may have different costs and benefits while the obvious advantage should be for higher rank individual while the low ranking individual should have anyway a net advantage when compared to individual life-style. However, current literature on social stress points to the negative consequences of social status (Sapolsky 2005, Goymann and Wingfield 2004). This should not let us conclude that dominance and subordination are stressful in themselves because these are naturally evolved social roles that may or not have costs depending on the context in which a given social status is gained (Bartolomucci et al., 2005). In particular dominance and subordination are not individual traits, but should be considered as states relative to a particular group of individual and the context in which the status is acquired (Benton et al., 1980; Lindberg 2001).

Social interactions often lead to negative consequences often referred to as stress which can be defined as anything which moves an organism from homeostasis. In the scientific literature two terms have been introduced. A *stressor* is anything that disrupts physiological balance. The *stress-response* is the physiological and behavioral reaction emitted by the organism to re-establish the balance. The balance to be maintained is what Claude Bernard called the "internal milieu" and in modern terms is referred to as *homeostasis* (Sapolsky, 1992). The stress-response is thus an adaptive reaction of a body to a challenge. However this short-term adaptive response may have negative effects if maintained over time or if repeatedly activated. The effect in this situation is an overload of the system with possible pathological implications (McEwen, 1998). The prototypical stress-response is today well understood (Sapolsky, 1992; Rabin, 1999; Sapolsky et al., 2000). Briefly, there are two major systems mediating most components of the stress response. The first is the hypothalamic–pituitary–adrenocortical (HPA) system, which stimulates the adrenal cortex to release glucocorticoids such as cortisol or corticosterone into the blood. The second is the sympathetic nervous system (SNS) that influences the stress-response through two different pathways working in parallel. One pathway is built up by the nerve endings that trigger the release of adrenaline from the chromaffin cells in adrenal medulla into the blood. The other pathway comprises the sympathetic nerve endings that innervate essentially every organ in the body (Sapolsky, 1992; Mason 1968a,b). The acute response, which the organism produces when challenged by an external and/or internal stimulus, is functionally implicated in the mobilization of energy needed for the behavioral response, and is adaptive in many ways. The acute rise of hormones and several neurotransmitters allow the organism to use all his resources to cope with the stress. However, when the stress-response is activated it has to be switched off because its mediators can highly damage the organism. Interestingly, some of the mediators involved in the acute response are also responsible for his termination in particular glucocorticoids (e.g. Sapolsky, 1992; Rabin, 1999; Sapolsky et al., 2000). The problem with the stress response is that it is adaptive in the short term but it can be highly maladaptive in the long term. For example a chronic elevation of glucocorticoids is well known to induce damage or at least a remodeling of the hippocampus, which is in comorbidity with the development of several psychopathologies (Sapolsky et al., 1986; McEwen, 1998, 2000; Fuchs et al., 2006).

As mentioned, sociality and stress are two major issues in the life of virtually all vertebrates. Researchers now accumulated a relevant number of experimental observations allowing to conclide that chronic stress derived by negative social interactions is a recurring factor in the lives of virtually all higher animal species. This suggests, that the social environment could be a relevant source of negative states of an individual and of the key factors in pathology development (Henry & Stephens, 1977; Altmann et al., 1995; Sachser et al., 1993, 1998; Sapolsky 1994; Johnston et al., 1996; von Holst, 1998; Kaplan & Manuck, 1999; Creel 2001). These landmark studies allow reaching a clear conclusion: social factors are powerful modulators of the stress response and more importantly they are so because they are real life events.

Most of recent evidences come from the social status effects on the stress response reproduction and HPA-axis and SNS functions. However, in the past 50 years the scientific credo definitely abandoned the view of organisms made by organs and systems each functionally separated by each other and in a way auto-regulating. Nowadays psychoneuro-immunology is a scientific modern discipline with his own realm of investigation, with his roots well fixed in historical disciplines such as endocrinology, neuroscience, immunology, psychology, and ethology (Ader et al., 1991, 2001). The experimental foundations of the discipline lie in several studies in which the bi-directional communication between the nervous system and the immune system were described and mechanicistically understood (Besedovsky & del Rey, 1996; Rabin, 1999). For example: from the periphery to the brain, activated immune cells can communicate their status/activity to the brain via cytokine receptors present in neurons, around the circumventricular organ and in the afferent braches of the vagus nerve (Goelher et al., 1997; Konsman et al., 2001); From the brain to the periphery, noradrenergic terminals innervate the spleen and other lymphoid organs and immunocytes express receptors for almost all neurohormones and neuropeptides known (Bellinger et al., 2001; Wybran et al., 1979; Sanders et al., 2001). Within the psychoneuroimmunological frameworks is possible to exactly investigate the interactions between the brain the neuroendocrine and immune system when stimulated/modulated by social factors.

EXPERIMENTAL EVIDENCES: RODENTS

Ever since the publication of the Bohus and Koolhaas paper in 1991, several studies have been published on this topic and the aim of this section is to integrate these new reports inside the Bohus and Koolhaas framework. Accordingly I will distinguish paper dealing with dyadic interactions; individual housing and group housing. Each part will distinguish the effects exerted on humoral immunity and cell-mediated immunity.

Dyadic Interactions

Within the category *dyadic interactions* I will consider all the models of social stress essentially based on the resident/intruder test which has been first introduced in the literature by Ginsburg & Allee (1942) and standardized later on by Miczek (1979). In this test, a resident male inhabiting a territory (alone or with a female) receives an intrusion in his territory by a younger and lighter conspecific. This is a widely used protocol and its effects are today described as the prototypical response to a social stress (Koolhaas et al., 1997b; Meerlo et al., 2002; Blanchard et al., 2002; Stefanski 2000). Other models such as the sensory contact model devised by Kudryavtseva to induce a depression-like state in mice (Kudryavtseva et al., 1991; Kudryavtseva 2000, 2003) or the chronic psychosocial stress which we recently developed (Bartolomucci et al., 2001, 2004, 2005; Bartolomucci 2005) may be considered as chronic extension of the classical resident intruder test where animals live chronically in sensory contact and interact on a daily base.

Ever since 1991 it was shown that subordinates and partially dominants showed impaired cellular immunity, such as reduced splenocytes proliferation, Th cell number and Th/Tcytotoxic ratio (Raab et al., 1986; Bohus et al., 1991). A depression of immune function has been confirmed and extended recently. For example, de Groot et al., (1999) proved that a single social defeat was able to transiently depress splenocytes proliferation, IL-2 and IFN-γ which returned to normal level afterwards.

Stefanski and Engler conducted a comprehensive investigation in rats (1998, 1999; see also Stefanski, 2000; Engler and Stefanski 2003; Engler et al., 2004). 2 hours after a resident/intruder test, intruder subordinates showed a strong increase in granulocytes and a marked lymphopenia in the whole blood. The percentage over the total lymphocytes count of Th and NK cells increased while B cells decreased. Proliferation to Concavalin A (ConA) was depressed as well. Two, as well as 7 days after the beginning of the procedure, the changes still persisted but a general reduction of the magnitude emerged. In the meanwhile NK cytotoxicity was strongly reduced. When the confrontation was stopped, several carry over effects emerged: granulocytosis, lower percentage of Th lymphocytes and lower proliferation. Those are the effects in the intruder becoming subordinates. The resident becoming dominants, shows a profile similar to subordinates until day 2, while a return to baseline level tended to emerged afterwards.

Kudryavtseva and coworkers (1998; Gryazeva et al., 2001) report similar immune effects in mice challenged with the sensory contact model. Indeed, they showed that repeated experience of social confrontation non-specifically increased the proportion of segmented neutrophils and lactate dehydrogenase activity in both participants in aggressive encounters, and decreased the numbers of CD4+ and CD8+ T-lymphocytes in the spleen. The proportion of CD4+ cells in defeated mice thymuses also decreased.

Azpiroz and coworkers observed a depression of NK activity in both dominant and subordinate mice under dyadic interactions (1994). A more recent investigation from the same group also showed a persistent decrease in mitogen induced proliferation (Vegas et al., 2006) which can be at least in part be reversed by fluoxetin treatment (Beitia et al., 2005). However, the observed immune impairments are often present when compared with undisturbed controls, while less or no difference emerged when compared with manipulated mice. This may resemble the findings by Bohus et al., (1993), who did not find main differences between socially defeated and foot-shock exposed rats.

Avitsur et al., (2002a) showed in pair fighting mice (daily defeat by an aggressive resident mice) that splenocytes developed a glucocorticoid resistance consisting in the resistance to the anti-proliferative effect of corticosterone (see below), as well as lower circulating lymphocytes accompanied by higher number of monocytes and neutrophils.

In a first study Bartolomucci et al. (2001) submitted mice to chronic psychosocial stress and showed that only Resident Subordinate mice showed a decreased splenocyte proliferation, IL-4 IL-10 and to a lower extent of IL-2 to ConA. In the follow up of this study (Bartolomucci et al., 2003a) mice were immunized with Keyhole Limpet Hemocyanine (KLH) seven days after stress-procedure onset. Splenocytes of KLH exposed mice were re-stimulated in vitro with KLH. Cell proliferation and production of IL-2, IL-4, IL-10 and IFN-γ were measured. As was for the previous study, Resident Subordinate mice showed a reduced KLH induced proliferation in vitro and lower IL-2 release when compared to controls. Results obtained in Intruder Subordinates mice prove that subordination per se does not affect the immune responses of Resident Subordinate mice. Indeed, Intruder Subordinate mice showed only minor immune-impairments. These results suggested that the loss of relevant resources, such as the territory, and not subordination in itself is the key factor determining why only certain stress-exposed individuals ultimately show malignancy and psychopathologies (Bartolomucci 2005).

As far as humoral immunity is concerned, several reports showed social defeat or victory to modulate Ig production. Particularly social defeat was associated with lower IgG in rats (Bohus et al., 1991), while augmentation of antibody titers was sometimes observed in dominant/winner mice (Fauman, 1987). Fleshner and coworkers (1989) also showed that a single exposure of an intruder rat previously immunized with KLH to an aggressive resident determined a long lasting decrease in anti-KLH-IgG. They also showed that the greater the amount of submissive posture displayed the lower the level of IgG. Consistent with the latter observations, Devonio et al., (2003) recently observed an increased anti SRBC immunity in dominants and a decrease in subordinates C57BL/6J mice. The role of the social status in mediating this immune response was proved by the inversion of the anti-SRBC response when a reversal of social status occurred. Dominant becoming subordinates showed a decrease in rosette forming cells, while subordinates becoming dominant showed an increase.

This result confirm previous observation from the same group reporting that dominant male CBA mice have increased anti-SRBC immunity, while subordinate but not dominant C57BL/6J males show a lower anti-SRBC immunity compared to controls (Devonio et al., 1993). A similar lower anti-SRBC response was shown, in socially defeated rats by Bohus and coworkers (1993).

De Groot et al. (2002) showed that anti-viral-IgG production is decreased after social defeat, however several subtleties emerged. First of all, the lower level of antibodies was observed in wounded but not in non-wounded subordinates. Second, the stress-induced suppression of antibody was strongest against inactivated rather than live viruses.

Bartolomucci et al. (2003a) first induced chronic psychosocial stress then, after 7 days of stress, mice were immunized with KLH (see above for experimental details). Results proved that Resident Subordinate and Dominant mice showed lower anti-KLH-IgG than controls. On the contrary, Intruder Subordinates showed no immune alteration.

Immunized syrian hamster males were introduced once or daily for 5 consecutive days in the cage of an aggressive unfamiliar hamster. They showed a significant decrease of the plasma level of anti-KLH-IgG (Jasnow et al., 2001).

In conclusion, consistent evidences exist in favor of a depression of cellular immunity mediated by dyadic models of social stress (both having intermittent daily based confrontation or continuous physical/sensory contact). In particular, reduced lymphocytes proliferation, decreased NK activity, lymphopenia and granulocitosis appear as the most robust immune alterations. The changes appear to be maximal acutely, but importantly, many changes still persist for up to 3 weeks and even after cessation of the confrontation. A straightforward conclusion emerged for humoral immunity too: it is consistently depressed in losers/subordinate individuals in dyadic models of social stress while winner/dominant showed inconsistent results having unaffected, decreased or even increased humoral immunity.

Individual Housing

Social isolation and lack of social support are considered deleterious for health, thus being regarded as one of the most relevant causes of diseases in animals including humans (Valzelli, 1973; Jessop and Bayer, 1989; Brown and Grunberg, 1995; Lyons et al., 1999). However, experimental evidences coming from behavioral and physiological studies are controversial, with individually housed animals showing higher, unaffected or lower cellular immune functions when compared with controls. It was proved by several independent studies with rats and mice that living in group when compared with individual housing is associated with lower proliferation and antigen presenting cells activity (Grewal et al., 1997; Lu et al., 1999), as well as lower mitogen stimulated cytokine release (Karp et al., 1994). The effects seem also strictly dependent on the interaction between social and genetic factors (Scislowska-Czarnecka et al., 2004). On the contrary, lower proliferation to ConA and cytokine release has been proved as well (Jessop & Bayer, 1989; Bartolomucci et al., 2003b).

These controversial finding may found explaination when we consider a few methodological questions. First of all, isolation in rodents determines a bi-phasic effect on immune functions. In rats after 1 week a lower proliferation to ConA and PHA emerged with values coming back to baseline and even increasing above baseline after 20 days of individual housing (Jessop & Bayer, 1989). In a recent study the link between individual housing and skin immune function was investigated in mice (Nakano 2004). Mice isolated from 2 to 90 days were first sensitized with haptens and then contact sensitivity was elicited. In agreement with results on cell proliferation (Jessop & Bayer, 1989), ear swelling was decreased in the first few days, than increased after 30 days and finally depressed after 90 days of individual housing. To further characterize this effect Nakano (2004) showed that the effect was paralleled by trafficking and antigen-presenting-cell functions of auricular drained lymph nodes langerhans cells but not of keratinocytes.

In addition to the temporal issue, Bartolomucci et al., (2003b) recently clarified that: individually housing mice for increasing time periods did not induce major immuno-endocrine effects compared to a stable sibling group housing despite, according to previous studies (Jessop & Bayer, 1989), a general lower level of ConA induced prolipheration and IL-2 emerged. On the contrary, when individually housed mice were exposed to a mild acute stress, i.e. forced exposure to a novel environment, they showed higher basal corticosterone and lower type 1 (IL-2) and type 2 (IL-4) cytokines as well as ConA induced-splenocytes proliferation compared to controls. Finally, when faced with a free choice between a novel environment and their home cage, individually housed mice showed reduced neophobic responses resulting in increased exploration of the novel environment, thus suggesting a low anxiety profile. Altogether, these results suggest that individual housing in itself does not dramatically impact baseline functions, but may impact cellular immune functions when individuals are further exposed to a stressful event.

Finally, a crucial point to evaluate the effects of individual housing regards the identity of the control group-housed individuals. Indeed, several findings showed that the lower the familiarity between animals in a group the higher the stress-induced effects (Bartolomucci et al., 2003b, 2004; Meehan and Henry, 1981; Vekovishcheva et al., 2000; Stefanski, 2001). Since in many of the above mentioned immune studies, individually housed mice were compared to non-siblings grouped subjects, it is likely that the often reported higher immune functions of individually housed mice might be due instead to the depressed immune function of rodents housed in groups of non-siblings.

Humoral immunity has also been investigated in individually housed rodents. Individually housed mice have been reported to produce a higher secondary anti-KLH IgG and IgM production when compared with mice housed in groups of 4, 6 and 12 animals per group (Karp et al., 1993). However, as was for cellular immunity the effects of individual housing are time dependent. Shanks et al., (1994) showed that when mice were isolated from group housing a transient depression of anti-SRBC-Ig emerged after 5 to 15 days but not for longer duration of individual housing and can be reversed by diazepam treatment; Liu and Wang (2005) confirmed this finding by showing that 10 days of individual housing determined a significant reduction of anti-SRBC-Ig when compared with group housed males. Klein & Nelson (1999), investigated sex and species differences in relation to housing condition in the polygamous meadow vole and the monogamous prairie voles: sex differences did not emerge when voles were singly housed but differences were only observed among meadow voles housed in pairs, in which females had higher anti-KLH IgM and IgG than males. Neither of these findings is related to circulating oestradiol, testosterone and corticosterone levels. The authors concluded that sex differences in humoral immunity are more pronounced among polygamous than monogamous species. Thus, social interaction is able to unmask sex differences in humoral immunity. A similar study was conducted in the siberian hamster by Demas and coworkers (2004). Male and female hamsters were either housed alone, in same sex or mixed sex pairs and immunized with KLH. Same sex peer reared males showed the lower level of anti-KLH IgM when compared with mixed sex pairs with isolated males showing an intermediate level. Instead, isolated females showed reduced IgM when compared with same sex peer-reared counterparts. Again same sex groups were the focus of the study by Heise and Van Acker (2000). Adult dominant female common voles showed lower immune response to SRBC than sisters. This finding was interpreted in terms of social stress induced immunosuppression.

In conclusion, evidences from individually housed rodents are less straightforward than from dyadic interaction models with evidences of depressed, unaffected or increased immune functions. The possible reason of these discrepancies may lie in different methodologies between studies.

Group Housing and Social Status

In this section I will discuss evidences coming from studies dealing with rank difference in a social group, social instability created by manipulating the individual group composition, social disruption within a group determined by intrusion of an aggressive conspecific and in a few case changes in stocking density such as crowding.

Bohus and coworkers (1991, 1993) aggregated male and females rats in colonies and investigated cellular immune functions. They observed striking differences between dominants, subdominants, subordinates and particularly outcast rats when compared to rats housed in conventional laboratory cages. Dominants and subdominants showed similar changes, with proliferation and several circulating lymphocytes subclasses (B, Th) being increased and few being decreased (Tcytotoxic). On the other hand, subordinate and outcast rats, which are rats loosing the dominant position, showed an almost complete suppression of all lymphocytes subclasses and lower prolipheration to ConA and Phytoagglutinin A (PHA).

In rabbits, grouping of unfamiliar individuals determined an increase of plasma corticosterone, an increase in LPS-induced IFN-γ release and a decrease in glucocorticoid-receptors (GR) capacity but not affinity in PBMC (Lupo et al., 2000). In a further study the same group showed that the dominance index (the higher it is the higher the rank in a hierarchy) positively correlate with GR level and negatively with IFN-γ release in PBMC of colony-housed rabbits (Muscetola et al., 2003).

Male mice challenged with the social disruption test (Padgett et al., 1998; Sheridan et al., 2000) showed the development of a glucocorticoid resistance, which consists in the resistance to the anti-proliferative effect of corticosterone. This state develops and is mediated by the increased level of circulating glucocorticoids and is regarded as a form of protection against the immune-suppressive effects of glucocorticoids while having clear pathophysiological effects. In several studies it was found that glucocorticoids resistance: needs more than one SDR cycle to occur (Avitsur et al., 2002b); develops preferentially in presence of injuries (Avitsur et al., 2001; Merlot et al., 2003; Engler et al., 2004) and in mice showing a subordinate behavioral profile (Avitsur et al., 2001); is mediated by the presence of macrophages, CD11b^{+}-positive monocytes but not B cells (Stark et al., 2001; Avitsur et al., 2002b); is paralleled, but not induced, by a higher release of IL-6 and TNF-α (Stark et al.,

2002; Avitsur et al., 2005). Recently, Engler et al. (2004, 2005) investigated immune cell trafficking associated with SDR. In the bone marrow B, T and NK cells strongly decreased while neutrophils and monocytes after an initial drop increased above control values. In the blood, B and T cells slightly decreased while NK, monocytes and neutrophils increased. Finally the spleen showed a mixed profile with monocytes, neutrophils, NK and B cells increasing and T cells decreasing. Therefore it is evident that after an initial depletion of immunocytes from the bone marrow to the circulation and peripheral lymphoid organs granulopoiesis strongly increased supporting ongoing granulocitosis. On the contrary, changes in lymphocytes number might be related either to the reduction in the number of recent thymic emigrants in the peripheral blood trafficking (Engler and Stefanski 2003) or to glucocorticoids induced trafficking/apoptosis (Engler et al., 2004).

Differential group hosing has also been found to affect humoral immunity. Animals entering an already established group and, more in general, changing housing condition will show a decreased antibody production (Fleshner et al., 1989; Edwards et al., 1980). Dominants in a group tended to have higher antibodies than subordinates (Beden and Brain, 1985).

The effect of altered immune functions observed in grouped rodents may even not be dependent upon direct social interaction but their urinary products. In fact, Cocke and coworkers (1993), showed odors of stressed conspecific to alter anti-KLH-IgM and IgG relative to home cage control and to animals exposed to the odor of non-stressed conspecifics (see also Fernandez, 2000).

In conclusion, it seems evident that social stress within groups of males and particularly in context of higher than normal aggression (because experimentally induced) is a potent modulator of cellular immunity. Individual differences, although poorly investigated, clearly emerged. In particular lower rank individuals show the highest social stress-induced cellular immune suppression. On the other hand, studies on humoral immunity are limited, therefore, it is still premature to draw a definitive conclusion.

Experimental Evidences: Other Animals

The effect of social stress on immunity has been mostly investigated in rodents. However, a number of studies have also been conducted in non-human primates, farm animals and birds. This section is included to provide the reader comparative evidences proving that social stress effect neuroimmune functions not only in rodents but also in almost all vertebrate species investigated so far.

Non-human Primates

Particularly macaques reared in different housing condition were the focus of the research while dyadic models are virtually missing from primate literature. Shapiro et al., (2000), reported a lower immune competence (in vitro proliferation to a number of gastrointestinal pathogens and cytokines production) for individually housed rhesus macaques when compared to pairs and group housed counterparts. Lilly et al., (1999) reached a similar conclusion reporting that red blood cells, hemoglobin, hematocrit, platelets, as well as white blood cells, the absolute number of T helper, T cytotoxic and total T and B cells were depressed in female rhesus after individual housing. In another set of experiments, removal of juvenile rhesus monkeys from their natal social group to indoor individual caging resulted in increased basal cortisol secretion and significant decrements in the frequency of lymphoid subpopulations such as that absolute numbers of total T cells declining (Gordon et al., 1992). The detrimental effect of social isolation is not restricted to rhesus because similar findings have been obtained also with *Cebus apella* (Kling et al., 1992). Not only social isolation may affect immunity, it can also increase mortality when social separation occur in the 90-day period before simian immunodeficiency virus (SIV) inoculation and in the 30-day period after inoculation (Capitanio and Lerche, 1998).

Differential group housing can be a tremendous event for primates too. Gust et al., (1993) in fact, showed that removal of adult rhesus monkeys from a large social group to peer housing resulted in no significant changes in basal cortisol levels or absolute numbers of T lymphocyte subsets 24 h later. However, the return of these males one year later to the same social group resulted in significant increases in cortisol levels and decrease in T-helper and T-cytotoxic cells 24 h later. On the contrary, the same authors showed that removal of naive juvenile rhesus monkeys from their natal social group to peer housing resulted in increased basal cortisol secretion and significant decrements in the absolute numbers of T-helper and T-cytotoxic lymphocyte subsets in the peripheral blood (Gust et al., 1992).

Similar changes can be obtained also with the formation of a new group of rhesus monkeys. Overall, the group showed a significant increase in cortisol and a significant decrease in the absolute number of total lymphocytes, T-helper and T-cytotoxic cells at 24 h post formation, but not thereafter. However, when partitioned into high and low dominance rank, differences in T-helper and T-cytotoxic cells were evident for up to 9 weeks with low ranking subjects showing significantly lower values (Gust et al., 1991).

Differential group housing may also affect monkey's health and not only immune functions. In the study already mentioned, Capitanio and Lerche, (1998), proved that housing relocation in the 90-day period before SIV inoculation and in the 30-day period after inoculation were associated with decreased survival. The effects are not limited to housing relocation and individual housing, because Capitanio et al., (1998) also showed that monkeys housed under unstable social conditions displayed more agonism and less affiliation, shorter survival to AIDS, and lower basal concentrations of plasma cortisol compared with stable animals. In part at variance with the studies by Capitanio and coworkers, Cohen et al., (1997) showed that in male cynomolgus monkeys randomly assigned to stable or unstable social conditions manipulation did not influence the probability of being infected by upper respiratory viruses. However, low social status was associated with a substantially greater probability of being infected.

This chapter mostly deals with the effects of social factors in adult life on neuroimmune functions. However it is worth mentioning that in a long series of studies Coe and coworkers were able to prove that disturbances during prenatal life can affect the baby's maturation including neuroimmune functions (Coe and Lubach 2005; Bailey et al,. 2004; Coet et al., 2002). Prenatally stressed monkeys evinced immature neuromotor reflexes at birth, greater emotionality during the first year of life, and a smaller hippocampus as juveniles. Immune responses of the infants were also affected: lymphocyte proliferation, natural killer activity and cytokine production were reduced.

Farm Animals

Social factors also affect neuroimmune functions of farm animals, The most investigated species being pigs.

A few studies investigated the modulating role of social factors in the early life of pig. For example, early social isolation determine a diminished plasma TNF-α response to LPS 1 day as well as 45 days after the isolation period as well as decreased lymphocyte proliferation in response to ConA and pokeweed mitogen one day after the isolation (Tuchscherer et al., 2004; Kanitz et al., 2004). Another social manipulation often investigated is the effect of mixing non-sibling pigs. Here the effect is strictly age-dependent. Indeed, when piglets are mixed in unfamiliar groups at weaning, they show a cortisol levels increase after mixing which returned to basal values within 24 h, while blood lymphocyte proliferation was not affected (Merlot et al., 2004). On the contrary altered immune functions are seen when mixing adult animals even if the direction of the effect is still controversial. For example, de Groot et al., (2001) showed that male but not female pig housed in mixed colonies showed a robust decrease in cellular immunity with dominant in mixed colonies having the lower immune response. Sutherland et al., (2006) while showing a clear effect of mixing report that that LPS-induced proliferation and NK cytotoxicity were greater in stressed pigs compared with controls. Furthermore, among stressed pigs, dominant pigs had a greater total white blood cell count, NK, and phagocytosis than the subordinate pigs.

Studies in other farm animals are more rare. Cunnick et al., (1994) investigate the effects of acute social isolation (30, 60, or 90 min) on young fowl immune system. The socially isolated birds showed an increase in body temperature and number of leukocytes in the blood but not in the spleen. In addition, isolation resulted in a transient increase in B-lymphocyte mitogen proliferation while a time-dependent decrease in T-lymphocyte mitogen proliferation was reported.

Finally, Hickey et al., (2003) showed that in calves disruption of the established social group at weaning (inclusive of social group disruption and maternal separation) determined an increase in the plasma cortisol concentration and neutrophil/lymphocyte ratio and decreased the leukocyte concentration and the in vitro interferon-γ response to ConA and KLH. The effects persisted for up to seven days after weaning.

Birds

A few studies also investigated the role of social factors on immune and neuro-endocrine functions (see also the effects on fowl discussed above). Edens et al., (1983) housed japanese quails either to isolation or subjected to a chronic resident-intruder test. The immune responsiveness to SRBC, as indicated by production of hemagglutinins of the intruder and residents, was significantly less than that of the isolates, and the intruder showed fewer antibodies than the residents.

More recently, Lindstrom et al. (2005) proved that house sparrows adjust their social status position to their physiological costs. Indeed, they showed that the cost of maintaining a social status position differed between individuals and were related to individual body size. Birds with small body size had increased costs in terms of increased stress responses and reduced cell-mediated immune responses while being experimentally kept as dominants, while birds with large body size had increased costs while they were kept as subordinates. Finally, birds with increased energetic and immunological costs as dominants obtained a low status position in the large group, while birds with increased costs as subordinates obtained a high status position in the large group.

A latter study is worth mentioning, Hawley et al. (2006) investigated the role of an experimentally stimulated social competition generated trough a reduction in feeding sites on house finches immune functions. They showed that finches under high social competition had significantly higher aggression rates, lower antibody responses, and lost more body mass. Within flocks, dominant individuals mounted stronger immune responses in both competition treatments.

Conclusion

The review of the available experimental studies published on social factors and immune functions clearly show how powerful its effect can be. Most of the studies where conducted in rodents while the comparative view, allowed by inclusion of studies in other vertebrate species, allows reaching a general conclusion: social factors does matter for vertebrate and strongly modulate their neuroimmune functions. This conclusion is important because due to the phylogenetical conservative nature of the social stress-response these studies can be considered as the foundation to extend the conclusion reached to human being.

References

Ader, R., Felten, D. L. & Cohen N. (1991). *Psychoneuroimmunology*, 2nd ed. San Diego: Academic Press.

Ader, R., Felten, D. L. & Cohen N. (2001). *Psychoneuroimmunology*, 3rd ed. San Diego: Academic Press.

Alexander, R. D. (1974). The evolution of social behavior. *Annual Review of Ecology and Systematic.*, *5*, 325-383.

Altmann, J. Sapolsky, R. M. & Licht, P. (1995). Baboon fertility and social status. *Nature*, *377*, 688-689.

Avitsur, R., Stark, J. L. & Sheridan, J. F. (2001). Social stress induces glucocorticoid resistance in subordinate animals. *Hormones and Behavior.*, *39*, 247-57.

Avitsur, R., Stark, J. L., Dhabhar, F. S. & Sheridan, J. F. (2002a). Social stress alters splenocyte phenotype and function. *Journal of Neuroimmunology.*, *132*, 66-71.

Avitsur, R., Stark, J. L., Dhabhar, F. S., Padgett, D. A. & Sheridan, J. F. (2002b). Social disruption-induced glucocorticoid resistance: kinetics and site specificity. *Journal of Neuroimmunology.*, *124*, 54-61.

Avitsur, R., Kavelaars, A., Heijnen, C. & Sheridan, J. F. (2005). Social stress and the regulation of tumor necrosis factor-alpha secretion. *Brain Behavior and Immunity.*, *19*, 311-317.

Azpiroz, A., Arregui, A., Fano, E., Garmendia, L. & Sanchez-Martin, J. R. (1994). Fighting experience and natural killer cell activity in male laboratory mice. *Aggressive Behavior.*, *20*, 67-72.

Bailey, M. T., Lubach, G. R. & Coe, C. L. (2004). Prenatal stress alters bacterial colonization of the gut in infant monkeys. *Journal of Pediatric Gastroenterology and Nutrition.*, *38*, 414-21.

Bartolomucci, A. (2005). Resource loss and stress-related disease: is there a link? *Medical Science Monitor.*, *11*, RA147-154.

Bartolomucci, A., Palanza, P., Gaspani, L., Limiroli, E., Panerai, A. E., Ceresini, G., Poli, M. D. & Parmigiani, S. (2001). Social status in mice: behavioral, endocrine and immune changes are context dependent. *Physiology and Behavior.*, *73*, 401-410.

Bartolomucci, A., Sacerdote, P., Panerai, A. E., Pederzani, T., Palanza, P. & Parmigiani, S. (2003a). Chronic psychosocial stress induced down-regulation of immunity depends upon individual factors. *Journal of Neuroimmunology.*, *151*, 58-64.

Bartolomucci, A., Palanza, P., Sacerdote, P., Ceresini, G., Chirieleison, A., Panerai, A. E. & Parmigiani, S. (2003b). Individual housing induces altered immuno-endocrine responses to psychological stress in male mice. *Psychoneuroendocrinology.*, *28*, 540-58.

Bartolomucci, A., Pederzani, T., Sacerdote, P., Panerai, A. E., Parmigiani, S. & Palanza, P. (2004). Behavioral and physiological characterization of male mice under chronic psychosocial stress. *Psychoneuroendocrinology.*, *29*, 899-910.

Bartolomucci, A., Palanza, P., Sacerdote, P., Panerai, A. E., Sgoifo, A., Dantzer, R. & Parmigiani, S. (2005). Social factors and individual vulnerability to chronic stress exposure. *Neuroscience and Biobehavioral Reviews.*, *29*, 67-81.

Beden, S. N. & Brain, P. F. (1985). The primary immune responses to sheep red blood cells in mice of differing social rank or from individual housing. *IRCS Medical Science.*, *13*, 364-365.

Beitia, G., Garmendia, L., Azpiroz, A., Vegas, O., Brain, P. F. & Arregi, A. (2005). Time-dependent behavioral, neurochemical, and immune consequences of repeated experiences of social defeat stress in male mice and the ameliorative effects of fluoxetine. *Brain Behavior and Immunity.*, *19*, 530-9.

Bellinger, D. L., Lorton, D., Lubahn, C. & Felten D. L. (2001). Innervation of lymphoid organs-association of nerves with cells of the immune system and their implications in desease. In: Ader, R., Felten, D. L., Cohen, N. (Eds.). *Psychoneuroimmunology* 3rd ed (Vol 1, 55-112). San Diego: Academic Press.

Benton, D., Dalrymple-Alford, J. C. & Brain P. F. (1980). Comparisons of measures of dominance in the laboratory mouse. *Animal Behaviour.*, *28*, 1274-1279.

Besedovsky, H. O. & Del Rey, A. (1996). Immuno-Neuro-Endocrine interactions: facts and hypotheses. *Endocrine Reviews.*, *17*, 64-102.

Blanchard, R. J., McKittrick, C. R. & Blanchard D. C. (2001). Animal models of social stress: effects on behavior and brain neurochemical systems. *Physiology and Behavior.*, *73*, 261-271.

Blanchard, D. C., McKittrick, C. R., Hardy, M. P. & Blanchard, R. J. (2002). Effects of social stress on hormones, brain, and behavior. In: D. W., Pfaff, A. P., Arnold, A. M., Etgen, S. E., Fahrbach, R. T., Rubin Eds., *Hormones, brain and behaviour* (Vol 1 pp. 735-772). San Diego: Academic Press.

Bohus, B. & Koolhaas, J. M. (2001). Psychoimmunology of social factors in rodents and other subprimate vertebrates. In: Ader, R., Felten, D. L., Cohen, N. (Eds.). *Psychoneuroimmunology* 3rd ed (Vol 1, pp. 807-830). San Diego: Academic Press.

Bohus, B., Koolhaas, J. M., de Ruiter, A. J. H. & Heijnen C. J. (1991). Stress and differential alterations in immune functions: conclusions from social stress studies in animals. *Nederland Journal of Medicine*, *39*, 306-315.

Bohus, B., Koolhaas, J. M., Heijnen, C. J. & de Boer, O. (1993). Immunological responses to social stress: dependence on social environment and coping abilities. *Neuropsychobiology*, *28*, 95-9.

Brown, K. J. & Grunberg, N. E. (1995). Effects of housing on male and female rats: crowding stresses males but calms females. *Physiology and Behavior.*, *58*, 1085-1089.

Capitanio, J. P. & Lerche N. W. (1998). Social separation, housing relocation, and survival in simian AIDS: a retrospective analysis. *Psychosomatic Medicine*, *60*, 235-44.

Capitanio, J. P., Mendoza, S. P., Lerche, N. W. & Mason, W. A. (1998). Social stress results in altered glucocorticoid regulation and shorter survival in simian acquired immune deficiency syndrome. *Proceedings of the National Academy of Science U S A*, *95*, 4714-9.

Cocke, R. C., Moynihan, J. A., Cohen, N., Grota, L. J. & Ader, R. (1993). Exposure to conspecific alarm chemosignals alters immune responses in BALB/c mice. *Brain Behavior and Immunity.*, *7*, 36-46.

Coe, C. L. & Lubach, G. R. (2005). Prenatal origins of individual variation in behavior and immunity. *Neuroscience and Biobehavioral Reviews*, *29*, 39-49.

Coe, C. L., Kramer, M., Kirschbaum, C., Netter, P. & Fuchs, E. (2002). Prenatal stress diminishes the cytokine response of leukocytes to endotoxin stimulation in juvenile rhesus monkeys. *Journal of Clinical Endocrinology and Metabolism*, *87*, 675-681.

Cohen, S., Line, S., Manuck, S. B., Rabin, B. S., Heise, E. R. & Kaplan, J. R. (1997). Chronic social stress, social status, and susceptibility to upper respiratory infections in nonhuman primates. *Psychosomatic Medicine*, *59*, 213-21.

Conradt, L. & Roper, T. J. (2003). Group decision-making in animals. *Nature.*, *421*, 155-8.

Creel, S. (2001). Social dominance and stress hormones. *Trends in Ecology & Evolution*, *16*, 491-497.

Cunnick, J. E., Kojic, L. D. & Hughes, R. A. (1994). Stress-induced changes in immune function are associated with increased production of an interleukin-1-like factor in young domestic fowl. *Brain Behavior and Immunity*, *8*, 123-36.

de Groot, J., van Milligen, F. J., Moonen-Leusen, B. W. M., Thomas, G. & Koolhaas J. M. (1999). A single social defeat transiently suppresses the anti-viral immune responce in mice. *Journal of Neuroimmunology*, *95*, 143-151.

de Groot, J., Ruis, M. A., Scholten, J. W., Koolhaas, J. M. & Boersma, W. J. (2001). Long-term effects of social stress on antiviral immunity in pigs. Physiology and Behavior, *73*, 145-58.

de Groot, J., Boersma, W. J. A., Scholten, J. W. & Koolhaas, J. M. (2002). Social stress in male mice impairs long-term antiviral immunity selectively in wounded subjects. *Physiology and Behavior*, *75*, 277-285.

Demas, G. E., Johnson, C. & Polacek, K. M. (2004). Social interactions differentially affect reproductive and immune responses of Siberian hamsters. *Physiology and Behavior 83*, 73-9.

Devonio, L., Algerina, E. & Pavina, T. (2003). Immunological consequences of the reversal of social status in C57BL76J mice. *Brain Behavior and Immunity*, *17*, 28-34.

Drummond, H. (2006). Dominance in vertebrate broods and litters. *Quarterly Review of Biology*, *81*, 3-32.

Edens, F. W., Thaxton, P., Morgan, G. W. & Parkhurst, C. R. (1983). Grouping in Japanese quail 2. Suppression of humoral immunity. *Poultry Science.*, *62*, 2479-85.

Edwards, E. A., Rahe, R. H., Stephens, P. M. & Henry, J. P. (1980). Antibody response to bovine serum albumine in mice: the effect of psychosocial environmental change. *Proceedings of the Society of Experimental Biology and Medicine*, *164*, 478-481.

Engler, H., Stefanski, V. (2003). Social stress and T cell maturation in male rats: transient and persistent alterations in thymic function. *Psychoneuroendocrinology*, *28*, 951-69.

Engler, H., Bailey, M. T., Engler, A. & Sheridan, J. F. (2004). Effects of repeated social stress on leukocyte distribution in bone marrow, peripheral blood and spleen. *Journal of Neuroimmunology*, *148*, 106-15.

Engler, H., Dawils, L., Hoves, S., Kurth, S., Stevenson, J. R., Schauenstein, K. & Stefanski, V. (2004). Effects of social stress on blood leukocyte distribution: the role of alpha- and beta-adrenergic mechanisms. *Journal of Neuroimmunology*, *156*, 153-62.

Engler, H., Engler, A., Bailey, M. T. & Sheridan, J. F. (2005). Tissue-specific alterations in the glucocorticoid sensitivity of immune cells following repeated social defeat in mice. *Journal of Neuroimmunology*, *163*, 110-9.

Fauman, M. A. (1987). The relation of dominant and submissive behavior to the humoral immune responce in Balb/c mice. *Biological Psychiatry*, *22*, 771-776.

Fernandes, G. A. (2000). Immunological stress in rats induces bodily alterations in saline-treated conspecifics. *Physiology and Behavior*, *69*, 221-230.

Fleshner, M., Laudenslager, M. L., Simons, L. & Maier, S. F. (1989). Reduced serum antibodies associated with social defeat in rats. *Physiology and Behavior*, *45*, 1183-1187.

Fuchs, E, Flugge, G & Czeh, B. (2006). Remodeling of neuronal networks by stress. *Frontiers in Bioscience.*, *11*, 2746-58.

Ginsburg, B. & Allee, W. C. (1942). Some effect of conditioning on social dominance and subordination in inbred strains of mice. *Physiological Zoology*, *15*, 485-506.

Goehler, L. E., Relton, J. K., Dripps, D., Kiechle, R., Tartaglia, N., Maier, S. F. & Watkins, L.R. (1997). Vagal paragangli bind biotinylated IL-1 antagonist. *Brain Research Bullettin*, *43*, 357-364.

Gordon, T. P., Gust, D. A., Wilson, M. E., Ahmed-Ansari, A., Brodie, A. R. & McClure H. M. (1992). Social separation and reunion affects immune system in juvenile rhesus monkeys. *Physiology and Behavior*, *51*, 467-72.

Goymann, W. & Wingfield, J. C. (2004). Allostatic load, social status and stress hormones: the cost of social status matter. *Animal Behaviour*, *67*, 591-602.

Grewal., I. S., Heilig, M., Miller, A. & Sercarz, E. E. (1997). Environmental regulation of T-cell function in mice: group housing of males affects accessory cell function. *Immunology.*, *90*, 165-168.

Gryazeva, N. I., Shurlygina, A. V., Verbitskaya, L. V., Mel'nikova, E. V., Kudryavtseva, N. N. & Trufakin V. A. (2001). Changes in various measures of immune status in mice subject to chronic social conflict. *Neuroscience and Behavioral Physiology*, 2001, *31*, 75-81.

Gust, D. A., Gordon, T. P., Wilson, M. E., Ahmed-Ansari, A., Brodie, A. R. & McClure, H. M. (1991). Formation of a new social group of unfamiliar female rhesus monkeys affects the immune and pituitary adrenocortical systems. *Brain Behavior and Immunity*, *5*, 296-307.

Gust, D. A., Gordon, T. P., Wilson, M. E., Brodie, A. R., Ahmed-Ansari, A. & McClure, H. M. (1992). Removal from natal social group to peer housing affects cortisol levels and absolute numbers of T cell subsets in juvenile rhesus monkeys. *Brain Behavior and Immunity*, *6*, 189-99.

Gust, D. A., Gordon, T. P. & Hambright, M. K. (1993). Response to removal from and return to a social group in adult male rhesus monkeys. *Physiology and Behavior*, *53*, 599-602

Hardy, C. A., Quay, J., Livnat, S. & Ader R. (1990). Altered T-Lymphocyte response following aggressive encounters in mice. *Physiology and Behavior*, *47*, 1245-1251.

Hawley, D. M., Lindstrom, K., Wikelski, M. (2006). Experimentally increased social competition compromises humoral immune responses in house finches. *Hormones and Behavior*, *49*, 417-24.

Heise, S. R. & Van Acker A. (2000). The effect of social environment on the immune response of female common voles in matriarchal laboratory groups. *Physiology and Behavior 71*, 289-96.

Hemelrijk, C. K. & Wantia, J. (2005). Individual variation by self-organisation. *Neuroscience and Biobehavioral Reviews 29*, 125-36.

Henry, J. P. & Stephens P. M. (1977). *Stress, health and the social environment*. Berlin: Springer Verlag.

Hickey, M. C., Drennan, M. & Earley, B. (2003). The effect of abrupt weaning of suckler calves on the plasma concentrations of cortisol, catecholamines, leukocytes, acute-phase proteins and in vitro interferon-gamma production. *Journal of Animal Science*, *81*, 2847-55.

Jasnow, A. M., Drazen, D. L., Huhman, K. L., Nelson, R. J. & Demas G. E. (2001). Acute and chronic social defeat suppresses humoral immunity of male Syrian hamsters (Mesocricetus auratus). *Hormones and Behavior*, *40*, 428-433.

Jessop, J. J. & Bayer, B. M. (1989). Time-dependent effects of isolation on lymphocyte and adrenocortical activity. *Journal of Neuroimmunology*, *23*, 143-147.

Johnson, E. O., Kamilaris, T. C., Carter, C. S., Calogero, A. E., Gold, P. W. & Chrusos G. P. (1996). The biobehavioral consequences of psychogenic stress in a small social primate (Callitrix jacchus jacchus). *Biological Psychiatry*, *40*, 317-337.

Kanitz, E., Tuchscherer, M., Puppe, B., Tuchscherer, A. & Stabenow, B. (2004). Consequences of repeated early isolation in domestic piglets (Sus scrofa) on their behavioural, neuroendocrine, and immunological responses. *Brain Behavior and Immunity*, *18*, 35-45.

Kaplan, J. R. & Manusk S. B. (1999). Status, stress, and atherosclerosis: the role of environment and individual behavior. *Annals of the N.Y. Academy Science*, *896*, 145-161.

Karp, J. D., Moynihan, J. A. & Ader, R. (1993). Effects of differential housing on the primary and secondary antibody responses of male C57BL/6 and BALB/c mice. *Brain Behavior and Immunity*, *7*, 326-333.

Klein, S. L. & Nelson, R. J. (1999). Social interactions unmask sex differences in humoral immunity in voles. *Animal Behavior*, *57*, 603.610.

Kling, A., Loyd, R., Tachiki, K., Prince, H., Klimenko, V. & Korneva, E. (1992). Effects of social separation on immune function and brain neurotransmitters in cebus monkey (C. apella). *Annals of the N.Y. Academy of Science*, *650*, 257-61.

Konsman, J. P., Parnet, P. & Dantzer, R. (2002). Cytokine-induced sickness behaviour: mechanisms and implications. *Trends in Neuroscience*, *25*, 154-159.

Koolhaas, J. M., Meerlo, P., de Boer, S. F., Strubbe, J. H. & Bohus B. (1997). The temporal dynamics of the stress response. *Neuroscience and Biobehavioral Reviews*, *21*, 775-782.

Kudryavsteva, N. N. (1991). A sensory contact model for the study of aggressive and submissive behavior in male mice. *Aggressive Behavior*, *17*, 285-291.

Kudryavtseva, N. N. (2000). The sensory contact model for the study of aggressive and submissive behaviors in male mice. *Aggressive Behavior*, *26*, 241-256.

Kudryavtseva, N. N. (2003). Use of the "partition" test in behavioral and pharmacological experiments. *Neuroscience and Behavioral Physiology*, *33*, 461-71.

Kudryavsteva, N. N. & Avgustinovich, D. F. (1998). Behavioral and physiological markers of experimental depression induced by social conflict (DISC). *Aggressive Behavior*, *24*, 271-286.

Lilly, A. A., Mehlman, P. T. & Higley, J. D. (1999). Trait-like immunological and hematological measures in female rhesus across varied environmental conditions. *American Journal of Primatology*, *48*, 197-223.

Lindberg, A. C. (2001). Group life. In L. J., Keeling, & H. W., Gonyou, eds. *Social behavior in farm animals* (37-57). Wallingford: CAB Publishing.

Lindstrom, K. M., Hasselquist, D. & Wikelski, M. (2005). House sparrows (Passer domesticus) adjust their social status position to their physiological costs. *Hormones and Behavior*, *48*, 311-20.

Liu, H. & Wang, Z. (2005). Effects of social isolation stress on immune response and survival time of mouse with liver cancer. *World Journal of Gastroenterology*, *11*, 5902-4.

Lu, Z. W., Hayley, S., Ravindran, A. V., Merali, Z. & Anisman, H. (1999). Influence of psychosocial, psychogenic and neurogenic stressors on several aspects of immune functioning in mice. *Stress*, *3*, 55-70.

Lupo, C., Fontani, G., Girolami, L., Lodi, L. & Muscetola, M. (2000). Immune and endocrine aspects of physical and social environmental variations in groups of male rabbits in seminatural conditions. *Ethology, Ecology and Evolution*, *12*, 281-289.

Lyons, D. M., Wang, O. J., Lindley, S. E., Levine, S., Kalin, N. H. & Schatzberg, A. F. (1999). Separation induced changes in squirrel monkey hypothalamic-pituitary-adrenal physiology resemble aspects of hypercortisolism in humans. *Psychoneuroendocrinology*, *24*, 131-142.

Mason, J. W. (1968a). A review of psychoendocrine research on the sympathetic–adrenal medullary system. *Psychosomatic Medicine*, *30*, 631-653.

Mason, J. W. (1968b). A review of psychoneuroendocrine research on the pituitary–adrenal cortical system. *Psychosomatic Medicine*, *30*, 576–609.

McEwen, B. S. (1998). Protective and damaging effects of stress mediators. *New England Journal of Medicine*, *338*, 171-179.

McEwen, B. S. (2000). The neurobiology of stress: from serendipity to clinical relevance. *Brain Research*, *886*, 172-189.

Meehan, W. P. & Henry J. P. (1981). Social stress and the role of attachment behavior in modifying aggression. In Brain, P. & Benton, D., Eds. *Multidisciplinary approach to aggression research*. (209-223). Amsterdam: Elsevier/North-Holland Biomedical Press.

Meerlo, P., Sgoifo, A. & Turek, F. W. (2002). The effect of social defeat and other stressors on the expression of circadian rhythms. *Stress*, *5*, 15-22.

Merlot, E., Moze, E., Dantzer, R. & Neveu, P. J. (2003). Importance of fighting in the immune effects of social defeat. *Physiology and Behavior*, *80*, 351-7.

Merlot, E., Meunier-Salaün, M. C. & Prunier, A. (2004) Behavioural, endocrine and immunenext term consequences of mixing in weaned piglets. *Applied Animal Behaviour Science.*, *85*, 247-257

Miczek, K. A. (1979). A new test for aggression in rats without aversive stimulation: differential effects of D-amphetamine and cocaine. *Psychopharmacology*, *60*, 253-259.

Muscettola, M., Massai, L., Lodi, L., Briganti, F., Fontani, G. & Lupo, C. (2003). IFN-gamma production in rabbits: behavioral and endocrine correlates. *Life Science*, *72*, 1331-43.

Nakano, Y. (2004). Stress-induced modulation of skin immune function: two types of antigen-presenting cells in the epidermis are differentially regulated by chronic stress. *British Journal of Dermatology*, *151*, 50-64.

Padgett, D. A., Sheridan, J. F., Dorne, J., Bernston, G. G., Candelora, J. & Glaser, R. (1998). Social stress and the reactivation of latent herpes simplex virus type 1. Proceedings of the *National Academy of Science USA*, *95*, 7231-7235.

Parmigiani, S., Palanza, P., Rodgers, R. J. & Ferrari PF. Selection, evolution of behavior and animal models in behavioral neuroscience. *Neuroscience and Biobehavioral Reviews*, *23*, 957-971.

Raab, A., Dantzer, R., Michaud, B., Mormede, P., Taghzouti, K., Simon, H. & Le Moal, M. (1986). Behavioral, Physiological and immunological consequences of social status and aggression in chronically coexisting resident-intruder dyads of male rats. *Physiology and Behavior*, *36*, 223-228.

Rabin, B. S. (1999). Stress, Immune functions and health. *The connection*. New York: Wiley & Sons.

Sachser, N. & Renninger, S. V. (1993). Coping with new social situations: the role of social rearing in guinea pig. *Ethology, Ecology and Evolution*, *5*, 64-74.

Sachser, N., Durschlag, M. & Hierzel, D. (1998). Social relationships and the management of stress. *Psychoneuroendocrinology*, *23*, 891-904.

Sanders, V. M., Kasprowicz, D. J., Kohm, A. P. & Swanson, M. A. (2001). Neurotreasmitter receptors on lymphocytes and other lymphoid cells. In: R., Ader, D. L., Felten, N., Cohen, (Eds.). *Psychoneuroimmunology*, 3rd ed. (vol 1, pp 161-196). San Diego: Academic Press.

Sapolsky, R. M. (1992). Neuroendocrinology of the stress-response. In J. B., Becker, S. M., Breedlove, D., Crews, Eds. *Behavioral endocrinology*. (p.287-324). Cambridge: The Mit Press.

Sapolsky, R. M. (1993). Endocrinology alfresco: psychoendocrine studies of wild baboons. *Recent Progress in Hormone Research*, *48*, 437-468.

Sapolsky, R. M. (1994). Individual differences and the stress response. *Seminars in Neuroscience*, *6*, 261-269.

Sapolsky, R. M. (2005). The influence of social hierarchy on primate health. *Science*, *308*, 648-52.

Sapolsky R. M., Krey, L. C. & McEwen, B. S. (1986). The neuroendocrinology of stress and aging: the glucocorticoid cascade hypothesis. *Endocrine Reviews*, *7*, 284-301.

Sapolsky, R. M., Romero, L. M. & Munck, A. U. (2000). How do glucocorticoids influence stress responses? Integrating permissive, suppressive, stimulatory, and preparative actions. *Endocrine Reviews*, *21*, 55-89.

Schapiro, S. J., Nehete, P. N., Perlman, J. E. & Sastry, K. J. (2000). A comparison of cell-mediated immune responses in rhesus macaques housed singly, in pairs, or in groups. *Applied Animal Behaviour Science*, *68*, 67-84.

Scislowska-Czarnecka, A., Chadzinska, M., Pierzchala-Koziec, K. & Plytycz, B. (2004). Long-lasting effects of social stress on peritoneal inflammation in some strains of mice. *Folia Biologica (Krakow)*, *52*, 97-104.

Shanks, N., Renton, C., Zalcman, S. & Anisman, H. (1994). Influence of change from grouped to individual housing on a T-cell-dependent immune response in mice: antagonism by diazepam. *Pharmacology, Biochemestry and Behavior*, *47*, 497-502.

Sheridan, J. F., Stark, J. L., Avitsur, R. & Padgett, D. A. (2000). Social disruption, immunity, and susceptibility to viral infection. Role of glucocorticoid insensitivity and NGF. *Annals of the NY Academy of Science*, *917*, 894-905.

Stark, J. L., Avitsur, R., Padgett, D. A., Campbell, K. A., Beck, F. M. & Sheridan, J. F. (2001). Social stress induces glucocorticoid resistance in macrophages. *American Journal of Physiology Regulatory Integrative and Comparative Physiology*, *280*, R1799-805.

Stark, J. L., Avitsur, R., Hunzeker, J., Padgett, D. A. & Sheridan, J. F. (2002). Interleukin-6 and the development of social disruption-induced glucocorticoid resistance. *Journal of Neuroimmunology*, *124*, 9-15.

Stefanski, V. (2000). Social stress in laboratory rats: hormonal responses and immune cell distribution. *Psychoneuroendocrinology*, *25*, 389-406.

Stefanski, V. (2001). Social rearing conditions before weaning influence numbers and proportions of blood immune cells in laboratory rats. *Developmental Psychobiology*, *39*, 46-52.

Stefanski, V. & Engler, H. (1998). Effects of acute and chronic social stress on blood cellular immunity in rats. *Physiology and Behavior*, *64*, 733-41.

Stefanski, V. & Engler, H. (1999). Social stress, dominance and blood cellular immunity. *Journal of Neuroimmunology*, *94*, 144-52.

Sutherland, M. A., Niekamp, S. R., Rodriguez-Zas, S. L. & Salak-Johnson, J. L. (2006). Impacts of chronic stress and social status on various physiological and performance measures in pigs of different breeds. *Journal of Animal Science*, *84*, 588-96.

Tuchscherer, M., Kanitz, E., Puppe, B., Tuchscherer, A. & Stabenow, B. (2004). Effects of postnatal social isolation on hormonal and immune responses of pigs to an acute endotoxin challenge. *Physiology and Behavior*, *82*, 503-11.

Valzelli, L. (1973). The «isolation syndrome» in mice. *Psychopharmacologia*, *31*, 305-320.

Vegas, O., Beitia, G., Sanchez-Martin, J.R., Arregi, A. & Azpiroz, A. (2004). Behavioral and neurochemical responses in mice bearing tumors submitted to social stress. *Behavioral Brain Research*, *155*, 125-34.

Vekovishcheva, O. Y., Sukhotina, I. A. & Zvartau, E. E. (2000). Co-housing in a stable hierarchical group is not aversive for dominant and subordinate individuals. *Neuroscience and Behavioral Physiology*, *30*, 1371-1324.

Von Holst, D. (1998). The concept of stress and its relevance for animal behavior. *Advances in the Study of Behavior*, *27*, 1-131.

Wybran, J., Appelboom, T., Famaey, J. P. & Govaerts, A. (1979). Suggestive evidence for receptors for morphine and methionine-enkephalin on normal human blood T lymphocytes. *Journal of Immunology*, *123*, 1068-1970

In: Veterinary Immunology and Immunopathology
Editors: L. Nermann, S. Meier

ISBN: 978-1-60876-342-9

Chapter 8

NEUROIMMUNOLOGY RESEARCH IN NON-HUMAN PRIMATES

Bert A. t Hart[1,2,*], Yolanda S. Kap[1,2], Anwar Jagessar[1], Sandra Amor[1] and Herbert P.M. Brok[1]

[1]Department of Immunobiology, Biomedical Primate Research Centre, Rijswijk – The Netherlands;

[2]Department of Immunology, Erasmus Medical Center, Rotterdam, The Netherlands.

ABSTRACT

Multiple sclerosis (MS) is one of the most intensively studied immune-based inflammatory diseases (IMID) of the human central nervous system. Experimental autoimmune encephalomyelitis (EAE) is the elected animal model of MS in which immunopathogenic mechanisms are investigated and the efficacy of new therapies can be tested. However, of the many new therapies that showed promising effects in EAE models only very few were found effective in patients. One possible explanation is the immunological gap between the laboratory mouse and rat strains in which EAE is modeled and the MS patient. In this publication we discuss how EAE models in non-human primates can bridge this gap. Not only is the immune system of non-human primates more closely related to humans, but the EAE models in these species also offer unique possibilities for preclinical research.

INTRODUCTION

Scientists working in the field of neuroimmunology investigate the interactions of the two most complex physiological systems in the human body, being the immune system and the nervous system. Neuroimmunology is a relatively new discipline in biomedical research as

[*]Correspondence concerning this article should be addressed to Bert A. 't Hart PhD, Department of Immunobiology, Biomedical Primate Research Centre, Lange Kleiweg 139 P.O. Box 3306, 2280 GH Rijswijk, The Netherlands. tel. : +31.15.2842691; fax. : +31.15.2843999; E-mail: hart@bprc.nl.

the main developments have taken place in the last three to four decades. It is hard to imagine nowadays that the awareness that certain neurological diseases are actually caused by an immunological attack on the nervous system first arose in the early 1960's. Well-known examples of these so-called autoimmune diseases are multiple sclerosis (MS), a debilitating disease of the central nervous system (CNS), Guillain-Barré syndrome (GBS) and Myasthenia gravis, the latter two being antibody-mediated diseases of the peripheral nervous system (PNS). However, only in the past decades it has become clear that also several neurodegenerative disorders, such as Alzheimer's disease or stroke, may have an immunological component and that the inflammation caused by the immune engagement contributes significantly to the disease outcome [1]. Since a few years it has become clear that also in chronic pain immunological processes are implicated [2]. The neuro-immune interaction is bi-directional, meaning that the immune system affects the CNS just as the CNS also affects immune functions (for review: [3-5]). The research of both forms of interaction has shown an equally impressive development in the past three decades. It has been known for years that the activated immune system signals to the nervous system, which in turn activates the hypothalamus-pituitary-adrenal (HPA) axis triggering the secretion of immunosuppressive corticosteroids. As lymphoid organs are innervated they receive potentiating signals directly from the nervous system. Via the same route the nervous system is informed about ongoing immune processes [6]. Intriguingly, research in the subdiscipline psychoneuroimmunology has shown that immune cells can produce neural hormones, such as endorphins or ACTH [7,8], and that non-neural cells of the nervous system (glia cells) can produce immunological hormones, i.e. cytokines and chemokines [9]. This allows communication between the two systems also via soluble factors.

Thus neural and immune system are tightly interconnected and may even be components of a large integrated feed-back network of adaptive responses to a pathogenic state [3]. Clearly an improved insight into the complexity of their integrated activity will have an enormous impact on future therapy development.

Animal Models in Neuroimmunology Research, a Case for Non-Human Primates

Immunologists investigating immunological effects on the nervous system use mainly experimental models in mice, either wild-type or strains engineered to over-express or delete certain genes of interest. That the research on neurological modulation of immunological processes has its roots in pharmacology may explain why most research has originally been done in rats, which can be well-trained for neurophysiological tasks [4]. This usage of two different animal species hampers the easy integration of research data from both disciplines.

It has become clear that immunological principles observed in laboratory mice cannot be easily transferred to the human immune system [10]. This is (partly) attributable to the fact that the immune system of a 12-week old mice from pathogen-free raised laboratory strains has experienced much less of the environmental stimuli that shape the human immune repertoire [11].

In the following paragraphs we will explain how non-human primates can help to integrate experimental data obtained in the three main species involved in Neuroimmunology research, i.e. mice, rats and humans. We will focus the discussion on our own field of interest, being the modeling of the neuroinflammatory disease multiple sclerosis (MS) in non-human primates. It is pertinent to emphasize, however, that the arguments are not confined to MS alone, but have a broader applicability to other IMID.

NON-HUMAN PRIMATES AS MODELS OF HUMAN NEURO-INFLAMMATORY DISEASE

The two key questions in MS research are a) what is the pathogenic sequence of events that underlies the development of the inflammatory lesion and b) which mechanisms underlay the lesions that cause progressive neurological deficit. We have developed experimental autoimmune encephalomyelitis (EAE) in non-human primates to investigate these issues. The past decade has shown a steep increase of our understanding of pathogenic mechanisms in MS [12,13]. Based on this knowledge a variety of new therapeutic approaches have been tested in animal models, mainly the experimental autoimmune encephalomyelitis (EAE) model in mice. However, although many treatments showed promising effects in animal models, most of these were ineffective in patients or were even found detrimental [14,15]. We have argued elsewhere that the immunological gap between the mouse EAE model and human MS may be too wide for the easy translation of a scientific principle into a therapeutic concept [16-18]. This is in our view the most important argument for the usage of non-human primate disease models.

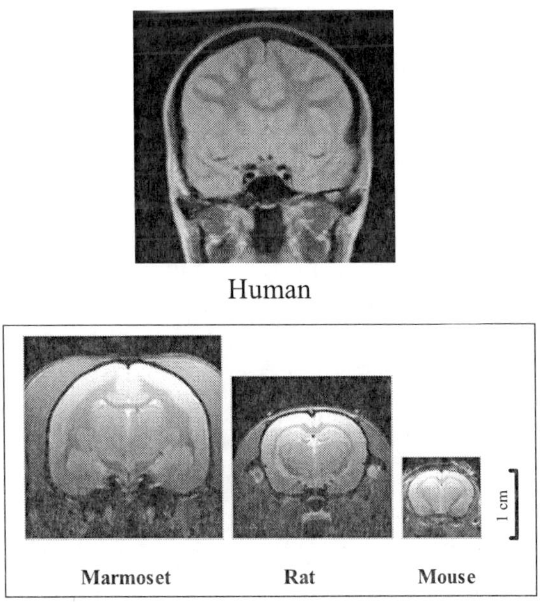

Figure 1. T2-weighted brain images of a healthy marmoset, rat and mouse compared to the human brain. The MR-image of the human brain was recorded at clinical field-strength (1 T) (a kind from Prof. dr. Barkhof, Amsterdam). The animal brain images were recorded at high-field (experimental animal scanner, 4.7 T horizontal bore, Varian, Palo Alto, California, USA) (a kind gift from Dr. Blezer, Utrecht). The white matter is well visible as a dark grey region, grey matter is lighter grey and ventricles are white.

Macaques are by far the most intensively used primate species in biomedical research and much is known about the physiological [19], genetic [20], and microbiological similarity to humans [21]. They have been used for the modeling of neurodegenerative diseases, including Parkinson's disease and Alzheimer's disease [22], as well as neuroinflammatory diseases, such as MS and ADEM [23-25]. New World monkeys are less frequently used, although the common marmoset is rapidly gaining popularity because of the small size and easiness in breeding and handling [26]. However, the in-depth analysis of immunopathogenic mechanisms is hampered by the limited knowledge of the marmoset's immune system and its genetic regulation as well as by the limited availability of reagents to perform analyses.

A particularly attractive aspect of marmosets for the modeling of CNS disorders is that, when adequately trained, they perform well in behavioral and cognitive tasks as well as neurophysiological tests [27-30]. Models of neurological diseases have been developed in the marmoset, including Parkinson's disease [31], Alzheimer's disease [32] and MS [33]. The possibility to combine physiological tests with well-developed brain imaging systems makes the marmoset a powerful model in neuroinflammation research [18,34]. In Figure 1 high resolution T2-weighted magnetic resonance images made of marmoset, rat and mouse brain are depicted, clearly showing the more human-like grey-white matter ratio in the marmoset.

Clinical and Pathological Features of the Marmoset EAE Model

By a single immunization of marmosets with CNS myelin from MS patients or recombinant human MOG_{1-125} (rhMOG) emulsified in complete adjuvant a chronic progressive neurological disease can be induced at 100% incidence. Reports from us and others demonstrate that the experimental disease shows striking pathological resemblance with MS [35,36]. Using similar MRI techniques as used for the diagnosis of MS, typical abnormalities in the brain can be visualized, resembling the human disease (Figure 2). Histological examination of these MRI-detectable changes shows that they represent sharply-edged focal areas of inflammation and demyelination with a variable degree of axonal injury and gliosis [33,37]. With serial MRI of rhMOG-immunized monkeys it can be shown that lesions can be detected long before neurological deficit is observed. Our observations in the marmoset model raise questions that are highly relevant to MS, such as:

- how can the 100% disease incidence be explained as the marmoset represents an outbred species?
- which event determines that clinically silent lesions, which can be observed during the asymptomatic period, become clinically manifest?
- can the onset of neurological defects be delayed or prevented by treating the clinically silent lesions?

We will discuss these aspects in the following paragraphs.

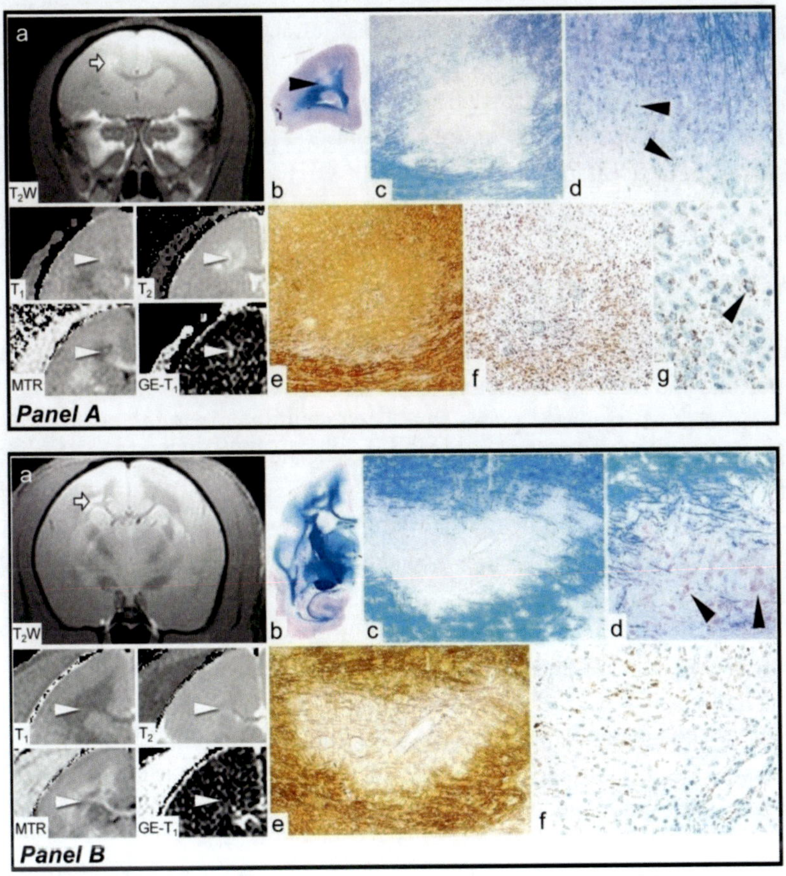

Figure 2. Appearance of brain lesions from a rhMOG-immunized marmoset on MRI (a) and histology (b to f). The lesion in panel A is diagnosed as a hyperintense area on T2W image in the white matter. That the lesion is inflammatory active can be deduced from the leakage of contrast agent through the blood brain barrier into the lesion (hyperintense appearance on GE-T_1 image) and the accumulation of oedema (hyperintense appearance on T_2 image). Immunological and histological staining confirm that this lesion is formed by demyelination (b,c) and contains many infiltrated inflammatory cells (lymphocytes, macrophages; d, f, g) but that axonal structures are left intact (e). The lesion in panel B is also diagnosed as a hyperintense area on T2W image in the white matter. This lesion is inflammatory inactive, and shows no hyperintensity on GE-T_1 and T_2 images. Both lesions are visible as a hypo-intense region on MTR images, which can be explained by increased water content (inflammation) as well as disappearance of macromolecules due to demyelination. Immunological and histological staining confirm that this lesion is formed by demyelination (b,c). The lesion contains few infiltrated inflammatory cells (lymphocytes, macrophages; d, f, g), but axonal structures are severely damaged (e).

Genetic Regulation of EAE Susceptibility in Marmosets

The strongest genetic influence on MS susceptibility in the human population is exerted by the major histocompatibility complex (MHC), a highly polymorphic genetic region that encodes for antigen presentation molecules. Of particular relevance to MS are MHC class I and II molecules. MHC class II molecules are constitutively expressed on B-cells and antigen presenting cells (APC) and are involved in the activation of CD4+ T-cell responses against extracellular antigens. MHC class I molecules are expressed on all nucleated cells and are involved in the activation of CD8+ T-cells against intracellular antigens, such as derived by viral antigens. The association of MS susceptibility with the MHC class II alleles HLA-DR15 (DRB1*1501) and –DQ6 (DQB2*0602) in the Caucasian population has been well established, but other linkages seem to exist in non-Caucasian populations. It is rather surprising that no clear association exists with MHC class I genes, while viruses are implicated as trigger of MS and CD8+ T-cells seem to play a prominent role in the MS pathogenesis [38].

The complexity of the marmoset's MHC region (indicated with *Caja*) is substantially reduced compared the human one (see: http://www.ebi.ac.uk/ipd/mhc/nhp). The human MHC class I region contains at least 18 genes, encoding the highly polymorphic classical *HLA-A*, *-B* and *–C* genes and the oligomorphic non-classical HLA-E, -F, -G, -H and –J genes. To date only three marmoset MHC class I genes have been identified, being the class Ia gene *Caja-G* and the class Ib genes *Caja-E* and *Caja-F* [39]. The human MHC class II region contains *HLA-DPA*, *-DPB*, *-DQA*, *-DQB*, *–DRA* and *-DRB* genes. *HLA-DPB*, *-DQA*, *-DQB* and *–DRB* genes were found to be highly polymorphic. In the marmoset the evolutionary equivalents of *HLA-DRA, -DRB, -DQA (2 –DQA1 alleles)* and *–DQB* (4 *–DQB1* and 2 *–DQB2* alleles) genes have been found, but *Caja-DP* sequences could not be identified [40]. The available genomic *Caja-DRB* sequences cluster into two polymorphic lineages *Caja-DRB1*03* (7 alleles) and *Caja-DRB*W16* (13 alleles) and the monomorphic lineage *Caja-DRB*W1201*. Analysis of cDNA from marmoset B-cell line RNA confirmed transcription of *Caja-DRB*W1201* and *Caja-DRB*W16* alleles, but transcripts of *Caja-DRB1*03* alleles were not found, indicating that these may represent null alleles. In some cases, however, transcripts of exon 2 segments from *Caja-DRB1*03* pseudogenes could be found as a recombination with *Caja-DRB*W16* [41].

The 100% EAE susceptibility maps to the MHC class II allele *Caja-DRB*W1201*, which was found present in each analyzed individual from a large cohort of monkeys from unrelated colonies [40,41]. This allele functions as the major restriction element for the activation of CD4+ T-cells specific for the MOG peptide 24-36 [42]. The capacity of CD4+ Th1 cells specific for this peptide to induce CNS inflammation was confirmed by adoptive transfer into naive recipients [43].

IMMUNE MECHANISMS IN THE CHRONIC DISEASE

In a meta-analysis of T- and B-cell responses of about 30 rhMOG-immunized monkeys we observed that T-cells from "rapid progressor" monkeys, which developed overt neurological disease within 9 weeks after EAE induction proliferate against a wider range of MOG peptides than T-cells from monkeys developing overt disease at a later time point (slow progressors) [44]. Interestingly T-cell responses against $MOG_{34\text{-}56}$ were confined to rapid progressor monkeys. This is a potentially important observation as this peptide contains dominant T-cell epitopes for rhMOG-immunized rhesus monkeys [45] and marmosets [42] and induces progressive EAE in both species ('t Hart, unublished observations). CD4+ T-cells immunized with this peptide have a T helper 1 (Th1) phenotype. Our hypothesis for the underlying mechanism is that the loading of certain MHC class II molecules with $MOG_{34\text{-}56}$ epitopes may trigger the activation of Th1 cells mediating chronic inflammation [46,47]. We have not been able (yet) to identify the MHC species that restrict the activation of CD4+ T-cells specific for this peptide and to determine whether these are over-represented in rapid progressor monkeys.

We observed no relation between the diversity of the antibody response and the rate of EAE progression in marmosets. This was not entirely unexpected. Although a role of anti-MOG antibodies in the induction of CNS demyelination has been proven [48,49], the main pathogenic antibody reactivity is thought to be mainly directed against non-linear/conformational epitopes [50].

EFFICACY OF ANTIBODY THERAPY TARGETING THE ACTIVATION OF TH1 CELLS

The data discussed thus far point at an important pathogenic role of MOG-induced T helper 1 cells. Hence, in a next set of experiments we have tested whether it is possible to intervene in the disease process with antibodies against interleukins (IL)-12 and -23. Both cytokines are intimately involved in the priming of T helper 1 cells [51]. IL-12 and -23 are hetero-dimeric cytokines with overlapping activity in the priming of Th1 cells. IL-12 and -23 are produced by APC when these are activated via engagement of Toll receptors (TLR) with pathogen-associated molecular patterns and/or the co-stimulatory molecule CD40 by CD154 molecules on activated T-cells and/or interferon-γ [52-54]. Studies in rodent EAE models confirm a key function of IL-23 and probably less so of IL-12 in the EAE pathogenesis. Moreover, expression of IL-12p40 was found associated with white matter inflammation in MS patients and EAE affected rodents. Several groups have documented the beneficial effect of anti-IL-12 antibody treatment on EAE development in mice [55-57]. The CD40 [58] and TLR-mediated {Visser, 2006 #1705} induction of IL-12p40 and the localization of IL-12p40 in the CNS white matter lesions of EAE-affected marmosets [59] have been described.

We have used a fully human IgG1k molecule against the shared p40 subunit of human IL-12 and IL-23 (IL-12p40). The antibody has been developed by Centocor Inc. (Malvern PA) for the treatment of T helper 1-mediated disorders, such as psoriasis [60], Crohn's disease and MS (not published). The antibody shows good cross-reactivity with marmoset IL-12p40 [61]. In a first experiment we have tested whether EAE development in human myelin-immunized monkeys could be prevented by administration of the antibody at a 10 mg/kg intravenous dose once weekly from psd 14 to the end of the observation period (day 86). It was observed in this experiment that 5 out of 5 placebo-treated animals developed EAE signs [mean day of onset psd 48 (21-63); mean disease duration to ethical end-point 8 days]. Only 1 of 5 antibody-treated monkeys showed signs of EAE from psd 64. Moreover, the severity of CNS pathology was substantially reduced by the antibody treatment and also the expression of inflammatory mediators within the CNS white matter was decreased [61]. Importantly, we observed in this study localization of the intravenously administered antibody in lymphoid organs as well as in the CNS white matter lesions. Essentially the same results were obtained by treatment with antagonist antibodies against human CD40 [58-62].

In the follow-up experiment we tested the therapeutic effect of the antibody when lesions were already present. This approach relies on the availability of MRI techniques with which not only the position and size of the lesions can be assessed using T2-weighted images, but also the inflammatory activity, using T2 relaxation time (RT) imaging, contrast-enhanced T1-RT imaging and magnetization transfer ratio imaging (MTR) (Figure 2) [18]. In a typical experiment rhMOG-immunized monkeys are scanned once every two or three weeks. We started treatment with the therapeutic antibody or a relevant control substance once formation of the typical lesions with sustained inflammatory activity was observed [63]. This occurred in this particular experiment at psd 53±22. Whether the treatment had a beneficial effect was assessed by measuring the effect on the same MRI parameters. With this treatment procedure we observed a complete suppression of the inflammatory activity and progressive enlargement of the lesions as well as a delayed time interval to the expression of overt neurological deficit (ataxia) from 13±5 to 46±23 days and of the time span to complete paralysis of the hind limbs from 31±10 to psd 64±20. The suppressed inflammatory activity of the lesions was confirmed at the histological level.

We have no direct mechanistic explanation for the delayed expression of overt neurological deficit. In a previous study we have shown that reversible axonal injury may occur in inflammatory active but not in inactive lesions, as evidenced from increased reactivity with antibodies against β-amyloid precursor protein and hypo-phosphorylated neurofilaments [64]. We therefore hypothesize that the delay of neurological disease by the anti-IL-12p40 antibody is actually caused by a suppression of the reversible axonal injury, while other pathological processes such as demyelination and irreversible axonal pathology may be less affected. Indeed, our histological data suggested that demyelination is less affected by the treatment than inflammation.

Perspectives for Modeling the Effect of External Factors on EAE in Marmosets

There is general consensus that the initiation and progression of MS is strongly influenced by the interaction of genetic and external risk factors, such as virus infections, stressful life events, smoking, dietary factors etc.. It is tempting to investigate the mechanism of these interactions in the marmoset EAE model.

Example 1: Virus Infection

Many MS investigators believe that MS may be caused by a virus infection. Although a long list of candidate viruses has been proposed a particularly strong case is made for herpes viruses, in particular avian herpes virus (Marek's disease virus, MDV) and the primate herpes viruses HHV-6 and EBV. For obvious ethical reasons it is impossible to test whether these viruses induce MS in humans. However, marmosets can be infected with all three viruses, i.e. MDV [65], HHV-6 (C.P. Genain, NIH Bethesda, personal communication at the 2005 meeting of the American Neurology Association) and EBV [66]. Hence, marmosets may serve as an attractive experimental model to study whether these viruses induce pathogenic aspects of MS.

Example 2: Stress

There is some evidence that stressful life-events may have an impact on the relapse rate in MS patients [67]. The common marmoset is an accepted model for modeling stress and the endocrinological consequences of stress and could therefore be used to model the effect of stress on the disease course [68,69].

Example 3: Vitamin-D

There is increasing interest for the possible relation between calcium malnutrition and vitamin D deficiency with MS [70]. It has been known for years that the common marmoset can serve as an interesting model for the clinical consequences of vitamin D deficiency, rickets type II in particular [71]. It would be interesting to investigate whether deprivation or dietary supplementation of vitamin D changes the incidence and course of EAE in the marmoset.

Example 4: Cortical Lesions

There is increasing awareness that axonal injury may be a more important pathological feature in MS than inflammation and demyelination and that lesions in the grey matter may have a bigger impact on the disease than white matter lesions. Both pathological aspects are present in the marmoset EAE model **{'t Hart, 2004 #984}{Pomeroy, 2005 #1567}** Preliminary data show significant pathomorphological differences between the grey and white mater lesions, suggesting that these are formed by distinct pathogenic mechanisms. Moreover, the model will be highly useful for the development of new MRI techniques for the visualization and characterization of cortical lesions in the intact animal.

Conclusions

We have discussed in this chapter that a neurological disease with strong resemblance to multiple sclerosis can be induced in the common marmoset. Because of the similarities with the human disease, this unique model can be used to study immunopathogenic mechanisms and their genetic regulation. Clinically relevant imaging techniques, especially MRI, have been developed with which lesions developing in the brain white matter can be visualized and characterized. Moreover, the model is highly suitable for the efficacy testing of new therapies that do not work in lower species. Finally, the model offers many possibilities to investigate the study mechanistic aspects of environmental factors for which a disease modifying role in MS is suspected.

REFERENCES

[1] Griffin, W. S. (2006). Inflammation and neurodegenerative diseases. *Am J Clin Nutr., 83,* 470S-474S.

[2] Watkins, L. R. & Maier, S. F. (2003). Glia: a novel drug discovery target for clinical pain. *Nature Immunol., 2*, 973-985.

[3] Ader, R. (2000). On the development of psychoneuroimmunology. *Eur J Pharmacol., 405*, 167-176.

[4] Exton, M. S., Herklotz, J., Westermann, J. & Schedlowski, M. (2001). Conditioning in the rat: an in vivo model to investigate the molecular mechanisms and clinical implications of brain-immune communication. *Immunol Rev., 184*, 226-235.

[5] Sternberg, E. M. (2006). Neural regulation of innate immunity: a coordinated nonspecific host response to pathogens. *Nat Rev Immunol., 6*, 318-328.

[6] Felten, D. L., Felten, S. Y., Carlson, S. L., Olschowka, J. A. & Livnat, S. (1985). Noradrenergic and peptidergic innervation of lymphoid tissue. *J Immunol 135*, 755s-765s.

[7] Heijnen, C. J., Kavelaars, A. & Ballieux, R. E. (1991). Beta-endorphin: cytokine and neuropeptide. *Immunol Rev 119*, 41-63.

[8] Harbour, D. V., Galin, F. S., Hughes, T. K., Smith, E. M. & Blalock, J. E. (1991). Role of leukocyte-derived pro-opiomelanocortin peptides in endotoxic shock. *Circ Shock., 35*, 181-191.

[9] Becher, B., Prat, A. & Antel, J. P. (2000). Brain-immune connection: immuno-regulatory properties of CNS-resident cells. *Glia., 29*, 293-304.

[10] Mestas, J. & Hughes, C. C. (2004). Of Mice and Not Men: Differences between Mouse and Human Immunology. *J Immunol., 172*, 2731-2738.

[11] Adams, A. B., Williams, M. A., Jones, T. R., Shirasugi, N., Durham, M. M., Kaech, S. M. Wherry, E. J., Onami, T., Lanier, J. G., Kokko, K. E., Pearson, T. C., Ahmed, R. & Larsen, C. P. (2003). Heterologous immunity provides a potent barrier to transplantation tolerance. *J Clin Invest., 111*, 1887-1895.

[12] Noseworthy, J. H., Lucchinetti, C., Rodriguez, M. & Weinshenker, B. G. (2000). Multiple sclerosis. *N Engl J Med., 343*, 938-952.

[13] Frohman, E. M., Racke, M. K. & Raine, C. S. (2006). Multiple sclerosis--the plaque and its pathogenesis. *N Engl J Med., 354*, 942-955.

[14] Wiendl, H. & Hohlfeld, R. (2002). Therapeutic approaches in multiple sclerosis: lessons from failed and interrupted treatment trials. *BioDrugs., 16*, 183-200.

[15] Hohlfeld, R. & Wiendl, H. (2001). The ups and downs of multiple sclerosis therapeutics. *Ann Neurol 49*, 281-284.

[16] 't Hart, B. A. & Amor, S. (2003). The use of animal models to investigate the pathogenesis of neuroinflammatory disorders of the central nervous system. *Curr Opin Neurol 16*, 375-383.

[17] 't Hart, B., Amor, S. & Jonker, M. (2004). Evaluating the validity of animal models for research into therapies for immune-based disorders. *Drug Discov Today., 9*, 517-524.

[18] 't Hart, B. A., Smith, P., Amor, S., Strijkers, G. J. & Blezer, E. L. A. (2005). MRI-guided immunotherapy development for multiple sclerosis in a primate. *Drug Discovery Today.*, *11*, 58 - 66.

[19] Roth, G. S., Mattison, J. A., Ottinger, M. A., Chachich, M. E., Lane, M. A. & Ingram, D. K. (2004). Aging in rhesus monkeys: relevance to human health interventions. *Science.*, *305*, 1423-1426.

[20] Miller, G., M. & Mardras, B. K. (2005). Similarities of non-human primates to humans: genetic variations and phenotypic associations common to rhesus monkeys and humans. In *The Laboratory Primate*. S. P. Wolfe-Coote, ed. Elsevier Science. 17-27.

[21] Lerche, N. W. (2005). Common viral infections of laboratory primates. In *The Laboratory Primate*. S. P. Wolfe-Coote, ed. *Elsevier Science.*, 75-85.

[22] Cork, L. C., Kitt, C. A., Struble, R. G., Griffin, J. W. & Price, D. L. (1987). Animal models of degenerative neurological disease. *Prog Clin Biol Res 229*, 241-269.

[23] Brok, H. P., Bauer, J., Jonker, M., Blezer, E., Amor, S., Bontrop, R. E., Laman, J. D. & 't Hart, B. A. (2001). Non-human primate models of multiple sclerosis. *Immunol Rev 183*, 173-185.

[24] Szabo, C. A. (2005). Primate models of Neurological disease. In *The Laboratory Primate*. S. P. Wolfe-Coote, ed. *Elsevier Science.*, 467-486.

[25] 't Hart, B. A., Bauer, J., Brok, H. P. & Amor, S. (2005). Non-human primate models of experimental autoimmune encephalomyelitis, Variations on a theme. *J Neuroimmunol 168*, 1-12.

[26] Mansfield, K. (2003). Marmoset models commonly used in biomedical research. *Comp Med.*, *53*, 383-392.

[27] Philippens, I. H., Melchers, B. P., Roeling, T. A. & Bruijnzeel. P. L. (2000). Behavioral test systems in marmoset monkeys. *Behav Res Methods Instrum Comput.*, *32,* 173-179.

[28] Roberts, A. C. & Wallis, J. D. (2000). Inhibitory control and affective processing in the prefrontal cortex: neuropsychological studies in the common marmoset. *Cereb Cortex.*, *10*, 252-262.

[29] Scott, L., Pearce, P., Fairhall, S., Muggleton, N. & Smith, J. (2003). Training nonhuman primates to cooperate with scientific procedures in applied biomedical research. *J Appl Anim Welf Sci.*, *6,* 199-207.

[30] Vitale, A. & Licata, E. (2004). Refinement techniques in experimental protocols involving Callitrichids. *Ann Ist Super Sanita.*, *40*, 237-240.

[31] Eslamboli, A. (2005). Marmoset monkey models of Parkinson's disease: which model, when and why? *Brain Res Bull.*, *68*, 140-149.

[32] Maclean, C. J., Baker, H. F., Ridley, R. M. & Mori, H. (2000). Naturally occurring and experimentally induced beta-amyloid deposits in the brains of marmosets (Callithrix jacchus). *J Neural Transm.*, *107*, 799-814.

[33] 't Hart, B. A., Laman, J. D., Bauer, J., Blezer, E. D., van Kooyk, Y., & Hintzen, R. Q. (2004). Modelling of multiple sclerosis: lessons learned in a non-human primate. *The Lancet Neurology.*, *3*, 589-597.

[34] Ferris, C. F., Snowdon, C. T., King, J. A., Sullivan, Jr., J. M., Ziegler, T. E., Olson, D. P., Schultz-Darken, N. J., Tannenbaum, P. L., Ludwig, R., Wu, Z., Einspanier, A.,

Vaughan, J. T. & Duong, T. Q. (2004). Activation of neural pathways associated with sexual arousal in non-human primates. *J Magn Reson Imaging.*, *19*, 168-175.

[35] 't Hart, B. A., Bauer, J., Muller, H. J., Melchers, B., Nicolay, K., Brok, H., Bontrop, R. E., Lassmann, H. & Massacesi, L. (1998). Histopathological characterization of magnetic resonance imaging- detectable brain white matter lesions in a primate model of multiple sclerosis: a correlative study in the experimental autoimmune encephalomyelitis model in common marmosets (Callithrix jacchus). *Am J Pathol.*, *153*, 649-663.

[36] Raine, C. S., Cannella, B., Hauser, S. L. & Genain, C. P. (1999). Demyelination in primate autoimmune encephalomyelitis and acute multiple sclerosis lesions: a case for antigen-specific antibody mediation. *Ann Neurol.*, *46*, 144-160.

[37] 't Hart, B. A., Vogels, J. T., Bauer, J., Brok, H. P. M. & Blezer, E. (2004). Non-invasive measurement of brain damage in a primate model of multiple sclerosis. *Trends Mol. Med.*, *10*, 85-91.

[38] Lassmann, H. & Ransohoff, R. M. (2004). The CD4-Th1 model for multiple sclerosis: a crucial re-appraisal. *Trends Immunol.*, *25*, 132-137.

[39] Cadavid, L. F., Shufflebotham, C., Ruiz, F. J., Yeager, M., Hughes, A. L. & Watkins, D. I. (1997). Evolutionary instability of the major histocompatibility complex class I loci in New World primates. *Proc Natl Acad Sci., U S A 94*, 14536-14541.

[40] Antunes, S. G., de Groot, N. G., Brok, H., Doxiadis, G., Menezes, A. A., Otting, N. & Bontrop, R. E. (1998). The common marmoset: a new world primate species with limited Mhc class II variability. *Proc Natl Acad Sci., U S A 95*, 11745-11750.

[41] Doxiadis, G. M., van der Wiel, M., Brok, H. P. M., de Groot, N. G., Otting, N., 't Hart, B. A., van Rood, J. J. & Bontrop, R. E. (2005). Reactivation of a conserved HLA-DR3-like pseudogene segment in a New World primate species by exon shuffling. *Proc Natl Acad Sci., U S A* accepted.

[42] Brok, H. P., Uccelli, A., Kerlero De Rosbo, N., Bontrop, R. E., Roccatagliata, L., de Groot, N. G., Capello, E., Laman, J. D., Nicolay, K., Mancardi, G. L., Ben-Nun, A. & 't Hart, B. A. (2000). Myelin/oligodendrocyte glycoprotein-induced autoimmune encephalomyelitis in common marmosets: the encephalitogenic T cell epitope pMOG24-36 is presented by a monomorphic MHC class II molecule. *J Immunol 165*, 1093-1101.

[43] Villoslada, P., Abel, K., Heald, N., Goertsches, R., Hauser, S. L. & Genain, C. P. (2001). Frequency, heterogeneity and encephalitogenicity of T cells specific for myelin oligodendrocyte glycoprotein in naive outbred primates. *Eur J Immunol.*, *31*, 2942-2950.

[44] Brok, H. P. M. (2002). Experimental Autoimmune Encephalomyelitis in the common marmoset; a novel model for multiple sclerosis. In *thesis*. Erasmus Medical Center, Rotterdam.

[45] Kerlero de Rosbo, N., Brok, H. P., Bauer, J., Kaye, J. F., 't Hart, B. A. & Ben-Nun, A. (2000). Rhesus monkeys are highly susceptible to experimental autoimmune encephalomyelitis induced by myelin oligodendrocyte glycoprotein: characterisation of immunodominant T- and B-cell epitopes. *J Neuroimmunol.*, *110*, 83-96.

[46] Yu, M., Kinkel, R. P., Weinstock-Guttman, B., Cook, D. J. & Tuohy, V. K. (1998). HLA-DP: a class II restriction molecule involved in epitope spreading during the development of multiple sclerosis. *Hum Immunol.*, *59*, 15-24.

[47] 't Hart, B. A., Brok, H. P., Amor, S. & Bontrop. R. E. (2001). The major histocompatibility complex influences the ethiopathogenesis of MS-like disease in primates at multiple levels. *Hum Immunol., 62*, 1371-1381.

[48] Genain, C. P., Nguyen, M. H., Letvin, N. L., Pearl, R., Davis, R. L., Adelman, M., Lees, M. B., Linington, C. & Hauser, S. L. (1995). Antibody facilitation of multiple sclerosis-like lesions in a nonhuman primate. *J Clin Invest., 96*, 2966-2974.

[49] Genain, C. P., Cannella, B., Hauser, S. L. & Raine, C. S. (1999). Identification of autoantibodies associated with myelin damage in multiple sclerosis. *Nat Med., 5*, 170-175.

[50] Von Budingen, H. C., Hauser, S. L., Ouallet, J. C., Tanuma, N., Menge, T. & Genain. C. P. (2004). Frontline: Epitope recognition on the myelin/oligodendrocyte glycoprotein differentially influences disease phenotype and antibody effector functions in autoimmune demyelination. *Eur J Immunol., 34*, 2072-2083.

[51] Gran, B., Zhang, G. X. & Rostami, A. (2004). Role of the IL-12/IL-23 system in the regulation of T-cell responses in central nervous system inflammatory demyelination. *Crit Rev Immunol., 24*, 111 - 128.

[52] Trinchieri, G. (2003). Interleukin-12 and the regulation of innate resistance and adaptive immunity. *Nat Rev Immunol., 3*, 133-146.

[53] O'Sullivan, B. & Thomas, R. (2003). CD40 and dendritic cell function. *Crit Rev Immunol., 23*, 83-107.

[54] Hunter, C. A. (2005). New IL-12-family members: IL-23 and IL-27, cytokines with divergent functions. *Nat Rev Immunol., 5*, 521-531.

[55] Constantinescu, C. S., Wysocka, M., Hilliard, B., Ventura, E. S., Lavi, E., Trinchieri, G., & Rostami, A. (1998). Antibodies against IL-12 prevent superantigen-induced and spontaneous relapses of experimental autoimmune encephalomyelitis. *J Immunol 161*, 5097-5104.

[56] Ichikawa, M., Koh, C. S., Inoue, A., Tsuyusaki, J., Yamazaki, M., Inaba, Y., Sekiguchi, Y., Itoh, M., Yagita, H. & Komiyama. A. (2000). Anti-IL-12 antibody prevents the development and progression of multiple sclerosis-like relapsing--remitting demyelinating disease in NOD mice induced with myelin oligodendrocyte glycoprotein peptide. *J Neuroimmunol., 102*, 56-66.

[57] Uyttenhove, C., Arendse, B., Stroobant, V., Brombacher, F. & Van Snick, J. (2004). Development of an anti-IL-12 p40 auto-vaccine: protection in experimental autoimmune encephalomyelitis at the expense of increased sensitivity to infection. *Eur J Immunol., 34*, 3572-3581.

[58] Laman, J. D., 't Hart, B. A., Brok, H., Meurs, M., Schellekens, M. M., Kasran, A., Boon, L., Bauer, J., Boer, M. & Ceuppens, J. (2002). Protection of marmoset monkeys against EAE by treatment with a murine antibody blocking CD40 (mu5D12). *Eur J Immunol., 32*, 2218-2228.

[59] 't Hart, B. A., van Meurs, M., Brok, H. P., Massacesi, L., Bauer, J., Boon, L., Bontrop, R. E. & Laman, J. D. (2000). A new primate model for multiple sclerosis in the common marmoset. *Immunol Today., 21*, 290-297.

[60] Kauffman, C. L., Aria, N., Toichi, E., McCormick, T. S., Cooper, K. D., Gottlieb, A. B., Everitt, D. E., Frederick, B., Zhu, Y., Graham, M. A., Pendley, C. E. & Mascelli, M. A. (2004). A phase I study evaluating the safety, pharmacokinetics, and clinical

response of a human IL-12 p40 antibody in subjects with plaque psoriasis. *J Invest Dermatol., 123*, 1037-1044.

[61] Brok, H. P., Van Meurs, M., Blezer, E., Schantz, A., x Peritt, A., Treacy, G., Laman, J. D., Bauer, J. & Hart. B. T. (2002). Prevention of experimental autoimmune encephalomyelitis in common marmosets using an anti-IL-12p40 monoclonal antibody. *J Immunol., 169*, 6554-6563.

[62] Boon, L., Brok, H. P., Bauer, J., Ortiz-Buijsse, A., Schellekens, M. M., Ramdien-Murli, S., Blezer, E., van Meurs, M., Ceuppens, J., de Boer, M., Hart, B. A. T. & Laman, J. D. (2001). Prevention of experimental autoimmune encephalomyelitis in the common marmoset (*Callithrix jacchus*) using a chimeric antagonist monoclonal antibody against human CD40 is associated with altered B cell responses. *J Immunol., 167*, 2942-2949.

[63] Blezer, E. L. A., Nicolay, K., Bauer, J., Brok, H. P. M. & 't Hart. B. A. (2006). Quantitative MRI-pathology correlations of brain white matter lesions developing in a non-human primate model of multiple sclerosis. *NMR Biomed.*, in the press.

[64] Mancardi, G., Hart, B. T., Roccatagliata, L., Brok, H., Giunti, D., Bontrop, R., Massacesi, L., Capello, E. & Uccelli, A. (2001). Demyelination and axonal damage in a non-human primate model of multiple sclerosis. *J Neurol Sci., 184*, 41-49.

[65] Sharma, J. M., Witter, R. L., Shramek, G., Wolfe, L. G., Burmester, B. R. & Deinhardt, F. (1972). Lack of pathogenicity of Marek's disease virus and herpesvirus of turkeys in marmoset monkeys. *J Natl Cancer Inst, 49*, 1191-1197.

[66] Cox, C., S. Chang, L. Karran, B. Griffin, & Wedderburn, N. (1996). Persistent Epstein-Barr virus infection in the common marmoset (Callithrix jacchus). *J Gen Virol 77* (Pt 6), 1173-1180.

[67] Mohr, D. C. & Pelletier, D. (2006). A temporal framework for understanding the effects of stressful life events on inflammation in patients with multiple sclerosis. *Brain Behav Immun.*, 20, 27-36.

[68] Abbott, D. H., Keverne, E. B., Bercovitch, F. B., Shively, C. A., Mendoza, S. P., Saltzman, W., Snowdon, C. T., Ziegler, T. E., Banjevic, M., Garland, T. Jr. & Sapolsky, R. M. (2003). Are subordinates always stressed? A comparative analysis of rank differences in cortisol levels among primates. *Horm Behav., 43*, 67-82.

[69] Saltzman, W., Prudom, S. L., Schultz-Darken, N. J., Wittwer, D. J. & Abbott, D. H. (2004). Social suppression of cortisol in female marmoset monkeys: role of circulating ACTH levels and glucocorticoid negative feedback. *Psycho-neuroendocrinology 29*, 141-161.

[70] Peterlik, M. & Cross. H. S. (2005). Vitamin D and calcium deficits predispose for multiple chronic diseases. *35*, 290-304.

[71] Suda, T., Takahashi, N., Shinki, T., Yamaguchi, A. & Tanioka. Y. (1986). The common marmoset as an animal model for vitamin D-dependent rickets, type II. *Adv Exp Med Biol., 196*, 423-435.

INDEX

A

B

C

D

E

F

G

H

I

J

K

L

M

N

O

P

Q

R

S

T

U

V